D0078772

Thermodynamics
Made Simple for Energy Engineers

Thermodynamics Made Simple for Energy Engineers

By S. Bobby Rauf

Library of Congress Cataloging-in-Publication Data

Rauf, S. Bobby, 1956-
Thermodynamics made simple for energy engineers / by S. Bobby Rauf.
p. cm.
Includes index.
ISBN-10: 0-88173-650-3 (alk. paper)
ISBN-10: 0-88173-651-1 (electronic)
ISBN-13: 978-1-4398-5201-9 (Taylor & Francis distribution : alk. paper)
1. Thermodynamics. 2. Engineers. I. Title.

TJ265.R25 2011
621.402'1--dc23

2011037041

Published by The Fairmont Press, Inc.
700 Indian Trail
Lilburn, GA 30047
tel: 770-925-9388; fax: 770-381-9865
http://www.fairmontpress.com

Distributed by Taylor & Francis Ltd.
6000 Broken Sound Parkway NW, Suite 300
Boca Raton, FL 33487, USA
E-mail: orders@crcpress.com

Distributed by Taylor & Francis Ltd.
23-25 Blades Court
Deodar Road
London SW15 2NU, UK
E-mail: uk.tandf@thomsonpublishingservices.co.uk

Printed in the United States of America
10 9 8 7 6 5 4 3 2 1

10: 0-88173-650-3 (The Fairmont Press, Inc.)
13: 978-1-4398-5201-9 (Taylor & Francis Ltd.)

Dedication

This book is dedicated to Dr. Stanley Peters, without whom, this wouldn't be!

Table of Contents

Preface

As the adage goes, "a picture is worth a thousand words;" this book maximizes the utilization of diagram, graphs and flow charts to facilitate quick and effective comprehension of the concepts of thermodynamics.

This book is designed to serve as a tool for building basic engineering skills in the filed of thermodynamics.

If your objective as a reader is limited to the acquisition of basic knowledge in thermodynamics, then the material in this book should suffice. If, however, the reader wishes to progress their knowledge and skills in thermodynamics to intermediate or advance level, this book could serve as a useful stepping stone.

In this book, the study of thermodynamics concepts, principles and analysis techniques is made relatively easy for the reader by inclusion of most of the reference data, in form of excerpts, within the discussion of each case study, exercise and self assessment problem solutions. This is in an effort to facilitate quick study and comprehension of the material without repetitive search for reference data in other parts of the book.

Certain thermodynamic concepts and terms are explained more than once as these concepts appear in different chapters of this text; often with a slightly different perspective. This approach is a deliberate attempt to make the study of some of the more abstract thermodynamics topics more fluid; allowing the reader continuity, and precluding the need for pausing and referring to chapters where those specific topics were first introduced.

Due to the level of explanation and detail included for most thermodynamics concepts, principles, computational techniques and analyses methods, this book is a tool for those energy engineers, engineers and non-engineers, who are not current on the subject of thermodynamics.

The solutions for end of the chapter self assessment problems are explained in just as much detail as the case studies and sample problem in the pertaining chapters. This approach has been adopted so that this book can serve as a thermodynamics skill building resource for not just energy engineers but engineers of all disciplines. Since all chapters and topics begin with the introduction of important fundamental concepts and principles, this book can serve as a "brush-up" review for even mechanical engineers whose current area of engineering specialty does not afford them the opportunity to keep their thermodynamics knowledge current.

In an effort to clarify some of the thermodynamic concepts for energy engineers whose engineering education focus did not include thermodynamics, analogies are drawn from non-mechanical engineering realms, on certain complex topics, to facilitate comprehension of the relatively abstract thermodynamic concepts and principles.

Each chapter in this book concludes with a list of questions or problems, for self-assessment, skill building and knowledge affirmation purposes. The reader is encouraged to attempt these problems and questions. The answers and solutions for the questions and problems are included under Appendix A of this text.

For reference and computational purposes, steam tables and Mollier (Enthalpy-Entropy) diagrams are included in Appendix B.

Most engineers understand the role units play in definition and verification of engineering concepts, principles, equations and analytical techniques. Therefore, most thermodynamic concepts, principles and computational procedures covered in this book are punctuated with proper units. In addition, for the reader's convenience, units for commonly used thermodynamic entities, and some conversion factors are listed under Appendix C.

Most thermodynamic concepts, principles, tables, graphs, and computational procedures covered in this book are premised on US/Imperial Units as well as SI/Metric Units. Certain numerical examples, case studies or self-assessment problems in this book are premised on ***either*** the SI unit realm ***or*** the US unit system. When the problems or numerical analysis are based on only one of the two unit systems, the given data and the final results can be transformed into the desired unit system through the use of unit conversion factors, in Appendix C.

Some of the Greek symbols, used in the realm of thermodynamics, are listed in Appendix D, for reference.

What readers can gain from this book:

- Better understanding of thermodynamics terms, concepts, principles, laws, analysis methods, solution strategies and computational techniques.
- Greater confidence in interactions with thermodynamics design engineers and thermodynamics experts.
- Skills and preparation necessary for succeeding in thermodynamics portion of various certification and licensure exams, i.e. CEM, FE, PE, and many other trade certification tests.
- A better understanding of the thermodynamics component of heat related energy projects.
- A compact and simplified thermodynamics desk reference.

Chapter 1

Introduction to Energy, Heat and Thermodynamics

INTRODUCTION

The term "thermodynamics" comes from two root words: **"thermo,"** which means heat, and "**dynamic,**" meaning energy in motion, or power. This also explains why the Laws of Thermodynamics are sometimes viewed as Laws of "Heat Power."

Since heat is simply thermal energy, in this chapter, we will review energy basics and lay the foundation for in depth discussion on heat energy and set the tone for discussion on more complex topics in thermodynamics.

ENERGY

The capacity of an, object, entity or a system to perform work is called energy. Energy is a scalar physical quantity. In the International System of Units (SI), energy is measured in Newton-meters (N-m) or Joules, while in the US system of units, energy is measured in ft-lbf, Btu's, therms or calories. In the field of electricity, energy is measured in watt-hours, (Wh), kilowatt-hours (kWh), Gigawatt-hours (GWh), Terawatt-hours (TWh), etc. Units for energy, such as ft-lbs and N-m, point to the equivalence of **energy** with **torque** (moment) and **work**. This point will be discussed later in this chapter.

Energy exists in many forms. Some of the more common forms of energy, and associated units, are as follows:

1) **Kinetic Energy**[1]; measured in ft-lbf, Btus, Joules, N-m (1 N-m = 1 Joule), etc. Where, Btu stands for British thermal units

2) **Potential Energy**[1]; measured in ft-lbf, Btus, Joules, N-m, etc.

3) **Thermal Energy**[1], or heat (Q); commonly measured in calories, Btus, Joules, therms, etc.

4) **Internal Energy**[1], (U); commonly measured in Btu's, calories or Joules.

5) **Electrical Energy**; measured in Watt-hours (Wh), killowatt-hours (kWh) and horsepower-hours (hp-hrs), etc.

6) **Gravitational Energy**; measured in ft-lbf, Joules, N-m, etc.

7) **Sound Energy**; measured in Joules.

8) **Light Energy**; measured in Joules.

9) **Elastic Energy**; measured in ft-lbf, Btus, Joules, N-m, etc.

10) **Electromagnetic Energy**; measured in Joules.

11) **Pressure Energy**[1]; measured in ft-lbf, Btus, Joules, N-m, etc.

[1]**Note:** These forms of energy are discussed in greater detail later in this chapter.

ROOT CONCEPTS AND TERMS THAT CONTRIBUTE TOWARD THE PRODUCTION OR TRANSFORMATION OF ENERGY

Force and Mass

Force or weight, in US or imperial units is measured in **lbf**. While, mass is measured in **lbm**. Mass, in lbm, can be converted to weight, in lbf, through multiplication by factor **g/g_c**, as follows:

Given: An object of mass, **m = 1 lbm**.

Weight = **Force = m.(g/g_c)**

= **(1-lbm) . (g/g_c)**

= (1-lbm) . (32.2 ft/sec^2)/(32.2 lbm-ft/lbf-sec^2).

= **1- lbf**

Force or weight, in SI, or metric units, is measured in Newton's, or "**N.**" Mass, in SI units, is measured in **kg.** Mass, available in kg, can be converted to weight or force, in Newton's, through multiplication by, simply, the gravitational acceleration "**g,**" which is equal to **9.81 m/sec^2.** Conversion of 1 kg of mass to its corresponding weight would be as follows:

Given: An object of mass, **m = 1 kg.**

Weight = **Force = m.g**
= (1-kg) . (9.81 m/sec^2)
= **9.81 kg-m/sec^2**

Since,

1 N = 1 kg-m/sec^2

Weight = **9.81 kg-m/sec^2 = 9.81 N**

Density and Weight Density

Density is defined as mass per unit volume. The symbol for density is: ρ (Rho).

Density of water, in the SI (Metric) realm, under STP conditions, is **1000 kg/m^3**. This density can be converted to **specific weight,** γ (gamma), **or weight density**, as follows:

$$\gamma = \rho \,.\, (g) \qquad \textbf{Eq. 1-1}$$
$$= (1000\ kg/m^3).\ (9.81\ m/sec^2)$$
$$= 9810\ N/m^3$$

Density ρ **of water**, in the US (Imperial) realm, under STP conditions, is **62.4 lbm/ft^3.** The density can be converted to **specific weight,** γ, **or weight density**, as follows:

$$\gamma = \rho \,.\, (g) = (62.4\ lbm/ft^3)\ .\ (g/g_c)$$
$$= (62.4\ lbm/ft^3).\ (32.2\ ft/sec^2)/(32.2\ lbm\text{-}ft/lbf\text{-}sec^2)$$
$$= 62.4\ lbf/ft^3$$

Specific Volume

Specific Volume is the ***inverse of density.*** The symbol for specific volume is "υ," (upsilon). The formula for specific volume, υ, is:

$\upsilon = 1/\rho$ **Eq. 1-2**

The **units** for specific volume are as follows:

- US (Imperial) Units: υ is measured in **cu-ft/lbm or ft^3/lbm.**
- SI (Metric) Units: υ is measured in **m^3/kg.**

Pressure

Pressure is defined as force applied per unit area. The symbol used for pressure is **p.** The formula for pressure is:

$p = F/A$ **Eq. 1-3**

Where,

F = Applied force
A = Area over which the force is applied

There are a number of units used, customarily, for measuring pressure. Common units for pressure, in the SI, or Metric system are N/m^2, Pa (Pascals), kPa, or MPa. **One Pascal** is equal to **1-N/m^2**. Common units for pressure, in the US, or Imperial, system are **psi**, or lbf/in^2, **psia** (psi-absolute) or **psig** (psi-gage). Other units utilized for pressure, and corresponding conversions are shown in **Table 1-1**.

Temperature

Temperature can be defined as a measure of the average kinetic energy of the particles in a substance, where such energy is directly proportional to the degree of hotness or coldness of the substance.

While temperature is one of the principal parameters of thermodynamics, it must be clear that temperature is ***not a direct*** measurement of heat, Q. Temperature is, however, is a parameter that is instrumental in determining the direction of flow of heat, Q. In that, heat travels from bodies at higher temperature to bodies at lower temperature. This role of temperature comports with the laws of thermodynamics.

From physics perspective, temperature is an indicator of the level of kinetic energy possessed by atoms and molecules in substances. In **solids**, at higher temperature, the atoms oscillate or vibrate at higher frequency. In atomic **gases**, the atoms, at higher temperatures, tend to exhibit faster translational movement. In molecular gases, the molecules, at higher temperatures, tend to exhibit higher rates of vibrational

Table 1-1. Units for Pressure and Associated Conversion Factors

Units for Pressure					
	Pascal	**bar**	**atmos.**	**torr**	**psi**
	(Pa)	**(bar)**	**(atm)**	**(Torr)**	**(psi)**
1 Pa	**1 N/m2**	10^{-5}	1.0197×10^{-5}	7.5006×10^{-3}	145.04×10^{-6}
1 bar	**100,000**	Approx. 106 dyn/cm2	1.0197	750.06	**14.50377**
1 atm	98,066.50	0.980665	Approx.1 kgf/cm2	735.56	14.223
1 atm	**101,325**	**1.01325**	**1.0332**	760	**14.696**
1 torr	133.322	1.3332×10^{-3}	1.3595×10^{-3}	Approx. 1 Torr; or 1 mmHg	19.337×10^{-3}
1 psi	6.894×10^{3}	68.948×10^{-3}	70.307×10^{-3}	51.715	**1 lbf/in2**

and rotational movement.

Even though, for a system in thermal equilibrium at a constant volume, temperature is thermodynamically defined in terms of its energy (***E***) and entropy (***S***), as shown in **Eq. 1-4**, unlike pressure, temperature is not commonly recognized as a derivative entity and, therefore, the units for temperature are not derived from the units of other independent entities.

$$T \equiv \frac{\partial E}{\partial S} \qquad \textbf{Eq. 1-4}$$

The universal symbol for temperature is: **T.** The unit for temperature, in the SI, or metric, realm is **°C** or degrees Celsius. In the Celsius temperature scale system, **0°C** represents the freezing point of water. The unit for temperature, in the US, or imperial, realm is **°F** or degrees Fahrenheit. On the Fahrenheit temperature scale system, **32°F** or degrees Fahrenheit represents the freezing point of water. The formulas used for conversion of temperature from metric to US realm, and vice versa, are as follows:

$$(°C \times {}^{9}/_{5}) + 32 = °F \qquad \textbf{Eq. 1-5}$$

$$(°F - 32) \times {}^{5}/_{9} = °C \qquad \textbf{Eq. 1-6}$$

Absolute Temperature

Unlike the Celsius temperature scale system, where 0°C represents the freezing point of water, the absolute temperature scale defines temperature independent of the properties of any specific substance. According to the laws of thermodynamics, absolute zero cannot be reached because this would require a thermodynamic system to be fully removed from the rest of the universe. **Absolute zero** is the theoretical temperature at which entropy would reach its **minimum value**. Absolute zero is defined as **0°K** on the Kelvin scale and as **−273.15°C** on the Celsius scale. This equates to **−459.67°F** on the Fahrenheit scale.

It is postulated that a system at absolute zero would possess finite quantum, mechanical, zero-point energy. In other words, while molecular motion would not cease entirely at absolute zero, the system would lack enough energy to initiate or sustain transference of energy to other systems. It would, therefore, be more accurate to state that ***molecular kinetic energy is minimal at absolute zero.***

According to the **Second Law of Thermodynamics** (discussed later in this text), at temperatures approaching the absolute zero, the ***change in entropy approaches zero***. This comports with the stipulation that as temperatures of systems or bodies approach absolute zero, the transference of heat energy diminishes. **Equation 1-7** is mathematical statement of the Second Law of Thermodynamics.

$$\lim_{T \to 0} \Delta S = 0 \qquad \textbf{Eq. 1-7}$$

Scientists, under laboratory conditions, have achieved temperatures approaching absolute zero. As temperature approaches absolute zero, matter exhibits quantum effects such as **superconductivity and superfluidity**. A substance in a state of superconductivity has ***electrical resistance approaching zero***. In superfluidity state, viscosity of a fluid approaches zero.

Table 1-2 shows factors for conversion of temperatures between Kelvin, Celsius, Fahrenheit and Rankin scales. This table also shows absolute temperature, freezing point of water, triple point for water and the boiling point of water. **Tables 1-3 and 1-4** list ***formulas*** for ***conversion of temperatures*** between Kelvin, Celsius, Fahrenheit and Rankin scales.

Table 1-2. Important Temperatures and Associated Conversion Factors

Important Temperatures and Conversion Table				
	Kelvin	**Celsius**	**Fahrenheit**	**Rankin**
Absolute Zero	0 °K	−273.15 °C	−459.67 °F	0 °R
Freezing Point of Water	273.15 °K	0 °C	32 °F	491.67 °R
Triple Point of Water	273.16 °K	0.01 °C	32.0 °F	491.69 °R
Boiling Point of Water	373.13 °K	99.98 °C	211.97 °F	671.64 °R

In the metric or SI system, the absolute temperature is measured in °K. The relationship between °C and °K is as follows:

$$T_{°K} = T_{°C} + 273°$$
$$\text{and, } \Delta T_{°K} = \Delta T_{°C}$$

In the US system, the absolute temperature is measured in °R. The relationship between °F and °R is as follows:

$$T_{°R} = T_{°F} + 460°$$
$$\text{and, } \Delta T_{°R} = \Delta T_{°F}$$

The absolute temperature system should be used for all thermodynamics calculations, unless otherwise required.

Law of Conservation of Energy

The law of conservation of energy states that energy can be converted from one form to another but ***cannot be created or destroyed***. This can be expressed, mathematically, as:

$\sum E = \sum \text{Energy} = \text{Constant}$

Table 1-3. Rankin Temperature Conversion Formulas

Rankin Temperature Conversion Formulas		
	From Rankin	**To Rankin**
Celsius	$[°C] = ([°R] - 492) \times 5/9$	$[°R] = ([°C] + 273) \times 9/5$
Fahrenheit	$[°F] = [°R] - 460$	$[°R] = [°F] + 460$
Kelvin	$[K] = [°R] \times 5/9$	$[°R] = [K] \times 9/5$

Table 1-4. Kelvin Temperature Conversion Factors

Kelvin Temperature Conversion Formulas		
	From Kelvin	**To Kelvin**
Celsius	$[°C] = [K] - 273$	$[K] = [°C] + 273$
Fahrenheit	$[°F] = [K] \times 9/5 - 460$	$[K] = ([°F] + 460) \times 5/9$
Rankin	$[°R] = [K] \times 9/5$	$[K] = [°R] \times 5/9$

FORMS OF ENERGY IN MECHANICAL AND THERMODYNAMIC SYSTEMS

Potential Energy

Potential energy is defined as energy possessed by an object by virtue of its height or elevation. Potential energy can be defined, mathematically, as follows:

$E_{potential} = m.g.h,$ {SI Units} **Eq. 1-8**

$E_{potential} = m.(g/g_c).h,$ US Units} **Eq. 1-8a**

When the change in potential energy is achieved through performance of work, **W**:

$W = \Delta E_{potential}$ **Eq. 1-9**

Kinetic Energy

Kinetic energy is defined as energy possessed by an object by virtue of its motion. Kinetic energy can be defined, mathematically, as follows:

$$\mathbf{E_{kinetic} = ½.m.v^2} \qquad \{\text{SI Units}\} \qquad \textbf{Eq. 1-10}$$

$$\mathbf{E_{kinetic} = ½.\ (m/g_c).\ v^2} \qquad \{\text{US Units}\} \qquad \textbf{Eq. 1-10a}$$

Where,

m = mass of the object in motion
v = velocity of the object in motion
$\mathbf{g_c}$ **= 32 lbm-ft/lbf-s^2**

When the change in kinetic energy is achieved through performance of work, **W**:

$$\mathbf{W = \Delta\ E_{kinetic}} \qquad \textbf{Eq. 1-11}$$

Energy Stored in a Spring[2]

Potential energy can be stored in a spring—or in any elastic object—by compression or extension of the spring. Potential energy stored in a spring can be expressed, mathematically, as follows:

$$\mathbf{E_{spring} = ½.k.x^2} \qquad \textbf{Eq. 1-12}$$

And,

$$\mathbf{W_{spring} = \Delta\ E_{spring}} \qquad \textbf{Eq. 1-13}$$

Where,

k = The spring constant
x = The contraction or expansion of the spring

[2]**Note:** In steel beam systems, beams act as springs, when loaded, to a certain degree. The deflection of a beam would represent the "**x**," in **Eq. 1-11.**

Pressure Energy

Energy stored in a system in form of pressure is referred to as pressure energy. For instance, energy stored in a compressed air tank is pressure energy. Pressure energy can be expressed, mathematically, as follows:

$$E_{pressure} = E_{flow} = m \cdot p \cdot \upsilon \qquad \textbf{Eq. 1-14}$$

Where,

m = mass of the pressurized system; this would be compressed air in a compressed air system
p = pressure in the system
υ = is the specific volume

Heat and Internal Energy of a System

If heat "**Q**" is added or removed from a system, in the absence of net work performed by or on the system, change in the internal energy "**U** "of a system would be:

$$U_f - U_i = \Delta Q \qquad \textbf{Eq. 1-15}$$

Where **Q** is positive when heat flows into a thermodynamic system and it is negative when heat exits a system.

Specific internal energy "**u**" is defined as ***internal energy per unit mass***. The units for internal energy are Btu/lbm, in the US System, and are kJ/kg, in the Metric or SI System.

Unit Conversions[3] Associated with Heat Energy:

Some of the common heat energy units and unit conversion formulas are listed below:

- Conversion of heat energy measured in **MMBtu's to Btu's**:
  — **1 MMBtu x (1000,000 Btu/MMBtu) = 10^6 Btu's**

- Conversion of heat energy measured in **Btu's to tons** and tons to Btu's:
  — **1 Btu x (8.333x 10.^{5} tons/Btu) = 0.00008333 tons**
  — **1 ton x (12,000 Btu/ton) = 12,000 Btu's**

- Conversion of heat energy measured in **Deca Therms to Btu's**:
  — **1 dT x (1,000,000 Btu/dT) = 1,000,000 Btu's or 1MMBtu**

- Conversion of heat energy measured in **Btu's to kWh and kWh to Btu's**:
  - **1 Btu x (2.928 x 10^{-4} kWh/Btu) = 0.0002928 kWh**
  - **1 kWh x (3413 Btu/kWh) = 3,413 Btu's**

[3]**Note:** These heat energy conversion formulas will be used in various analysis and example problems through this text.

Molar Internal Energy

Molar internal energy "***U***" is defined as internal energy per mole. The units for internal energy ***U*** are Btu / lbmole, in the US system, and are kJ / kmole, in the Metric or SI System.

CASE STUDY 1-1: ENERGY AND ENERGY UNIT CONVERSION

As an energy engineer, you are to analyze substitution of coal, as heating fuel, in lieu of nuclear energy derived from complete conversion, of **2.5 grams** of a certain mass. The nuclear reaction is similar to the Uranium fission reaction shown in Figure 1-1. If the heating value of coal is **13,000 Btu/lbm**, how many U.S. tons of coal must be burned in order derive the same amount of energy?

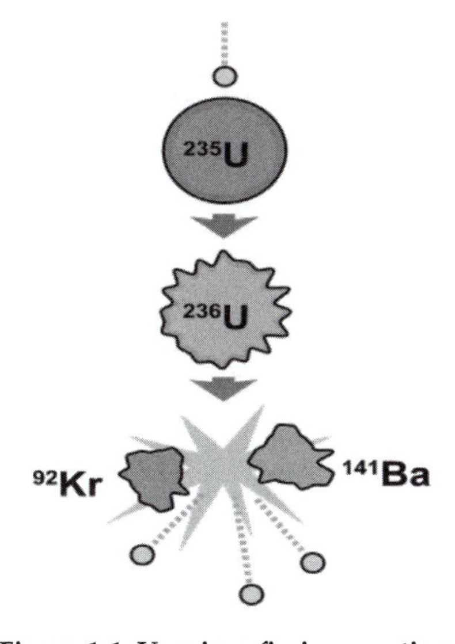

Figure 1-1. Uranium fission reaction

Solution

Given or known

c = Speed of light = **3×10^8 m/s**

m = **Mass of material to be converted to Energy**: 2.5 g, or 0.0025kg

Energy content of coal: 1 lbm of coal contains **13,000 Btu's** of energy

Mass conversion factor; lbm to US tons, and vice and versa: 2000 lbm / ton

Energy unit conversion between Joules and Btu's: 1055 Joules / Btu

According to Einstein's Equation:

$$E = m.\ c^2 \qquad \textbf{Eq. 1-16}$$

By applying **Eq. 1-16,** Energy derived from 2 grams of given mass would be:

$$E = m.\ c^2 = (0.0025\ kg) \times (3 \times 10^8\ m/s)^2 = 2.25 \times 10^{14}\ Joules$$

This energy can be converted into Btu's as follows:

$$E = (2.25 \times 10^{14}\ Joules)/(1055\ Joules/Btu)$$
$$= 2.13 \times 10^{11}\ Btu's.$$

Since 1 lbm of coal contains 13,000 Btu's of heat, the number of lbs of coal required to obtain **2.13×10^{11} Btu's** of heat energy would be:

$$E = (2.13 \times 10^{11}\ Btu's)/(13,000\ Btu's/lbm)$$
$$= 1.64 \times 10^{07}\ lbm$$

Since there are 2000 lbm per ton:

$$E = (1.64 \times 10^{07}\ lbm)/(2000\ lbm/ton)$$
$$= 8,203\ US\ tons\ of\ coal.$$

Conclusion: Energy derived from fission of **2.5 grams** of fissile material is equivalent to the energy derived from **8,203 tons of coal**.

Work

As we will see, through the exploration of various topics in this text, work can be viewed as a vehicle for converting energy contained in various types of fuels to mechanical or electrical energy. In this section, we will elaborate on aspects of work that will be applied in the discussion and analysis of thermodynamic systems.

Work in a Mechanical System:

In a mechanical system, work performed by an external force is referred to as **external work**. While, work performed by an internal force is referred to as **internal work**. Units such as Btu's or kilocalories are not, customarily, used to measure mechanical work.

In a mechanical system, **work is positive** when it is the result of force acting the direction of motion. **Work is negative** when it the result of a force opposing motion. Work attributed to friction is an example of negative work. Where, friction "$\mathbf{F_f}$" is defined, mathematically, as:

$$\mathbf{F_f = \mu_f \cdot N} \qquad \textbf{Eq. 1-17}$$

Where,

μ_f = Coefficient of friction
N = Normal force applied by the surface against the object

Frictional force, $\mathbf{F_f}$, can either be **static frictional force** or **dynamic frictional force.** As stipulated by **Eq. 1-17**, $\mathbf{F_f}$ is directly proportional to the normal force **N**. The coefficient of friction for static friction is, typically, higher as compared to the coefficient of friction for dynamic friction. This is ostensible from the fact that greater force is required to set an object in motion as opposed to the force required to maintain the object in motion.

Mathematical Equations for Work

Work can be performed, defined and computed in several ways. Some of the diferences stem from the realm or frame of reference that work occurs in. Listed below are some of the scenarios in which work can

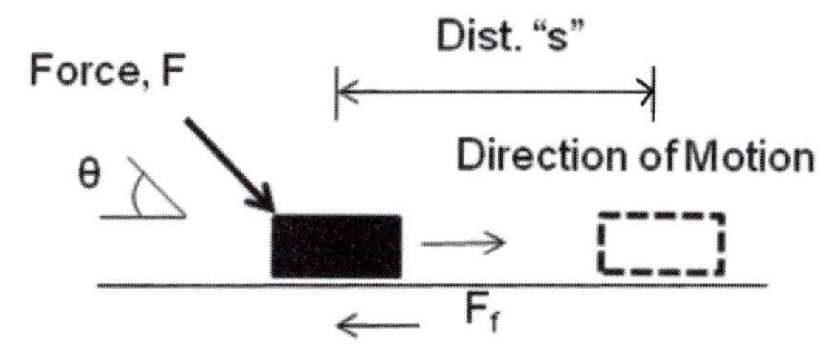

Figure 1-2. Illustration of mechanical work, in a system with friction

occur, and pertinent formulas:

In a rotational system, with ***variable torque***:

$$\mathbf{W_{variable\ torque}} = \qquad \textbf{Eq. 1-18}$$

In a rotational system, with ***constant torque***:

$$\mathbf{W_{constant\ torque}} = \tau \cdot \theta \qquad \textbf{Eq. 1-19}$$

Where,

- **T** = τ = Torque
- θ = Angular distance traversed in the same direction as the torque, τ.

In a linear system, with ***variable force***:

$$\mathbf{W_{variable\ force}} = \qquad \textbf{Eq. 1-20}$$

In a linear system, with ***constant force***, where, force and distance are ***colinear***:

$$\mathbf{W_{constant\ force} = F \cdot s.} \qquad \textbf{Eq. 1-21}$$

General equation for work performed by a constant frictional force where, force and distance are colinear:

$$\mathbf{W_{friction} = F_f \cdot s} \qquad \textbf{Eq. 1-22}$$

When work is performed by a force, **F**, that is applied at an angle, θ, with respect to the direction of motion – as shown in Figure 1-2 - it can be defined, mathematically, as follows:

$$\mathbf{W = Work = (F.Cos\theta - F_{f)} \cdot s} \qquad \textbf{Eq. 1-23}$$

Or,

$$\mathbf{W = (F.Cos\theta) \cdot s - F_f \cdot s} \qquad \textbf{Eq. 1-24}$$

Where,

s = Distance over which the force is applied

In Equations 1-23 and 1-24, the mathematical term **"(F.Cosθ) . s"** constitutes ***positive work*** performed by the force **F** in the direction of motion, and **(F_f . s)** constitutes ***negative work***, performed against the direction of motion. Note that component "**F.Cosθ**," in Equations 1-23 and 1-24, represents the ***horizontal component of force*** contributed by the diagonally applied force **F**.

Work performed by **gravitational force** is defined, mathematically, as:

$W_{g\,(SI)} = m.\ g.\ (h_f - h_i)$, in the Metric Unit Systems **Eq. 1-25**

$W_{g(US)} = m.\ (g/g_c)\ .\ (h_f - h_i)$, in the US Unit Systems **Eq. 1-26**

Where,

h_f = The final elevation of the object
h_i = The initial elevation of the object

Work performed in the case of a linear spring **expansion or contraction** is represented, mathematically, as:

$W_{spring} = ½\ .\ k\ .\ (x_f - x_i)^2$ **Eq. 1-27**

Where,

k = The spring constant
x_i = The initial length of the spring
x_f = The final length of the spring

Work Performed in a Thermodynamic System

In the thermodynamics domain, work constitutes the phenomenon of changing the energy level of an object or a system.

The term "**system**," in thermodynamics, is often used to represent the **medium**. For instance, in the case of an open thermodynamic system—such as steam powered turbine—steam is considered as a system performing work on the surroundings, i.e., the turbine.

In a thermodynamic system, **work is positive** when an object or system performs work on the surroundings. Example: If the vanes of an air compressor are considered to constitute the system, then the work performed on air, by the vanes in an air compressor, would be positive. **Work is negative** when the surroundings perform work on the object. Inflating of a raft or an inner tube constitutes negative work as the air (environment or surrounding) performs work on the walls of the raft or tube (the system) during the inflation process.

Specific Heat

Specific heat is defined as the amount of the heat, **Q**, required to change the temperature of mass "**m**" of a substance by **ΔT**. The symbol for specific heat is "**c**."

The mathematical formula for specific heat of solids and liquids is:

$$\mathbf{c = Q/(m.\ \Delta T)} \qquad \textbf{Eq. 1-28}$$

Or,

$$\mathbf{Q = m.\ c.\ \Delta T} \qquad \textbf{Eq. 1-29}$$

Where,

- **m** = Mass of the substance; measured in **kg**, in the SI system, and in **lbm** in the US system
- **Q** = The heat added or removed; measured in **Joules** or **kJ** in the SI System, or in **Btu's** in the US system
- ΔT = The change in temperature, measured in °**K** in the SI Systems, or in °**R** in the US System

The units for **c** are **kJ/(kg. °K), kJ/(kg. °C), Btu/(lbm. °F) or Btu/(lbm. °R).**

The thermodynamic equation involving **specific heats of gases** are as follows:

$\mathbf{Q = m.\ c_v.\ \Delta T}$, when volume is held constant. **Eq. 1-30**

$\mathbf{Q = m.\ c_p.\ \Delta T}$, when pressure is held constant. **Eq. 1-31**

Approximate specific heat, $\mathbf{c_p}$, for selected liquids and solids are listed in **Table 1-5**.

The next case study, **Case Study 1-2,** is designed to expand our exploration of energy related analysis methods and computational techniques. Some of the energy, work and heat considerations involved in this case study lay a foundation for more complex energy work and thermodynamics topics that lie ahead in this text. This case study also provides us an opportunity to experience the translation between the SI (Metric)

Table 1-5. Approximate Specific Heat, c_p, for Selected Liquids and Solids, in kJ/kg °K, cal/gm °K, Btu/lbm °F, J/mol °K

Substance	c_p in kJ/kg °K	c_p in kcal/kg °K or BTU/lbm °F	Molar C_p J/mol °K
Aluminum	0.9	0.215	24.3
Bismuth	0.123	0.0294	25.7
Copper	0.386	0.0923	24.5
Brass	0.38	0.092	N/A
Gold	0.126	0.0301	25.6
Lead	0.128	0.0305	26.4
Iron	0.460	0.11	N/A
Silver	0.233	0.0558	24.9
Tungsten	0.134	0.0321	24.8
Zinc	0.387	0.0925	25.2
Mercury	0.14	0.033	28.3
Ethyl Alcohol	2.4	0.58	111
Water	4.186	1	75.2
Ice at -10 °C	2.05	0.49	36.9
Granite	0.79	0.19	N/A
Glass	0.84	0.2	N/A

Table 1-6. Densities of Common Materials

Metal	g/cm^3	lb/in^3	lb/ft^3	lb/gal
Water	1.00	0.036	62	8.35
Aluminum	2.7	0.098	169	22.53
Zinc	7.13	0.258	445	59.5
Iron	7.87	0.284	491	65.68
Copper	8.96	0.324	559	74.78
Silver	10.49	0.379	655	87.54
Lead	11.36	0.41	709	94.8
Mercury	13.55	0.49	846	113.08
Gold	19.32	0.698	1206	161.23

unit system and the US (Imperial) unit system. As we compare the solutions for this case study in the US and SI unit systems, we see that choosing one unit system versus another, in some cases does involve the ***use of different formulas.*** This difference in formulas for different unit systems is evidenced in the potential and kinetic energy components of the energy conservation equations.

Furthermore, this case study helps us understand the vital and integrated role that work, kinetic energy and potential energy play in application of law of conservation of energy in thermodynamic system analysis. In this case study we start off with energy and work considerations and conclude our analysis with the quantitative assessment of ***thermodynamic impact*** on ***steel, air*** and ***water***, key substances involved in the overall process.

CASE STUDY 1-2. ENERGY CONSERVATION, ENERGY CONVERSION AND THERMODYNAMICS

At a foundry, a solid rectangular block of carbon steel, density **7,850 kg/m^3 (491 lbm/ft^3)**, from Table 1-6, is released to a downward inclined ramp as shown in **Figure 1-3**. The volume of the block is **1.0 m^3 (35.32 ft3)** and its release velocity, at the top of the ramp is **1.5 m/s (4.92 ft/s)**. The force of friction between the block, the inclined surface and the flat conveyor bed is **400 N (89.924 lbf).** The block is stopped on the flat section through compression of a shock absorbing spring system before it settles on a roller conveyor operating at a linear speed of **2 m/s (6.562 ft/s)**. Assume that the frictional force stays constant through the entire path of the block.

a) Determine the velocity of the steel block when it enters the horizontal segment of the travel, i.e. point "**w.**"

b) Employing the law of energy conservation and principles of energy conversion, calculate the value of the spring constant for the shock absorbing spring system.

c) How much energy is stored in the shock absorbing spring upon complete compression?

d) What would the steady state speed of the block be after it settles onto the roller conveyor?

e) If the spring type shock absorbing system is replaced by a compressed air cylinder of **1.0 m³ (35.32 ft³)** uncompressed volume, at room temperature of **20°C (68 °F)** and **standard atmospheric pressure**, what would be the rise in temperature of the cylinder air immediately after the steel block's impact? The final, compressed, volume is **0.75 m³ (26.49 ft³)**, and the pressure gage on the cylinder reads **2 bar (29 psia)**.

f) If the conveyor, at the bottom of the incline, is a belt driven roller conveyor and the rate of flow of blocks onto the conveyor is **one per 10 seconds**, determine the horsepower rating of the conveyor motor. Assume the conveyor belt to be directly driven off the conveyor motor shaft and that there is no slip between the belt and the rollers. Assume the motor efficiency to be **90%**.

g) The conveyor transports the blocks to a cooling/quenching tank. The temperature of the blocks, when they are dropped into the quenching tank, is **100°C (212 °F)**. The initial temperature of the water in the quench tank is **20°C (68 °F)** and volume of water is **6.038 m³ (213.23 ft³)**. The final, equilibrium, temperature of the water and the block is **30°C (86 °F)**. Determine the amount of heat extracted by the quench water per block.

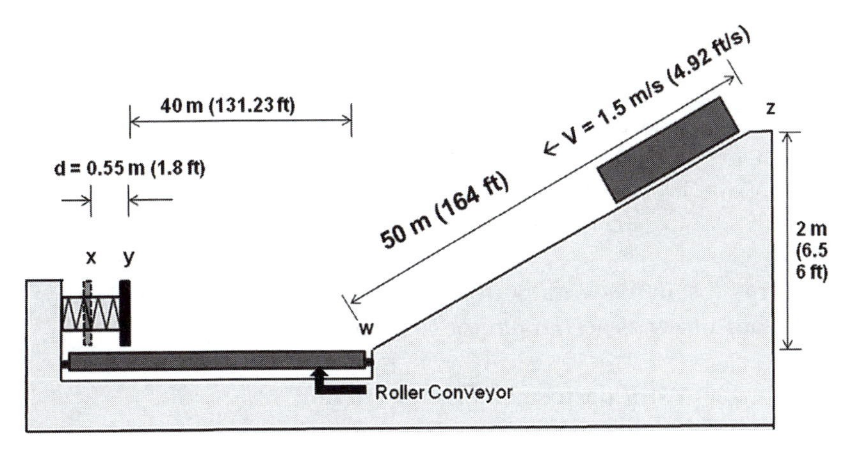

Figure 1-3. Case Study 1-2

Solution, Case Study 1-2—SI (Metric) version

Given, known or derived values:

Volume of the steel block = $\mathbf{V_{steel\ block} = 1.0\ m^3}$

Density of the steel block = $\mathbf{\rho_{steel} = 7850\ kg/m^3}$

Mass of the steel block = $\mathbf{m_{steel\ block}}$

$$= \rho_{steel} \cdot V_{steel\ block}$$
$$= (7850\ kg/m^3) \times (1.0\ m^3)$$
$$\therefore m_{steel\ block} = 7850\ kg$$

Velocity of block at point "z" = $\mathbf{V_z = 1.5\ m/s}$

Elevation at point "z" = $\mathbf{h_z = 2.0\ m}$

Spring constant for the shock absorbing spring = **k**

Frictional force, $\mathbf{F_f = N.\ \mu_f = 400\ N}$

Density of air at STP: $\mathbf{1.29\ kg/m^3}$

At STP:

Temperature = **0 °C**

Pressure = **1 bar, 101.33kPa,or 101,330 Pa**

a) Determine the velocity, $\mathbf{v_w}$, of the steel block when it enters the horizontal segment of the travel, i.e. point **w**:

Solution Strategy: The velocity variable $\mathbf{v_w}$ is embedded in the kinetic energy of the block, $\mathbf{1/2\ .\ m.\ v_w^2}$. So, if we can determine the amount of **kinetic energy** possessed by the block at point **w**, we can derive the required velocity $\mathbf{v_w}$. To find the kinetic energy at point **w**, we can apply the law of conservation of energy at points **z** and **w** as shown below:

Let total energy at point z = $\mathbf{E_{z\text{-}total}}$

Then,

$$E_{z\text{-}total} = E_{z\text{-}kinetic} + E_{z\text{-}potential}$$
$$E_{z\text{-}total} = 1/2\ .\ m.\ v_z^2 + m\ .\ g.\ h_z$$
$$E_{z\text{-}total} = 1/2\ .\ (7850\ kg)\ .\ (1.5\ m/s)^2 + (7850\ kg)\ .\ (9.81\ m/s^2)\ .\ (2\ m)$$
$$\therefore E_{z\text{-}total} = 162{,}848\ J$$

The energy lost in the work performed against friction, during the block's travel from **z** to **w**, is accounted for as follows:

$\mathbf{W_{f\text{-}wz}}$ **= Work performed against friction**

$$= (Dist.\ w\text{-}z)\ .\ (F_f)$$
$$= (50\ m)\ .\ (400\ N) = 20{,}000\ J$$

Therefore, the energy left in the block when it arrives at point **w**, at the bottom of the ramp, would be as follows:

$$\mathbf{E_{w\text{-}total} = E_{z\text{-}total} - W_{f\text{-}wz}}$$
$$\mathbf{= 162{,}848\ J - 20{,}000\ J}$$
$$\mathbf{= 142{,}848\ J}$$

Since the block is at "ground" elevation when it arrives at point **w**, the potential energy at point **w** would be "**zero.**"

$$\therefore\ \mathbf{E_{w\text{-}total} = 1/2\ .\ m.\ v_w^2}$$

Or,

$$\mathbf{v_w = \{2\ .\ (E_{w\text{-}total})/m\}^{1/2}}$$
$$\mathbf{= \{2\ .\ (142{,}848\ J)/7850\ kg\}^{1/2}}$$
$$\mathbf{= 6.03\ m/s}$$

b) Employing the law of energy conservation and principles of energy conversion, calculate the value of the spring constant for the shock absorbing spring system.

Solution Strategy: The unknown constant **k** is embedded in the formula for the **potential energy stored** in the spring after it has been fully compressed, upon stopping of the block. This potential energy is equal to the work performed on the spring, i.e., $\mathbf{W_{spring} = 1/2\ .\ k\ .\ x^2}$. So, if we can determine the amount of **work performed on the spring,** during the compression of the spring, we can derive the required value of **k.**

To derive the value of $\mathbf{W_{spring}}$, we will apply the law of conservation of energy to the travel of the block from point **z** to point **x.**

Based on the dimensions in Figure 1-3:

Distance x-z = (0.55m + 50 m + 40m)
= **90.55 m**

$\mathbf{W_{f\text{-}xz}}$ = Work performed against friction over Dist. x-z
= **(Dist. x-z) .** $\mathbf{F_f}$
= (90.55 m) . (400 N)
= **36,220 J**

Therefore, the total energy at point **z** could be stated as:

$$E_{z\text{-total}} = W_{spring} + W_{f-xz}$$

Or,

$$W_{spring} = E_{z\text{-total}} - W_{f-xz}$$
$$= 162{,}848\ J - 36{,}220\ J$$
$$= \mathbf{126{,}\ 628\ J}$$

Energy stored in the spring is quantified as:

$$W_{spring} = 1/2 \cdot k \cdot x^2$$

Since W_{spring} has been determined to be equal to **126, 628 J,**

$$126{,}\ 628\ J = ½ \cdot k \cdot (0.55)^2$$

Therefore,

$$k = 2 \cdot W_{spring}/x^2$$
$$= 2 \cdot (126{,}628\ J)/(0.55)^2$$
$$= 837{,}212\ N/m$$

c) How much energy is stored in the shock absorbing spring upon complete compression?

Solution:

Energy stored in the spring is equal to the work performed on the spring. The work performed on the spring, as computed in part (b) above, is:

$$W_{spring} = E_{z\text{-total}} - W_{f-xz}$$
$$= 162{,}848 - 36{,}220$$
$$= 126{,}628\ J$$

d) What would the steady state speed of the block be after it settles onto the roller conveyor?

Solution/answer:

After the block settles into a steady state condition on the conveyor, it assumes the speed of the conveyor, i.e. **2 m/s.**

e) Rise in the temperature of the compressed air in the shock absorbing cylinder:

Solution:
The rise in the cylinder's air temperature can be determined after calculating the final temperature of the air through the application of the ideal gas law. Ideal gas laws can be applied in this case because air, for most practical purposes, is assumed to act as an ideal gas.

According to ideal gas law:

$$(P_1 . V_1)/T_1 = (P_2 . V_2)/T_2 \qquad \textbf{Eq. 1-32}$$

Or, through rearrangement of Eq. 1-32:

$$T_2 = (P_2 . V_2 . T_1)/(P_1 . V_1) \qquad \textbf{Eq. 1-33}$$

Given or known:

P_1 = 1 Bar = 101.33 kPa
V_1 = 1.0 m^3
T_1 = 20 °C => 273 + 20C = 293 °K
P_2 = 2 bar or 202.66 kPa
V_2 = 0.75 m^3

Then, by applying Eq. 1-33:

T_2 = {(202.66 kPa) . (0.75 m^3) . (293 °K)} / {(101.33 kPa) . (1.0 m^3)}
T_2 = 439.5 °K i.e. 166.5 °C;

Therefore, the rise in the cylinder air temperature would be:

= 166.5 °C - 20°C
= 146.5 °C

Note: In Chapter 8, this part, (e), will be extended, as an illustration of isothermal process, to determine the amount of heat that must be removed to maintain the air temperature at 20°C.

f) If the conveyor, at the bottom of the incline, is a belt driven roller conveyor and the rate of flow of blocks onto the conveyor is **one per 10 seconds**, determine the horsepower rating of the conveyor motor. Assume the conveyor belt to be directly driven off the conveyor motor shaft and that there is no slip between the belt and the rollers.

Assume the motor efficiency to be **90%**.

Solution:

Apply the **power**, **velocity** and **force** formula to determine the power requirement, as follows:

$$\mathbf{P = F \cdot v} \qquad \textbf{Eq. 1-34}$$

Where,

- **P** = Power required to move the steel blocks
- F = Force required to move the block or the force required to move the conveyor belt with the block on the rollers
- v = Velocity of the belt; i.e., **2.0 m/s,** as given.

While the velocity **v** is given, the force **F** is unknown and must be derived. Force can be defined in terms of mass flow rate $\dot{m}$ and the change in velocity **Δv**, as stated in **Eq. 1-35**:

$$\mathbf{F = \dot{m} \cdot \Delta v} \qquad \textbf{Eq. 1-35}$$

Based on the derived mass of the block as 7850 kg and the fact that 1 block is moved every 10 seconds:

$$\dot{m} = \textbf{mass flow rate} = \textbf{7850 kg/10 secs} = \textbf{785 kg/s}$$

And, based on the given conveyor speed of 2.0 m/s:

$$\begin{aligned}
\Delta v &= \text{Change in the velocity of the block} \\
&= \mathbf{v_f - v_{\dot{c}}} \\
&= \mathbf{2.0\ m/s - 0} \\
\therefore \mathbf{\Delta v} &= \mathbf{2.0\ m/s}
\end{aligned}$$

{Note: This change in the velocity is in the direction of the roller conveyor}

Therefore, applying **Eq. 1-35,** the force required to move the block would be:

$$\begin{aligned}
\mathbf{F} &= \mathbf{\dot{m} \cdot \Delta v} \\
&= \mathbf{(785\ kg/s) \cdot (2.0 m/s - 0)} \\
&= \mathbf{1570\ N}
\end{aligned}$$

{Note: (kg/s . m/s) => kg . m/s^2 = > m . a = F in Newtons

Then, by applying **Eq. 1-34:**

$$\mathbf{P = F \,.\, v}$$
$$\mathbf{= (1570\ N)\,.\,(2\ m/s)}$$
$$\mathbf{= 3140\ W}$$

Since there are 746 watts per hp, the computed power of **3140 W,** in hp, would be:

$$\mathbf{P = (3140\ W)/(746\ W/hp)}$$
$$\mathbf{= 4.21\ hp}$$

Therefore, choose a standard **5 hp** motor.

Note: The efficiency of the motor is not needed in the motor size determination. Motor is specified on the basis of the brake horsepower required by the load; which in this case, is 4.21hp.

g) The conveyor transports the blocks to a cooling / quenching tank. The temperature of the blocks, when they are dropped into the quenching tank, is **100°C**. The initial temperature of the water in the quench tank is **20°C** and volume of water is **6.038 m^3**. The final, equilibrium, temperature of the water and the block is **30°C**. How much heat is extracted by the quench water per block?

Solution:

Given or known:

$c_{\text{cast iron}}$ = **0.460 kJ/kg. °K** {From **Table 1-5**}

$m_{\text{steel block}}$ = **7850 kg**, as determined earlier

$T_{\text{block - i}}$ = 100 °C = 273 + 100C = **373 °K**

$T_{\text{block - f}}$ = 30 °C = 273 + 30°C = **303 °K**

$\therefore \Delta T_{\text{block}}$ = 303 °K - 373 °K= **- 70°K**

According to **Eq. 1-29:**

$$\mathbf{Q = m.\ c.\ \Delta T}$$

Therefore,

$$Q_{\text{lost by the block}} = \mathbf{(m_{\text{block}})\ .\ (c).\ (\Delta T_{\text{block}})}$$
$$Q_{\text{lost by the block}} = (7850\ \text{kg})\ .\ (0.460\ \text{kJ / kg. °K}).\ (-\ 70\text{K})$$
$$\mathbf{= -\ 252{,}770\ kJ}$$

Since,

$\mathbf{Q}_{\text{absorbed by water}} = -\ \mathbf{Q}_{\text{lost by the block}},$
$\mathbf{Q}_{\text{absorbed by water}} = \mathbf{-\ (-\ 252{,}770\ kJ)}$
$\mathbf{= +\ 252{,}770\ kJ}$

Solution, Case Study 1-2—US (Imperial) Unit Version
Given, known or derived values:

Volume of the steel block = $\mathbf{V_{steel\ block} = 35.32\ ft^3}$
Density of the steel block = $\mathbf{\rho_{steel} = 491\ lbm/ft^3}$
Mass of the steel block = $\mathbf{m_{steel\ block}}$
$\mathbf{= steel \cdot V_{steel\ block}}$
$\mathbf{= (491\ lbm/ft^3) \times (35.32\ ft^3)}$
$\therefore\ \mathbf{m_{steel\ block} = 17{,}342\ lbm}$
Velocity of block at point "**z**" = $\mathbf{V_z = 4.92\ ft/s}$
Elevation at point "**z**" = $\mathbf{h_z = 6.56\ ft}$
Spring constant for the shock absorbing spring = **k**
Frictional force, $\mathbf{F_f = N.\ \mu_f = 89.92\ lbf}$
Density of air at STP: $\mathbf{0.0805\ lbm/ft^3}$, at STP: Temperature = 32 °F, Pressure = 1 atm or 14.5 psia

a) Determine the velocity, $\mathbf{v_w}$, of the steel block when it enters the horizontal segment of the travel, i.e. point **w**:

Solution Strategy: The velocity variable $\mathbf{v_w}$ is embedded in the kinetic energy of the block: $\mathbf{1/2\ .\ m/g_c.\ v_w^2}$. So, if we can determine the amount of **kinetic energy** possessed by the block at point **w**, we can derive the required velocity $\mathbf{v_w}$. To find the kinetic energy at point **w**, we can apply the law of conservation of energy at points **z** and **w** as shown below:

Let total energy at point z = $\mathbf{E_{z\text{-}total}}$

Then,

$\mathbf{E_{z\text{-}total} = E_{z\text{-}kinetic} + E_{z\text{-}potential}}$
$\mathbf{E_{z\text{-}total} = 1/2\ .\ (m/g_c)\ .\ v_z^2 + m\ .\ (g/g_c)\ .\ h_z}$ **Eq. 1-36**
$\mathbf{E_{z\text{-}total}} = 1/2\ .\ \{(17{,}342\ lbm/(32\ lbm\text{-}ft/lbf\text{-}s^2)\}\ .\ (4.92\ ft/s)^2 + (17{,}342\ bm)\ .\ (32\ ft/s^2/32\ lbm\text{-}ft/lbf\text{-}s^2)\ .\ (6.56\ ft)$
$\therefore\ \mathbf{E_{z\text{-}total} = 120{,}282\ ft\text{-}lbf}$

The energy lost in the work performed against friction, during the block's travel from **z** to **w**, is accounted for as follows:

$$\mathbf{W_{f\text{-}wz}} = \textbf{Work performed against friction}$$
$$= \mathbf{(Dist.\ w\text{-}z)\ .\ (F_f)}$$
$$= (164\ \text{ft})\ .\ (89.92\ \text{lbf})$$
$$= \mathbf{14{,}747\ ft\text{-}lbf}$$

Therefore, the energy left in the block when it arrives at point **w**, at the bottom of the ramp, would be as follows:

$$\mathbf{E_{w\text{-}total} = E_{z\text{-}total} - W_{f\text{-}wz}}$$
$$= 120{,}282\ \text{ft-lbf} - 14{,}747\ \text{ft-lbf}$$
$$= \mathbf{105{,}535\ ft\text{-}lbf}$$

Since the block is at "ground" elevation when it arrives at point **w**, the potential energy at point **w** would be "**zero**."

$$\therefore\ \mathbf{E_{w\text{-}total} = 1/2\ .\ (m/g_c)\ .\ v_w^2} \qquad \textbf{Eq. 1-37}$$

Or,

$$\mathbf{v_w = \{2\ .\ (g_c)\ .\ (E_{w\text{-}total})/m\}^{1/2}}$$
$$= \{2\ .\ (32\ \text{lbm-ft/lbf-s}^2)\ .\ (105{,}535\ \text{ft-lbf})/17{,}342\ \text{lbm}\}^{1/2}$$
$$= \mathbf{19.74\ ft/s}$$

b) Employing the law of energy conservation and principles of energy conversion, calculate the value of the spring constant for the shock absorbing spring system.

Solution Strategy: The unknown constant **k** is embedded in the formula for the **potential energy stored** in the spring after it has been fully compressed, upon stopping of the block. This potential energy is equal to the work performed on the spring, i.e., $\mathbf{W_{spring} = 1/2\ .\ k\ .\ x^2}$. So, if we can determine the amount of **work performed on the spring,** during the compression of the spring, we can derive the required value of **k.**

To derive the value of $\mathbf{W_{spring}}$, we will apply the law of conservation of energy to the travel of the block from point **z** to point **x.**

Based on the dimensions in **Figure 1-3**:

Distance x-z $= (1.8\ \text{ft} + 131.23\ \text{ft} + 164\ \text{ft})$
$= \mathbf{297\ ft}$

$\mathbf{W_{f-xz}}$ = Work performed against friction over Dist. x-z
= **(Dist. x-z) . $\mathbf{F_f}$**
= (297 ft) . (89.92 lbf)
= **26,706 ft-lbf**

Therefore, the total energy at point **z** could be stated as:

$\mathbf{E_{z\text{-}total} = W_{spring} + W_{f-xz}}$

Or,

$\mathbf{W_{spring} = E_{z\text{-}total} - W_{f-xz}}$
= 120,282 ft-lbf – 26,706 ft-lbf
= **93,576 ft- lbf**

Energy stored in the spring is quantified as:

$\mathbf{W_{spring} = 1/2 . k . x^2}$

Since $\mathbf{W_{spring}}$ has been determined to be equal to **93,576 ft- lbf,**
93,576 ft- lbf = ½ . k . (1.8 ft) ^{2}

Therefore,
$\mathbf{k = 2 . W_{spring}/x^2}$
= **2 . (93,576 ft- lbf)/(1.8 ft)** ^{2}
= **57,763 lbf/ft**

Ancillary Exercise: This value of **k = 57,763 lbf/ft,** in US units, is within 0.7% of the value of **k = 837,212 N/m,** derived in SI units. The reader is encouraged to perform the unit conversions necessary to prove the practical equivalence between the **k** values calculated in US and SI units.

c) How much energy is stored in the shock absorbing spring upon complete compression?

Solution:
Energy stored in the spring is equal to the work performed on the spring. The work performed on the spring, as computed in part (b) above, is:

$\mathbf{W_{spring} = E_{z\text{-}total} - W_{f-xz}}$
= **120,282 ft-lbf – 26,706 ft-lbf**
= **93,576 ft- lbf**

d) What would the steady state speed of the block be after it settles onto the roller conveyor?

Solution/answer:
After the block settles into a steady state condition on the conveyor, it assumes the speed of the conveyor, i.e. **6.56 ft/s.**

e) Rise in the temperature of the compressed air in the shock absorbing cylinder:

Solution:
The rise in the cylinder's air temperature can be determined after calculating the final temperature of the air through the application of the ideal gas law. Ideal gas laws can be applied in this case because air, for most practical purposes, is assumed to act as an ideal gas.

According to ideal gas law:

$$(P_1 \cdot V_1)/T_1 = (P_2 \cdot V_2)/T_2 \qquad \textbf{Eq. 1-32}$$

Or, through rearrangement of Eq. 1-32:

$$T_2 = (P_2 \cdot V_2 \cdot T_1)/(P_1 \cdot V_1) \qquad \textbf{Eq. 1-33}$$

Given or known:

$P_1 = 1$ Atm $= 14.5$ psia
$V_1 = 35.32$ ft^3
$T_1 = \Rightarrow 461 + 68$ °F $= 529$ °R
$P_2 = 2$ Atm. $= 29$ psia
$V_2 = 26.49$ ft^3

Then, by applying **Eq. 1-33:**

$T_2 = \{(29 \text{ psia}) \cdot (26.49 \text{ ft}^3) \cdot (529\ °\text{R})\} / \{(14.5 \text{ psia}) \cdot (35.32 \text{ ft}^3)\}$
$T_2 = \mathbf{793.5}$°R

Or,

$T_2 = 793.5$°R - 461 = **332.5 °F**

Therefore, the rise in the cylinder air temperature would be:

= 332.5 °F - 68 °F
= **264.5** °F

f) If the conveyor, at the bottom of the incline, is a belt driven roller conveyor and the flow of blocks onto the conveyor is one per 10 seconds, determine the horsepower rating of the conveyor motor. Assume the conveyor belt to be directly driven off the conveyor motor shaft and that there is no slip between the belt and the rollers. Assume the motor efficiency to be 90%.

Solution:
Apply the **power, velocity** and **force** formula to determine the power requirement, as follows:

$$\mathbf{P = F \cdot v} \qquad \textbf{Eq. 1-34}$$

Where,

P = Power Required to move the steel blocks
F = Force required to move the block or the force required to move the conveyor belt with the block on the rollers
v = Velocity of the belt; i.e. **6.562 ft/s,** as given.

While the velocity **v** is given, the force **F** is unknown and must be derived. Force can be defined in terms of mass flow rate $\dot{m}$ and the change in velocity Δv, as stated in **Eq. 1-38**:

$$\mathbf{F = (\dot{m}/g_c) \cdot \Delta v} \qquad \textbf{Eq. 1-38}$$

Based on the derived mass of the block as 7850 kg and the fact that 1 block is moved every 10 seconds:

$$\dot{m} = \textbf{mass flow rate} = \textbf{17,342 lbm/10 secs} = \textbf{1,734 lbm/sec}$$

And, based on the given conveyor speed of **6.562 ft/s:**

Δv = **Change in the velocity of the block**
$= v_f - v_{\dot{c}}$
$= 6.562 \text{ ft/s} - 0$
$\therefore \Delta v = 6.562 \text{ ft/s}$ {Note: This change in the velocity is in the direction of the roller conveyor}

Then, by applying **Eq. 1-38,** the force required to move the block would be:

$\mathbf{F = (\dot{m}/g_c) \,.\, \Delta v}$
$= \{(1{,}734 \text{ lbm/sec})/(32 \text{ lbm-ft/lbf-s}^2)\} \,.\, (6.562 \text{ ft/s})$
$= \mathbf{356 \text{ lbf}}$

Then, by applying **Eq. 1-34:**
$\mathbf{P = F \,.\, v}$
$= (356 \text{ lbf}) \,.\, (6.562 \text{ ft/s})$
$= \mathbf{2336 \text{ ft-lbf/s}}$

In hp, the computed power of **2336 ft-lbf/s** would be:
$\mathbf{P = (2336 \text{ ft-lbf/s})/(550 \text{ ft-lbf/s/hp})}$
$= \mathbf{4.25 \text{ hp}}$

Therefore, choose a standard **5 hp** motor.

Note: The efficiency of the motor is not needed in the motor size determination. Motor is specified on the basis of the brake horsepower required by the load; which in this case, is 4.21hp.

g) The conveyor transports the blocks to a cooling / quenching tank. The temperature of the blocks, when they are dropped into the quenching tank, is **212 °F**. The initial temperature of the water in the quench tank is **68 °F** and volume of water is **213.23 ft^3**. The final, equilibrium, temperature of the water and the block is **86 °F**. Determine the amount of heat extracted by the quench water per block?

Solution:
Given or known:

$c_{\text{cast iron}} = \mathbf{0.11 \text{ Btu/lbm °F}}$ or **0.11 Btu/lbm °R** {From **Table 1-5**}
$m_{\text{steel block}} = \mathbf{17{,}342 \text{ lbm}}$, as determined earlier.
$T_{\text{block - i}} = 212 \text{ °F} \Rightarrow 461 + 212 \text{ °F} = \mathbf{673 \text{ °R}}$
$T_{\text{block - f}} = 86 \text{ °F} \Rightarrow 461 + 86 \text{ °F} = \mathbf{547 \text{ °R}}$
$\therefore \Delta T_{\text{block}} = \mathbf{673 \text{ °R} - 547 \text{ °R}} = \mathbf{126 \text{ R}}$

And,
$\Delta T_{\text{block}} = \mathbf{212 \text{ °F} - 86 \text{°F}} = 126 \text{ °F}$

According to **Eq. 1-29**:
$\mathbf{Q = m. \, c. \, \Delta T}$ **Eq. 1-29**

Therefore,

$$Q_{\text{lost by the block}} = \mathbf{(m_{block}) \cdot (c) \cdot (\Delta T_{block})}$$

$$Q_{\text{lost by the block}} = \mathbf{(17{,}342\ lbm) \cdot (0.11 Btu/lbm\ °R) \cdot (-126\ °R)}$$

$$= \mathbf{-\ 240{,}360\ Btu}$$

Since,

$$Q_{\text{absorbed by water}} = -\ Q_{\text{lost by the block}},$$

$$Q_{\text{absorbed by water}} = \mathbf{-\ (-\ 240{,}360\ Btu)}$$

$$= \mathbf{+\ 240{,}360\ Btu}$$

Chapter 1—Self Assessment Problem & Question

1. Determine the amount of heat extracted by the quench water, per block, in Case Study 1-2, ***using the temperature rise of the water*** when the steel block is dropped into the quenching tank. The temperature of the block is **100°C** when it enters the quench water. The initial temperature of the water in the quench tank is **20°C** and volume of water is **6.038 m³**. The final, equilibrium, temperature of the water and the block is **30°C.**

Chapter 2

Thermodynamics and Power

INTRODUCTION

This text focuses not only on the important concepts, theories, principles and analyses techniques associated with thermodynamics but also demonstrates their practical applications through case studies that illustrate the flow of energy from thermal form to utilities such as electricity. Through some of the later chapters in this text we will learn how energy is harvested from fuels, fossil or non-fossil, transferred to a medium like water and packaged in form of the medium's enthalpy—or enthalpy of the superheated steam. In subsequent chapters of this text we will get an opportunity to study the concepts of enthalpy, entropy and work in greater depth. We will also learn how the superheated steam transfers its heat energy to turbines and how the turbines transform the thermal energy (enthalpy) into electrical energy. However, in order to be able to understand practical thermodynamic systems in a comprehensive fashion in the chapters ahead, in this chapter, we will examine what happens to the energy after it is transformed from enthalpy to work performed by the turbine. Since thermodynamic systems are constructed and installed for applications and purposes that extend beyond the boilers and turbines, knowledge and appreciation of flow of energy downstream of the turbines is essential for ensuring that investment in comprehensive power generating systems, as a whole, is productive and effective.

This brief chapter prepares us to better understand the flow of energy beyond the turbines, through examination of energy as it transforms and flows from superheated steam enthalpy form to electrical power delivered onto to the electrical power grid.

Before we embark on the exploration of flow of energy, let's review the concepts of power and efficiency.

POWER AND EFFICIENCY

Power

The concept of power was introduced briefly in the last chapter. We introduced the fact that ***power is rate of performance of work,*** or **P = Work/Elapsed Time** and that one of the ways power can be calculated is through the mathematical relationship **Power = Force x Velocity.** The counterpart of the last power formula, in the ***rotational motion*** realm, would be **Power = Torque x Rotational Velocity.** Other aspects of power, forms of power and formulas for power will be introduced and discussed, in depth, in subsequent chapters of this text.

Units for Power

US/Imperial Unit System: hp, ft-lbf/sec, ft-lbf/min, Btu/sec
SI or Metric: Watts, kW, MW, GW, TW (10^{12} W)

Common Power Conversion Factors in the SI System

1 J/s = 1 N-m/s = 1 W
1 kJ/s = 1 kW
1000 kW = 1 MW
1.055 kJ/s = 1 Btu/s
One hp = 746 Watts = 0.746 kW = 550 ft-lbf/sec

Since the units for power and energy are often confused, let's also examine common units for energy so that the similarities and differences between the units for power and energy can be observed and noted.

Units for Energy

US/Imperial Unit System: ft-lbf, Btu

SI or Metric Unit System

N-m, Joules or J, Wh, kWh, MWh, GWh, TWh (10^{12} Wh)

Common Energy Conversion Factors

1 J = 1 N-m
1 W x 1h = 1 Wh
1 kW x 1h = 1 kWh
1000 kW x 1h = 1 MWh
1 Btu = 1055 J = 1.055 kJ

1 Btu = 778 ft-lbf
1 hp x 1hour = 1 hp-hour

Efficiency

Efficiency is defined, generally, as the ratio of output to input. The output and input could be in form of power, energy or work. Efficiency assumes a more specific definition when considered in the context of a specific form of energy, work or power. The concept of efficiency, when applied in the thermodynamics domain, can involve power, energy or work.

In thermodynamics, when ***power*** is the subject of analysis, efficiency is defined as follows:

Efficiency = η = (Output Power)/(Input Power)
Efficiency in percent = = (Output Power)/(Input Power) x 100

Where,
η (Eta) is a universal symbol for efficiency

Also, in thermodynamics, when energy is the subject of analysis, efficiency is defined as follows:

Efficiency = η = (Output Energy)/(Input Energy)
Efficiency in percent = = (Output Energy)/(Input Energy) x 100

Although work is not used as commonly in the computation of efficiency, in thermodynamics, where applicable, the efficiency calculation based on work would be as follows:

Efficiency = η = (Work Performed by The System)/(Work Performed on System)

Efficiency in Percent = η = (Work Performed by The System)/(Work Performed on System) x 100

As obvious from the definitions of efficiency above, since energy cannot be created, efficiency is always less than 1, or less than 100%. The decimal result for efficiency is often converted to, and stated as, a percentage value.

In the following section, we will explore the relationship between power and efficiency in steam, mechanical and electrical systems, and develop better understanding of the flow of power in steam type electrical power generating systems.

Power—Steam, Mechanical and Electrical

The power delivered by steam to the turbine blades, $\mathbf{P_{steam}}$, in a simplified scenario—with no heat loss, no kinetic head loss, no potential head loss and zero frictional head loss—scenario can be represented by the mathematical relationship stated in form of **Eq. 2-1**. In the context of flow of energy from steam to electricity, functional relationship between electrical power, $\mathbf{P_{Electrical}}$, generator efficiency $\eta_{Generator}$, steam turbine efficiency $\eta_{Turbine}$, and $\mathbf{P_{steam}}$ can be expressed in form of **Eq. 2-2.**

$$P_{steam} = (h_i - h_f) \cdot \dot{m} \qquad \text{Eq. 2-1}$$

$$P_{Electrical} = (P_{steam}) \cdot (\eta_{Turbine}) \cdot (\eta_{Generator}) \qquad \text{Eq. 2-2}$$

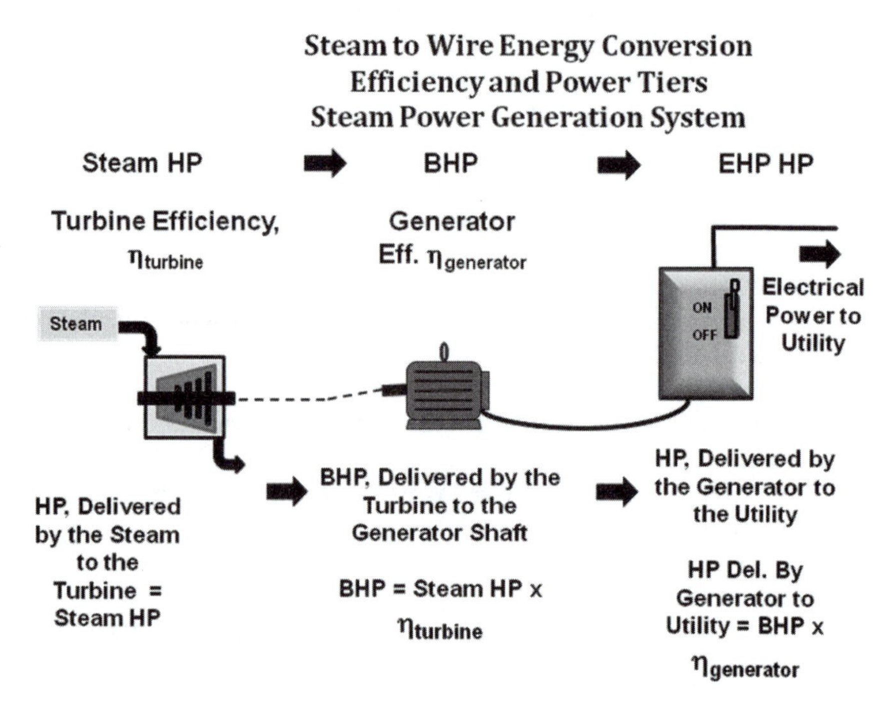

Figure 2-1. Steam to Wire Power Flow—Steam Power Generation System

The flow of power and energy from steam to electricity is depicted, in a power flow diagram, in **Figure 2-1**. This diagram is, essentially a pictorial illustration of the **Eq. 2-2**. The power flow diagram in Figure 2-1 also forms the crux of the scenario analyzed in Case Study 2-1, below.

CASE STUDY 2-1. STEAM TO ELECTRICITY CONVERSION

As an energy engineer, you are charged with the task to estimate the heat content or enthalpy, $\mathbf{h_i}$, of the superheated steam that must be fed to a steam turbine in order to supply **10 MW** (Mega Watt) of electrical power to the electrical grid. Assume that there is no heat loss in the turbine system and that difference between the enthalpies on the entrance and exit ends of the turbine is converted completely into work, with the inefficiency of the turbine accounted for. All of the data available and pertinent to this project is listed below:

- Electrical Power Generator Efficiency: **90%**
- Steam Turbine Efficiency: **70%**
- Mass flow rate for steam, $\dot{\mathbf{m}}$**: 25 kg/s (55 lbm/s)**
- Estimated exit enthalpy, $\mathbf{h_f}$, of the steam: **2875 kJ/kg (1239 Btu/lbm)**

Solution

Solution Strategy

In order to determine the estimated enthalpy, $\mathbf{h_i}$, of the incoming steam, we need to start with the stated output **(10 MW)** of the generator and work our way upstream to derive the energy delivered to the vanes of the turbine. The assumption that there is no heat loss in the turbine system and that the difference between the enthalpies on the entrance and exit ends of the turbine is converted completely into work, with the inefficiency of the turbine, implies that the energy lost by the steam is equal to the net energy delivered to the turbine vanes. Also, note that net energy delivered to the turbine vanes is reduced or derated according to the given efficiency of the turbine.

Solution in SI/Metric Units

Since, $1\text{J}/\text{s} = 1\text{W}$ and $1\ \text{kJ}/\text{s} = 1\text{kW}$,

Power output of the generator = 10 MW = 10,000kW
= **10,000kJ/s**

Brake horsepower delivered by the turbine to the generator, through the turbine shaft, is determined as follows:

BHP = Generator Output/Generator Efficiency
= 10,000kJ/s/0.9
= **1.11 x 10^4 kJ/s or 11,111 kJ/s**

Power delivered by the steam to the turbine vanes is determined as follows:

P_{steam} = BHP/Turbine Efficiency
= (1.11 x 10^4 kJ/s)/0.7
= **1.5873 x 10^4 kJ/s or 15,873 kJ/s**

Of course, we could obtain the same result, in one step, by rearranging and applying Eq. 2-2 as follows:

$$P_{Electrical} = (P_{steam}) \cdot (\eta_{Turbine}) \cdot (\eta_{Generator}) \quad \textbf{Eq. 2-2}$$
$$P_{steam} = P_{Electrical}/\{(\eta_{Turbine}) \cdot (\eta_{Generator})\}$$
$$P_{steam} = (10{,}000 \text{ kJ/s})/\{(0.9) \cdot (0.7)\}$$
$$= 15{,}873 \text{ kJ/s}$$

Since the difference in the turbine entrance and exit enthalpies, in this scenario, is equal to the energy delivered to the turbine vanes:

$$P_{steam} = (h_i - h_f) \cdot \dot{m} \quad \textbf{Eq. 2-1}$$
15,873 kJ/s = (h_i - 2875 kJ/kg) . 25 kg/s
h_i = (15,873 kJ/s)/(25 kg/s) + 2875 kJ/kg
h_i = 3,509 kJ/kg

Solution in US/Imperial Units

Since, 1J/s = 1W and 1 kJ/s = 1kW,

Power output of the generator = 10 MW = 10,000kW
= 10,000kJ/s

Since 1.055 kJ = 1.0 Btu,

Power output of the generator = (10,000kJ/s) . (1/1.055kJ/Btu)
= **9,479 Btu/s**

Brake horsepower delivered by the turbine to the generator, through the turbine shaft, is determined as follows:

BHP = Generator Output/Generator Efficiency
= (9,479 Btu/s)/0.9
= **10,532 Btu/s**

Power delivered by the steam to the turbine vanes is determined as follows:

P_{steam} = **BHP/Turbine Efficiency**
= (10,532 Btu/s)/0.7
= **15,046 Btu/s**

Alternatively, we could obtain the same result, in one step, by rearranging and applying Eq. 2-2 as follows:

$P_{Electrical} = (P_{steam}) \cdot (\eta_{Turbine}) \cdot (\eta_{Generator})$ **Eq. 2-2**
$P_{steam} = P_{Electrical}/\{(\eta_{Turbine}) \cdot (\eta_{Generator})\}$
P_{steam} = (9,479 Btu/s)/{(0.9) . (0.7)}
= **15,046 Btu/s**

Since the difference in the turbine entrance and exit enthalpies, in this scenario, is equal to the energy delivered to the turbine vanes:

$P_{steam} = (h_i - h_f) \cdot \dot{m}$ **Eq. 2-1**
15,046 Btu/s = (h_i - 1239 Btu/lbm) . (55 lbm/s)
h_i = (15,046 Btu/s)/(55 lbm/s) + 1239 Btu/lbm
h_i = **1512 Btu/lbm**

Chapter 2—Self Assessment Problems & Questions

1. As an energy engineer, you are charged with the task to estimate the amount of electrical power produced, in MW, by a steam based power generating plant. Assume that there is no heat loss in the turbine system and that difference between the enthalpies on the entrance and exit ends of the turbine is completely converted into work, less the inefficiency of the turbine. All of the data available, pertinent to this project, is listed below:

- Electrical Power Generator Efficiency: **87%**
- Steam Turbine Efficiency: **67%**
- Mass flow rate for steam, $\dot{m}$: **20 kg/s (44 lbm/s)**
- Exit enthalpy, h_f, of the steam: **2900 kJ/kg (1249 Btu/lbm)**
- Incoming superheated steam enthalpy, h_i: **3586 kJ/kg (1545 Btu/lbm)**

2. Consider the scenario described in Problem (1). Your client has informed you that the power generating plant output requirement has now **doubled**. Based on the concepts and principles learned in Chapter 2, what is the most suitable alternative for doubling the power output if the exit enthalpy, h_f, of the steam must be kept constant at the original 2900 kJ/kg (1249 Btu/lbm) level?

a. Double the mass flow rate, $\dot{m}$, only.
b. Double the incoming superheated steam enthalpy, h_i only.
c. Double the efficiency of the turbine.
d. Double the efficiency of the generator.
e. Increase mass flow rate, $\dot{m}$, incoming superheated steam enthalpy, h_i and increase the efficiency specification on the turbine.

Chapter 3

Study of Enthalpy and Entrophy

INTRODUCTION

Similar to the last chapter, the goal in this brief chapter is to continue the introduction of basic, yet critical, concepts in the field of thermodynamics. In this chapter, we will introduce the concept of ***entropy*** and we will expand on the concept of ***enthalpy***. As we progress through this text, you will notice that the discussion on entropy will be limited, reflecting the somewhat limited role of entropy in practical thermodynamics. On the other hand, our continued exploration of enthalpy, in this chapter, and the ones heretofore, is indicative of the instrumental and ubiquitous role of enthalpy in the study of thermodynamics. We received a brief, preliminary, introduction to enthalpy in the last chapter—in the context of energy flow in power generating realm. In this chapter, we will expand on enthalpy in preparation for its examination in more complex thermodynamic scenarios.

Enthalpy

Enthalpy is defined as the total heat content or total useful energy of a substance. The symbol for enthalpy is "**h**." Enthalpy is also considered to be the sum of internal energy "**u**" and flow energy (or flow work) **p.V**. This definition of enthalpy can be expressed, mathematically, as follows:

$$\mathbf{h = u + p.v} \qquad \textbf{Eq. 3-1}$$

Where,

h = Specific enthalpy, measured in **kJ/kg** (SI Units) or **Btu/lbm** (US Units)

u = Specific internal energy, measured in **kJ/kg** (SI Units) or **Btu/lbm** (US Units)

p = Absolute Pressure measured in **Pa** (SI Units), or **psf** (US Units) v = Specific volume measured in $\mathbf{m^3/kg}$ (SI Units), or $\mathbf{ft^3/lbm}$ (US Units)

p.V = Flow Energy, Flow Work or p-V work, quantified in **kJ/kg** (SI Units) or **Btu/lbm** (US Units)

In practical saturated or superheated steam systems, internal energy, **u,** specific enthalpy, **h**, and specific volume, υ, can be assessed through saturated steam tables and superheated steam tables, respectively. The terms saturated steam and superheated steam are defined in depth later in this text. Chapters 5 and 6 cover classifications of steam and associated steam tables in detail. Reference steam tables, in US and SI form, are included in Appendix B of this text.

In order to maintain consistency of units in practical thermodynamic situations, where computation is performed in **US units**, a more suitable form of the enthalpy equation **Eq. 3-1** would be as follows:

$$h = u + p.V/J \qquad \textbf{Eq. 3-2}$$

Where,

h = Enthalpy, measured in **Btu's**
u = Internal energy, measured in **Btu**
p = Absolute Pressure measured in **psf** or $\mathbf{lbf/ft^2}$
V = Volume measured in $\mathbf{ft^3}$
J = Joule's constant; value of **J is 778 ft-lbf/Btu**

Note that in SI unit system, an alternate version of enthalpy equation Eq. 3-1 is not necessary because units in Eq. 3-1 are congruent.

Enthalpy can also be quantified in molar form. In molar form, enthalpy is referred to as ***molar enthalpy*** and represented by the symbol "**H**". The units for molar enthalpy **H** are **Btu/lbmole**, in the US system, and are **kJ/kmole**, in the Metric or SI System. Where a mole of a substance is defined or calculated through division of the mass of that substance by the atomic weight of the substance, if it is a solid, or by the molecular weight, if it is a liquid or gas.

The mathematical equation for **molar enthalpy "H,"** is as follows:

$$\mathbf{H = U + p.V} \qquad \textbf{Eq. 3-3}$$

Where,

U = **Molar Internal Energy,** can be expressed in **Btu/lbmol** (US Units) or **kJ/kmol** (SI Units)

p = Absolute pressure measured in **Pa** (SI Units), **psf** (US Units) or **lbf/ft^2**

V = Molar specific volume measured in **m^3/kmol** (SI Units), or **ft^3/lbmole** (US Units)

EXAMPLE 3-1

Calculate the absolute enthalpy, ***h***, in **Btu's**, for **1 lbm of vapor** under the following conditions:

h = Enthalpy, measured in **Btu's =?**
u = **1079.9 Btu/lbm**
p = **14.14 psia**
V = **27.796 ft^3**
J = Joule's constant; value of **J is 778 ft-lbf/Btu**

Solution

The pressure is given in psia, or **lbf/in^2**. In order to streamline the pressure for application in **Eq. 3-2**, we must convert in into **lbf/ft^2**.

Therefore,

p = **(14.14 lbf/in^2).(144 in^2/ft^2)**
= **2,036 lbf/ft^2**

Then, by applying **Eq. 3-2**, and by substitution of known and derived values:

$h = u + p.V/J$ **Eq. 3-2**
h = **1079.9 Btu/lbm + (2,036 lbf/ft2). (27.796 ft^3)/778 ft-lbf/Btu**
h = **1152.67 Btu**

Entropy

Entropy is defined as the non-work producing form of energy. It is also regarded as the energy that is not available for performing useful work within a certain environment. The symbol for entropy is "**s.**" Some facts, principles and laws associated with entropy are summarized below:

- Increase in entropy is referred to as entropy production.

  The total absolute entropy of a system is said to be equal to the sum of all absolute entropies that have occurred over the life of the system.

  $$s_{total} = \sum \Delta s_i \qquad \textbf{Eq. 3-4}$$

  Where, Δs_i represents change in enthalpy at each object or in each substance. Application of this entropy principle will be demonstrated through Case Study 3-1

- According to the **third law of thermodynamics**, the absolute entropy of a perfect crystalline solid, in thermodynamic equilibrium, approaches zero as the temperature approaches absolute zero.

  $$s = 0$$

- In an ***isothermal*** (constant temperature) process, the entropy production, Δs, is a function of the energy transfer rate:

  $$\Delta s = q/T_{abs} \qquad \textbf{Eq. 3-5}$$

Where,

s = entropy in **kJ/kg.°K** (SI Units System), or in **Btu/lbm.°R** (US Unit System)

q = Heat transferred in **kJ/kg**, (SI Units) or **Btu/lbm** (US Units)

T_{abs} = Absolute Temperature of the object or substance, in °**K** (SI Units System), or in °**R** (US Unit System)

CASE STUDY 3-1. ENTROPY ANALYSIS

In a certain solar system there are four (4) planets oriented in space as shown in Figure #2. Their temperatures are indicated in the diagram, in °K as well as in °R. As apparent from the orientation of these planets in Figure 3-1, they are exposed to each other such that heat transfer can occur freely through radiation. All four (4) planets are assumed to be massive enough to allow for the interplanetary heat transfer to be isothermal for each of the planets.

a) Will heat transfer occur, through radiation, from planet Z to planets X and Y?

b) If the 3,000 kJ/kg of radiated heat transfer occurs from planet X to planet Y, what would be the entropy changes at each of the two planets?

c) Can convectional heat transfer occur between any of two planets in this solar system?

d) If certain radiated heat transfer between Planets Y and Z causes an entropy change of **–11.77 kJ/kg°K** at Planet Y and an entropy change of **12.66 kJ/kg.°K** at Planet Z, what would be the overall, resultant, entropy of this planetary system?

e) Can planet X be restored to its original state? If so, how?

Solution—Case Study 3-1:

a) Will heat transfer occur, through radiation, from planet Z to planets X and Y?

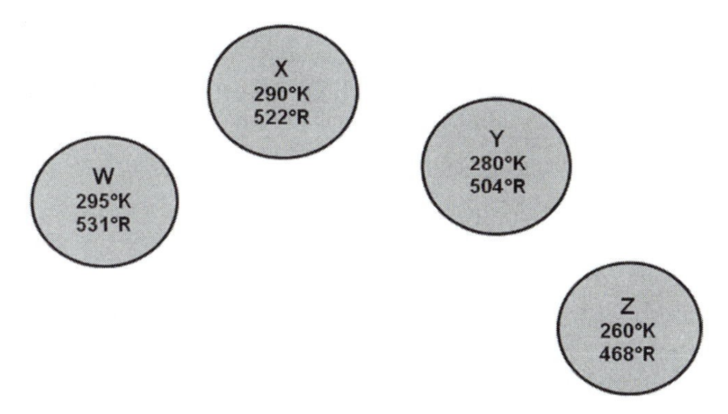

Figure 3-1. Case Study 3-1, Entropy

Solution/Answer:

Heat flows ***from a body at higher temperature to one that is at a lower temperature***. The temperature of Planet Z is lower than the temperature of planets X and Y. Therefore, **NO radiated heat transfer will occur from planet Z to planets X and Y.**

b) If the **3000 kJ/kg** of radiated heat transfer occurs from planet X to planet Y, what would be the entropy changes at each of the two planets?

Solution/Answer:

In an ***isothermal*** (constant temperature) process, the entropy production, **Δs**, is a function of the energy transfer rate, and its relationship with heat **q** and absolute temperature, $\mathbf{T_{abs}}$ is represented by **Eq. 3-5:**

$$\Delta s = q/T_{abs} \qquad \textbf{Eq. 3-5}$$

$\therefore \Delta s_X = (-\,3{,}000\ kJ/kg)/(290°K)$
$= -\,10.34\ kJ/kg.°K$ {Due to heat loss by Planet X}

And,

$\Delta s_Y = (+3{,}000\ kJ/kg)/(280\ °K)$
$= +\,10.71\ kJ/kg.°K$ {Due to heat gain by Planet Y}

c) Can convectional heat transfer occur between any of two planets in this solar system?

Solution/Answer:

Convectional heat transfer is dependent on bulk movement of a fluid (gaseous or liquid) and, therefore, it can ***only*** occur in liquids, gases and multiphase mixtures. Since, the system in this problem is a planetary system, the medium between the bodies is devoid of fluids needed for convectional heat transfer. Heat transfer in this planetary system occurs through radiation, primarily.

Therefore, the answer is *NO*.

d) If the heat transfer from part (b) occurs simultaneous to a certain radiated heat transfer between Planets Y and Z—where the entropy change of - 11.77 kJ/kg. °K is recorded at Planet Y and an entropy change of12.66 kJ/kg.°K is recorded at Planet Z—what would be the overall, resultant, entropy of this planetary system?

Solution/Answer:

Overall $\Delta s_{\text{Planetary System}} = \sum(\Delta s_i)$

$\therefore$ Overall $\Delta s_{\text{Planetary System}} = \Delta s_X + \Delta s_Y + \Delta s_{YZ} + \Delta s_Z$

Or,

$\Delta s_{\text{Planetary System}}$ = -10.34 kJ / kg.°K + 10.71 kJ / kg.°K - 11.77 kJ / kg.°K + 12.66 kJ / kg. °K

$\therefore$ Overall $\Delta s_{\text{Planetary System}}$ = + 1.2643 kJ/kg. °K

e) Can planet X be restored to its original state? If so, how?

Solution/Answer:

Planet X ***can*** be restored to its original state; through absorption of **3,000 kJ/kg of (specific) heat energy**. Note: Conditions stipulated in part (d) do not pertain to part (e).

Chapter 3—Self Assessment Problems & Questions

1. Calculate the volume **1 kg** of vapor would occupy under the following conditions:

h = **2734 kJ**
u = **2550 kJ**
p = **365.64 kPa = 365.64 kN/m^2**
V = ?

2. In a certain solar system there are four (4) planets oriented in space as shown in Figure 3-1. As apparent from the orientation of these planets, they are exposed to each other such that heat transfer can occur freely through radiation. All four (4) planets are assumed to be massive enough to allow for the interplanetary heat transfer to be an isothermal phenomenon for each of the planets. ***Perform all computation in the US Unit System.***

a. If the **1,300 Btu/lbm** of radiated heat transfer occurs from **planet X to planet Y,** what would be the entropy changes at each of the two planets?

b. If a certain radiated heat transfer between Planets Y and Z causes an entropy change of **-2.9 Btu/lbm.°R** at Planet **Y** and an entropy change of **3.1 Btu/lbm.°R** at Planet **Z**, what would be the overall, resultant, entropy of this planetary system?

3. If the mass of vapor under consideration in problem 1 were tripled to 3 kg, what would be the impact of such a change on the volume?

4. Would **Eq. 3-2** be suitable for calculation of enthalpy if all available data are in **SI (Metric) units**?

Chapter 4

Understanding Mollier Diagram

INTRODUCTION

Mollier diagram is named after Richard Mollier (1863-1935), a German professor who pioneered experimental research on thermodynamics associated with water, steam and water-vapor mixture. Mollier diagram is a graphical representation of functional relationship between enthalpy, entropy, temperature, pressure and quality of steam. Mollier is often referred to as **Enthalpy-Entropy Diagram** or **Enthalpy-Entropy Chart**. The enthalpy-entropy charts in Appendix B are Mollier Diagrams. It is used commonly in the design and analysis associated with power plants, steam turbines, compressors, and refrigeration systems.

Mollier diagram is available in two basic versions: The SI/Metric Unit version and the US/Imperial Unit version. **Figure 4-1** depicts the SI/Metric version of the Mollier diagram. The US and SI versions of the Mollier diagram are included in Appendix B. The abscissa (horizontal or x-axis in a Cartesian coordinate system) and ordinate (vertical or y-axis in a Cartesian coordinate system) scales represent **entropy** and **enthalpy**, respectively. Therefore, Mollier diagram is also referred to as the **Enthalpy-Entropy Chart**.

The constant pressure and constant temperature lines in the Mollier diagram are referred to as isobars and isotherms, respectively. In addition, the graph includes lines representing constant steam quality, "**x**," in the bottom half of the diagram. The bold line, spanning from left to right, in the lower half of Mollier diagram is the **saturation line**. The saturation line, labeled as **x = 1,** represents the set of points on Mollier diagram where the steam is 100% vapor. All points above the saturation line are in the **superheated steam realm**. All points below the saturation line represent a **mixture of liquid and vapor phases.** The concept of quality is explained and illustrated in Chapter 5.

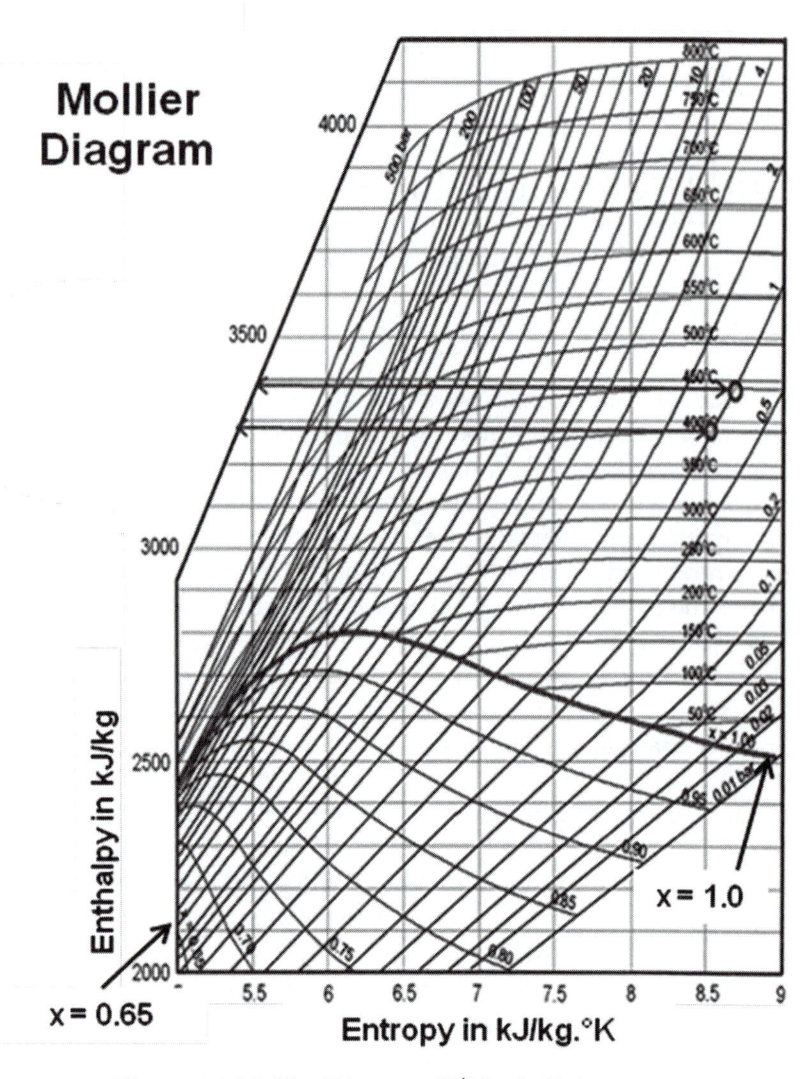

Figure 4-1. Mollier Diagram, SI/Metric Units

A comparison of the Mollier diagram and the psychrometric chart reveals convincing similarity between these two versatile and commonly applied thermodynamics tools. Some schools of thought explain the process of transformation of the Mollier diagram to the psychrometric chart on the basis of geometric manipulation. This relationship between Mollier diagram and the psychrometric chart is apparent from the fact that both involve critical thermodynamic properties such as **enthalpy, temperature, sensible heat, latent heat and quality.**

A comparison of the **Mollier diagram** and the **steam tables** also reveals a marked similarity and equivalence between the two. This equivalence is illustrated through **Example 5-6** in Chapter 5. The reader would be better prepared to appreciate the illustration of relationship between the Mollier diagram and the steam tables after gaining a clear comprehension of the saturated and superheated steam tables in Chapter 5. Example 5-6 demonstrates the interchangeability of the Mollier diagram and the **Superheated Steam Tables** as equivalent tools in deriving the enthalpy values associated with the change in the temperature of superheated steam. This equivalence between Mollier diagram and the steam tables is further reinforced by the fact that ***both*** involve critical thermodynamic properties of steam such as **enthalpy, entropy, temperature and pressure.**

APPLICATION OF MOLLIER DIAGRAM

A common application of Mollier diagram involves determination of an unknown parameter among the key Mollier diagram parameters such as, ***enthalpy, entropy, temperature pressure and quality***. Typical applications of Mollier diagram are illustrated through the example problems that follow.

EXAMPLE 4-1

Determine if steam at 450°C and 1 bar is saturated or superheated. Find the enthalpy and entropy of this steam.

Solution:

See the Mollier diagram in **Figure 4-2.** Identify the point of intersection of the 450°C line (or 450°C isotherm) and the constant pressure line (or isobar) of 1 bar. This point of intersection of the two lines is labeled **A**. As explained above, this region of the Mollier diagram is the superheated steam region.

Therefore, the steam at 450°C and 1 bar is superheated.

Enthalpy Determination

To determine the enthalpy at point **A**, draw a straight horizontal line from point A to the left till it intersects with the diagonal enthalpy line. This horizontal line intersects the enthalpy line at an enthalpy value of, approximately, 3380 kJ/kg.

Therefore, h_A, or enthalpy at point A, is 3380 kJ/kg.

Entropy Determination

To determine the entropy at point A, draw a straight vertical line from point **A** to the bottom, until it intersects with the entropy line. The vertical line intersects the entropy line at, approximately, 8.7kJ/kg.°K.

Therefore, s_A, or entropy at point A, is 8.7kJ/kg.°K.

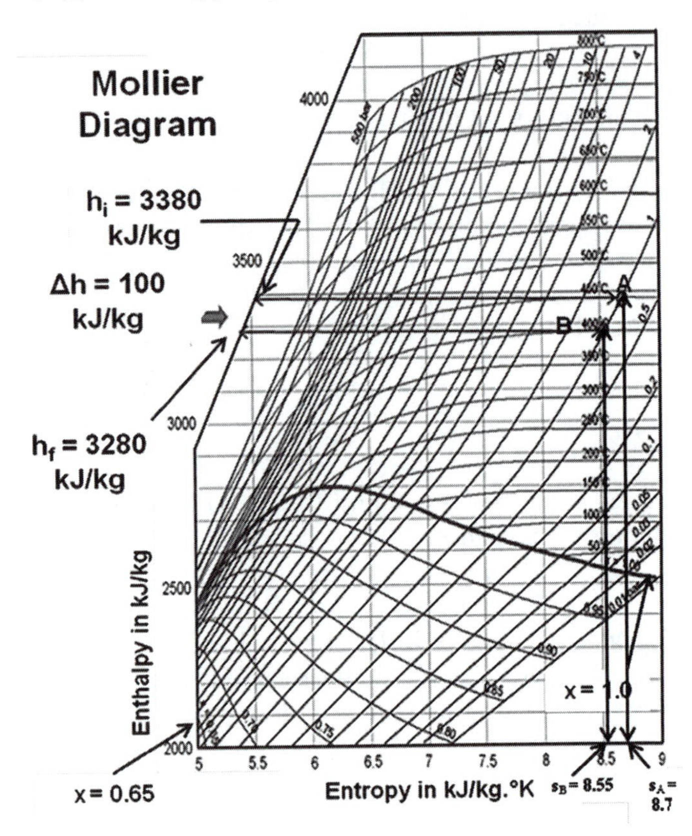

Figure 4-2. Mollier Diagram, SI/Metric Units

EXAMPLE 4-2

Determine the amount of heat that must be removed from a system, on per kg basis, in order to reduce the temperature of steam from **450°C,** at **1 Atm.** to **400°C,** at **1 Atm.**

Solution:

To determine the amount of heat that must be removed from the steam in order to cool the steam from **450°C,** at **1 Atm.** to **400°C,** at **1 Atm,** we must assess the enthalpies at those two points.

The first point, at **450°C and 1 Atm,** was labeled as point **A** in Example 4-1. The enthalpy, $\mathbf{h_i}$ at point A was determined to be **3380 kJ/kg.**

The enthalpy, $\mathbf{h_f}$, at the second point—referred to as point **B**—as shown on the Mollier diagram in Figure 4-2, is **3280 kJ/kg.**

Therefore, the amount of heat that must be removed from the system in order to lower the temperature from **450°C to 400°C, at 1 Atm,** would be:

$$\Delta h = h_i - h_f$$

$$\therefore \Delta h = h_i - h_f = 3380 \text{ kJ/kg} - 3280 \text{ kJ/kg} = 100 \text{ kJ/kg}$$

In other words, **100 kJ of heat must be removed from each kg of steam** in order to cool it from **450°C,** at **1 Atm.** to **400°C,** at **1 Atm.**

Chapter 4—Self Assessment Problems & Questions

1. Using the Mollier diagram, find the entropy of steam at 400°C and 1 Atm.

2. Heat is removed from a thermodynamic system such that the temperature drops from **450°C,** at **1 Atm** to **150°C,** at **1 Atm.** Determine the following:

a) The new, or final, Enthalpy
b) The new entropy
c) The state of steam at **150°C** and **1 Atm**

Chapter 5

Saturated and Superheated Steam Tables

INTRODUCTION

In this text, as we study various topics of thermodynamics, we will utilize and focus on two main categories of steam tables: (1) The Saturated Steam Tables and (2) The Superheated Steam Tables.

Appendix B of this text includes the compact version of the saturated steam tables and the superheated steam tables. These tables are referred to as the compact version because they do not include certain properties or attributes that are customarily included only in the detailed or comprehensive version. Characteristics or properties included in most comprehensive version of the saturated steam tables, but omitted in Appendix B steam tables, are as follows:

1) Internal energy "**U**."
2) The heat of vaporization "$\mathbf{h_{fg}}$."

Internal energy, absolute and specific, is not required in most common thermodynamic analysis. And, heat of vaporization, $\mathbf{h_{fg}}$ for water—as explained in Chapter 6—is a derivative entity. In that, $\mathbf{h_{fg}}$ can be derived from $\mathbf{h_L}$ and $\mathbf{h_V}$ as stipulated by **Eq. 5-1** :

$$\mathbf{h_{fg} = h_V - h_L} \qquad \textbf{Eq. 5-1}$$

EXAMPLE 5-1

Using the saturated liquid enthalpy value for $\mathbf{h_L}$ and the saturated vapor enthalpy value for $\mathbf{h_V}$, at 1 MPa and 180°C, as listed in the saturated steam table excerpt in **Table 5-1**, verify that $\mathbf{h_{fg}}$ **= 2015 kJ/kg.**

Solution:
As stated in **Eq. 5-1**:

$$h_{fg} = h_V - h_L$$

As read from **Table 5-1**:

$$h_V = 2777 \text{ kJ/kg, and}$$
$$h_L = 762.68 \text{ kJ/kg}$$

$$\therefore h_{fg} = h_V - h_L$$
$$= 2777 - 762.68$$
$$= 2014.32 \text{ kJ/kg}$$

The value for h_{fg}, at 1.0 MPa and 180°C, as listed in **Table 5-1**, is **2015 kJ/kg**, versus the derived value of **2014 kJ/kg**. The difference between the calculated value of h_{fg}, at 1.0 MPa and 180°C and the value listed in **Table 5-1** is only 0.05% and is, therefore, negligible. Hence, we can say that **Eq. 5-1** stands verified as a tool or method for deriving the heat of vaporization h_{fg} from the compact version of steam tables included in Appendix B.

The saturated and superheated steam tables in Appendix B are presented in the US/Imperial unit realm as well as the SI/Metric realm. Note that in this chapter—as well as other chapters in this text—for the

Table 5-1. Properties of Saturated Steam, by Pressure, SI Units

Properties of Saturated Steam By Pressure Metric/SI Units									
Abs. Press. MPa	Temp. °C	Specific Volume m^3/kg		Enthalpy kJ/kg			Entropy kJ/kg		Abs. Press. MPa
		Sat. Liquid v_L	Sat. Vapor v_V	Sat. Liquid h_L	Evap. h_{fg}	Sat. Vapor h_V	Sat. Liquid s_L	Sat. Vapor s_V	
0.010	45.81	0.0010103	14.671	191.81	2392.8	2583.9	0.6492	8.1489	0.010
0.10	99.61	0.0010431	1.6940	417.44	2258.0	2674.9	1.3026	7.3588	0.10
0.20	120.21	0.0010605	0.88574	504.68	2201.9	2706.2	1.5301	7.1269	0.20
1.00	179.89	0.0011272	0.19435	762.68	2015.3	2777.1	2.1384	6.5850	1.00

readers' convenience, saturated steam table excerpts include the heat of vaporization, $\mathbf{h_{fg}}$, values. See **Tables 5-1, 5-3, 5-4, and 5-5.**

Also, for illustration of various numerical examples, and thermodynamics discussion in general, excerpts from the superheated steam tables in Appendix B, are included in this chapter in form of **Tables 5-2, 5-6 and 5-7.**

SATURATED STEAM TABLES

Saturated water and steam tables, as presented in Appendix B, are categorized as follows:

A. Saturated water and steam tables, by temperature, in US Units
B. Saturated water and steam tables, by pressure, in US Units
C. Saturated water and steam tables, by temperature, in SI/Metric Units
D. Saturated water and steam tables, by pressure, in SI/Metric Units

A. Saturated water and steam tables, by temperature, in US Units

As apparent from the inspection of the four categories of saturated steam tables above, two distinguishing factors between these categories of tables are temperature and pressure. First category of tables, listed under bullet A, represents saturated water and steam data by ***temperature***, in US Units. In other words, this set of tables is used when temperature is the determining factor, or when the current or future state of the saturated water or saturated steam is premised on, or defined by, the temperature. So, if saturated water or saturated steam is said to exist at a given temperature, the following properties can be identified:

a) **Saturation pressure,** in psia, at the given temperature, in °F.

b) **Specific volume,** $\boldsymbol{\nu_L}$, in ft³/lbm, of saturated liquid, at the given temperature and saturation pressure.

c) **Specific volume,** $\boldsymbol{\nu_v}$, in ft³/lbm, of saturated vapor, at the given temperature and saturation pressure.

d) **Specific enthalpy,** $\mathbf{h_L}$, in Btu/lbm, of saturated liquid, at the given temperature and saturation pressure.

e) **Specific enthalpy, h_v,** in Btu/lbm, of saturated vapor, at the given temperature and saturation pressure.

f) **Specific entropy, s_L,** in Btu/lbm-°R, of saturated liquid, at the given temperature and saturation pressure.

g) **Specific entropy, s_v,** in Btu/lbm-°R, of saturated vapor, at the given temperature and saturation pressure.

B. Saturated water and steam tables, by pressure, in US Units

The second category of tables represents saturated water and steam data by ***pressure,*** in US Units. In other words, this set of tables is used when pressure is the determining factor, or when the current or future state of the saturated water or saturated steam is defined by the pressure. So, if saturated water or saturated steam is said to exist at a given pressure, the following properties can be identified:

a) **Saturation temperature,** in °F, at the given pressure, in psia.

b) **Specific volume, ν_L,** in ft^3/lbm, of saturated liquid, at the given pressure and saturation temperature.

c) **Specific volume, ν_v,** in ft^3/lbm, of saturated vapor, at the given pressure and saturation temperature.

d) **Specific enthalpy, h_L,** in Btu/lbm, of saturated liquid, at the given pressure and saturation temperature.

e) **Specific enthalpy, h_v,** in Btu/lbm, of saturated vapor, at the given pressure and saturation temperature.

f) **Specific entropy, s_L,** in Btu/lbm-°R, of saturated liquid, at the given pressure and saturation temperature.

g) **Specific entropy, s_v,** in Btu/lbm-°R, of saturated vapor, at the given pressure and saturation temperature.

C & D

Saturated steam tables categorized as C and D above are similar to categories A and B, with the exception of the fact that the temperature, pressure, specific volume, enthalpy and entropy are in the **metric unit system.**

SUPERHEATED STEAM TABLES

Superheated steam tables, as presented in Appendix B, are categorized as follows:

a) Superheated steam tables in US Units
b) Superheated steam tables in SI/Metric Units

Unlike the saturated steam tables, regardless of the unit system, the superheated steam tables differ from the saturated steam tables as follows:

a) Superheated steam tables, such as the ones included under Appendix B, provide only the **specific volume**, **enthalpy** and **entropy,** for a given set of temperature and pressure conditions.

b) Retrieval of specific values of enthalpy and entropy from the superheated steam tables requires knowledge of the exact temperature and pressure.

c) When the exact temperature and pressure for a given superheated steam condition are not available or listed in the superheated steam tables, **single or double interpolation** is required to identify the specific volume, enthalpy and entropy.

SINGLE AND DOUBLE INTERPOLATION OF STEAM TABLE DATA

Interpolation is often required when the retrieving data from tables such as the Saturated Steam Tables or the Superheated Steam Tables. Interpolation, is needed when the given pressure or temperature don't coincide with the standard pressure and temperature values on the given tables.

Example 5-2 offers an opportunity to study the interpolation method, in the US unit realm. Even though the interpolation method is being illustrated on the basis of steam tables in this chapter, this technique can be employed for interpolation of other types of tabular data, as well.

EXAMPLE 5-2

Calculate the enthalpy of 450 psia and 950°F superheated steam.

Solution:

As you examine the superheated steam tables for these parameters, in Appendix B, you realize that exact match for this data is not available in the table. See **Tables 5-2 and 5-3** for excerpts from the superheated steam tables in Appendix B.

While the given pressure of 450 psia is listed, the stated temperature of 950°F is not listed. Therefore, the enthalpy for 450 psia and 950°F superheated steam and must be derived by applying interpolation to the enthalpy data listed in the tables for 900°F and 1,000°F.

The formula for ***single interpolation***, applied between the stated or available enthalpy values for 900°F and 1000°F, at 450 psia, is as follows:

$$\mathbf{h}_{950\,°F,\,450\,psia} = ((\mathbf{h}_{1000\,°F,\,450\,psia} - \mathbf{h}_{900\,°F,\,450\,psia})/(1000°F - 900°F)).(950-900) + \mathbf{h}_{900\,°F,\,450\,psia}$$

By substituting enthalpy values and other given data from superheated steam table excerpt, shown in **Table 5-2**:

$$\mathbf{h}_{950\,°F,\,450\,psia} = ((1522.4\ Btu/lbm - 1468.6\ Btu/lbm)/(1000°F - 900°F)).(950-900) + 1468.6\ Btu/lbm = 1496\ Btu/lbm$$

Note: The available enthalpy values are circled in **Table 5-2**.

Example 5-3 offers an opportunity to study the **double interpolation method**. As is the case with single interpolation method, even though the double interpolation method is being illustrated on the basis of steam tables in this chapter, this technique can be employed for ***double interpolation*** of other types of tabular data, as well.

Table 5-2. Superheated Steam Table Excerpt, US/Imperial Units

Properties of Superheated Steam					
US/Imperial Units					
Abs.		**Temp.**	**Note: v is in ft³/lbm, *h* is in Btu/lbm**		
Press.		**°F**		**and *s* is in BTU/(lbm-°R)**	
psia					
(Sat. Temp. °F)		500	700	900	1000
260	*v*	2.062	2.5818	3.0683	3.3065
(404.45)	*h*	1262.5	1370.8	1475.2	1527.8
	s	1.5901	1.6928	1.7758	1.8132
360	*v*	1.446	1.8429	2.2028	2.3774
(434.43)	*h*	1250.6	1365.2	1471.7	1525
	s	1.5446	1.6533	1.7381	1.7758
450	*v*	1.1232	1.4584	1.7526	1.8942
(456.32)	*h*	1238.9	1360	1468.6	1522.4
	s	1.5103	1.6253	1.7117	1.7499
600	*v*			1.3023	1.411
(486.25)	*h*			1463.2	1518
	s			1.577	1.7159

EXAMPLE 5-3

Calculate the enthalpy of 405 psia and 950°F superheated steam.

Solution:

As you examine the superheated steam tables for these parameters, in Appendix B, you realize that exact match for this data is not available. See **Table 5-2** for an excerpt of the superheated steam tables in Appendix B.

In this example, neither the given pressure of 405 psia nor the stated temperature of 950°F is listed in Appendix B superheated steam tables. Therefore, the enthalpy for 405 psia and 950°F superheated steam and must be derived by applying double interpolation to the enthalpy data listed in the Table 5-2 for 360 psia[1], 450 psia, 900°F and 1,000°F.

[1]**Note:** Since the enthalpy data for 400 psia is available in Appendix B, double interpolation could be performed on 400 psia and 450 psia points ***yielding the same results***. The lower pressure point of 360 psia is chosen in this example simply to maintain continuity with the superheated steam table excerpt in **Table 5-2**.

The double interpolation approach, as applied here, will entail three steps.

The first step involves determination of $\mathbf{h}_{900\,°F,\,405\,psia}$, the enthalpy value at 405 psia and 900°F. The enthalpy values available and used in this first interpolation step are circled in Table 5-2. The following formula sums up the mathematical approach to this first step:

$$\mathbf{h}_{900\,°F,\,405\,psia} = ((\mathbf{h}_{900\,°F,\,360\,psia} - \mathbf{h}_{900\,°F,\,450\,psia})/(450\text{ psia} - 360\text{ psia})).(450\text{ psia} - 405\text{ psia}) + \mathbf{h}_{900\,°F,\,450\,psia}$$

Substituting enthalpy values and other given data from superheated steam table excerpt, shown in **Table 5-2**:

$$\mathbf{h}_{900\,°F,\,405\,psia} = ((1471.7\text{ Btu/lbm} - 1468.6\text{ Btu/lbm})/(450\text{ psia} - 360\text{ psia})).(450\text{ psia} - 405\text{ psia}) + 1468.6\text{ Btu/lbm}$$

$$= 1470\text{ Btu/lbm}$$

Second interpolation step involves determination of $\mathbf{h}_{1000\,°F,\,405\,psia}$, the enthalpy value at 405 psia and 1000°F. The enthalpy values available and used in this interpolation step are circled in Table 5-2. The following formula sums up the mathematical approach associated with this interpolation step:

$$\mathbf{h}_{1000\,°F,\,405\,psia} = ((\mathbf{h}_{1000\,°F,\,360\,psia} - \mathbf{h}_{1000\,°F,\,450\,psia})/(450\ \text{psia} - 360\ \text{psia})).(450\ \text{psia} - 405\ \text{psia}) + \mathbf{h}_{1000\,°F,\,450\,psia}$$

Substituting enthalpy values and other given data from superheated steam table excerpt, shown in **Table 5-2**:

$$\mathbf{h}_{1000\,°F,\,405\,psia} = ((1525\ \text{Btu/lbm} - 1522.4\ \text{Btu/lbm})/(450\ \text{psia} - 360\ \text{psia})).(450\ \text{psia} - 405\ \text{psia}) + 1522.4\ \text{Btu/lbm} = 1524\ \text{Btu/lbm}$$

The final step in the double interpolation process, as applied in this case, involves interpolating between $\mathbf{h}_{1000\,°F,\,405\,psia}$ and $\mathbf{h}_{900\,°F,\,405\,psia}$, the enthalpy values derived in the first two steps above, to obtain the desired final enthalpy $\mathbf{h}_{950\,°F,\,405\,psia}$.

The formula for this final step is as follows:

$$\mathbf{h}_{950\,°F,\,405\,psia} = ((\mathbf{h}_{1000\,°F,\,405\,psia} - \mathbf{h}_{900\,°F,\,405\,psia})/(1000°F - 900°F)).(950°F - 900°F) + \mathbf{h}_{900\,°F,\,405\,psia}$$

Substituting enthalpy values derived in the first two steps above:

$$\mathbf{h}_{950\,°F,\,405\,psia} = ((1524\ \text{Btu/lbm} - 1470\ \text{Btu/lbm})/(1000°F - 900°F)).(950°F - 900°F) + 1470\ \text{Btu/lbm} = 1497\ \text{Btu/lbm}$$

EXAMPLE 5-4

Determine the enthalpy of saturated water at 20°C and 1 Bar.

Solution:

The saturation temperature at 1 Bar, 1 Atm, or 101 kPa, as stated in the saturated steam tables in Appendix B, is 99.6°C or, approximately, 100°C. The saturated water in this problem is at 20°C; well below the saturation temperature. Therefore, the water is in a subcooled state.

In the subcooled state, saturated water's enthalpy is determined by its temperature and not the pressure. Hence, the enthalpy of saturated water at 20°C must be retrieved from the temperature based saturated steam tables.

From Appendix B, and as circled in Table 5-3, the enthalpy of saturated water at 20°C is **83.92 kJ/kg [2].**

[2]**Note**: Since the water is referred to as "***saturated*** ***water***" and is clearly identified to be subcooled, the enthalpy value selected from the tables is $\mathbf{h_L}$ and not $\mathbf{h_V}$.

QUALITY OF STEAM CONSIDERATION IN THERMODYNAMIC CALCULATIONS:

In thermodynamics, there are myriad scenarios where water exists, simultaneously, in liquid and vapor forms. In such conditions, the concept of ***quality*** of steam plays a vital role. Quality, as described earlier is the ratio of the mass of vapor and the total mass of vapor and liquid. Mathematically, quality is defined as follows:

$$Quality = x = \frac{m_{vapor}}{m_{vapor} + m_{liquid}}$$

Where,

$\mathbf{x}$ = Quality, or quality factor

$\mathbf{m_{vapor}}$ = mass of vapor in the liquid and vapor mixture

$\mathbf{m_{liquid}}$ = mass of liquid in the liquid and vapor mixture

$\mathbf{m_{liquid} + m_{vapor}}$ = Total mass of the liquid and vapor mixture

When quality of steam is less than one (1), or less than100%, determination of enthalpy—and other parameters that define the state of water under those conditions—requires consideration of the proportionate amounts of saturated water (liquid) and vapor. For instance, if the

Table 5-3. Properties of Saturated Steam, by Temperature, SI Units

Properties of Saturated Steam By Temperature Metric/SI Units										
Temp. °C	Abs. Press. MPa	Specific Volume m^3/kg		Enthalpy kJ/kg			Entropy kJ/kg		Temp. °C
		Sat. Liquid v_L	Sat. Vapor v_V	Sat. Liquid h_L	Evap. h_{fg}	Sat. Vapor h_V	Sat. Liquid s_L	Sat. Vapor s_V	
20	0.002339	0.0010018	57.7610	83.920	2454.1	2537.5	0.2965	8.6661	20
50	0.012351	0.0010121	12.0280	209.34	2382.7	2591.3	0.7038	8.0749	50
100	0.101420	0.0010435	1.6719	419.10	2257.0	2675.6	1.3070	7.3541	100
200	1.554700	0.0011565	0.1272	852.39	1940.7	2792.1	2.3308	6.4303	200

quality of steam is 50%, determination of total enthalpy would entail 50% of the enthalpy contribution from saturated vapor and 50% from saturated water (liquid). This principle is formulated mathematically through equations 5-2, 5-3, 5-4 and 5-5, and illustrated through the Example 5-5.

The basic formulae for computing enthalpy, entropy, internal energy and specific volume when quality of steam is less than 100% are as follows:

$$\mathbf{h_x = (1\text{-}x).\ h_L + x.\ h_V} \quad \textbf{Eq. 5-2}$$

$$\mathbf{s_x = (1\text{-}x).s_L + x.s_v} \quad \textbf{Eq. 5-3}$$

$$\mathbf{u_x = (1\text{-}x).u_L + x.u_v} \quad \textbf{Eq. 5-4}$$

$$\boldsymbol{\nu_x = (1\text{-}x).\nu_L + x.\nu_v} \quad \textbf{Eq. 5-5}$$

EXAMPLE 5-5

Determine the enthalpy and specific volume for 100 psia steam with a quality of 55%.

Solution:

Given:

Quality, x = **0.55**
Absolute Pressure = **100 psia**

From saturated steam tables in Appendix B, and the excerpt in Table 5-4, the values of enthalpies and specific volumes, at 100 psia, are:

$\mathbf{h_L}$ = **298.57** Btu/lbm
$\mathbf{h_V}$ = **1187.5** Btu/lbm
$\boldsymbol{\nu_L}$ = **0.017736** ft^3/lbm
$\boldsymbol{\nu_v}$ = **4.4324** ft^3/lbm

Apply equations **5-2** and **5-5**:

$$\mathbf{h_x = (1\text{-}x).\ h_L + x.\ h_V} \quad \textbf{Eq. 5-2}$$
$$\boldsymbol{\nu_x = (1\text{-}x)\ .\nu_L + x\ .\nu_v} \quad \textbf{Eq. 5-5}$$

Then,

$$\mathbf{h_x} = \mathbf{(1- 0.55) \,.\, (298.57}\ \text{Btu/lbm}\mathbf{) + (0.55) \,.\, (1187.5}\ \text{Btu/lbm}\mathbf{)}$$

$$\mathbf{h_x} = \mathbf{787.48}\ \text{Btu/lbm}$$

And,

$$\boldsymbol{\nu_x} = \mathbf{(1- 0.55) \,.\, (0.017736}) + \mathbf{(0.55) \,.\, (4.4324}\ \text{ft}^3/\text{lbm})$$

$$\boldsymbol{\nu_x} = \mathbf{2.446}\ \text{ft}^3/\text{lbm}$$

EXAMPLE 5-6

Prove the equivalence of the Mollier Diagram and the Steam Tables by verifying the results of Example 4-1, Chapter 4, through the use of Steam Tables in Appendix B.

Solution:

The solution from Example 4-1, as restated with the aid of Figure 5-1, is as follows:

h_i = Enthalpy at 450°C and 1 Atm., as read from the Mollier Diagram
= **3380 kJ/kg**

h_f = Enthalpy at 400°C and 1 Atm., as read from the Mollier Diagram
= 3280 kJ/kg

Then, using the Mollier Diagram, the amount of heat that must be removed from the system in order to lower the temperature from 450°C to 400°C, at 1 Atm, would be:

$$\Delta h = h_i - h_f = 3380\ \text{kJ/kg} - 3280\ \text{kJ/kg} = 100\ \text{kJ/kg}$$

Now, lets determine the amount of heat to be removed using the steam tables, in Appendix B:

h_i = Enthalpy at 450°C and 1 Atm., from Appendix B
= 3382.8 kJ/kg

h_f = Enthalpy at 400°C and 1 Atm., from Appendix B
= 3278.5 kJ/kg

Table 5-4. Properties of Saturated Steam, by Pressure, US Units

Properties of Saturated Steam By Pressure US/Imperial Units									
Abs. Press. psia	Temp. °F	Specific Volume ft³/lbm		Enthalpy Btu/lbm			Entropy Btu/(lbm.°R)		Abs. Press. psia
		Sat. Liquid v_L	Sat. Vapor v_V	Sat. Liquid h_L	Evap. h_{fg}	Sat. Vapor h_V	Sat. Liquid s_L	Sat. Vapor s_V	
1.0	101.69	0.016137	333.51	69.728	1036	1105.4	0.1326	1.9776	1.0
4.0	152.91	0.016356	90.628	120.89	1006.4	1126.9	0.2198	1.8621	4.0
14.0	209.52	0.016697	28.048	177.68	972.0	1149.4	0.3084	1.7605	14.0
100	327.82	0.017730	4.4324	298.57	889.2	1187.5	0.4744	1.6032	100

Table 5-5. Properties of Saturated Steam, by Temperature, US Units

Properties of Saturated Steam By Temperature US/Imperial Units										
Temp. °F	Abs. Press. psia	Specific Volume ft^3/lbm		Enthalpy Btu/lbm			Entropy Btu/(lbm.°R)		Temp. °F
		Sat. Liquid v_L	Sat. Vapor v_V	Sat. Liquid h_L	Evap. h_{fg}	Sat. Vapor h_V	Sat. Liquid s_L	Sat. Vapor s_V	
50	0.17813	0.016024	1702.9	18.066	1065.2	1083.1	0.0361	2.1257	50
100	0.95044	0.016131	349.87	68.037	1037.0	1104.7	0.1296	1.9819	100
210	14.1360	0.016701	27.796	178.17	971.6	1149.5	0.3092	1.7597	210
250	29.8430	0.017001	13.816	218.62	945.6	1164.0	0.3678	1.7000	250

Table 5-6. Properties of Superheated Steam, SI Units

Properties of Superheated Steam Metric/SI Units						
Abs. Press. MPa (Sat. T, °C)		Temp. °C				
		150	300	500	650	800
0.05 (81.33)	n	3.889	5.284	7.134		
	h	2780.1	3075.5	3488.7		
	s	7.9401	8.5373	9.1546		
0.1 (99.61)	n	1.9367	2.6398	3.5656		
	h	2776.6	3074.5	3488.1		
	s	7.6147	8.2171	8.8361		
1.0 (179.89)	n		0.2580	0.3541	0.4245	0.4944
	h		3051.7	3479.0	3810.5	4156.2
	s		7.1247	7.7640	8.1557	8.5024
2.5 (223.99)	n		0.0989	0.13998	0.1623	0.1896
	h		3008.8	3462.1	3799.7	4148.9
	s		6.6438	7.3234	7.7056	8.0559
3.0 (233.86)	n		0.0812	0.1162	0.1405	0.1642
	h		2994.3	3457.0	3797.0	4147.0
	s		6.5412	7.2356	7.6373	7.9885
4.0 (250.36)	n		0.0589	0.0864	0.1049	0.1229
	h		2961.7	3445.8	3790.2	4142.5
	s		6.3638	7.0919	7.4989	7.8523

Table 5-7. Properties of Superheated Steam, US Units. V = specific volume in ft^3/lbm; h = enthalpy in Btu/lbm; s = entropy in Btu/lbm-°R

Properties of Superheated Steam **US/Imperial Units**						
Abs. Press. psia (Sat. T, °F)		**Temp. °F**				
		200	300	500	1000	1500
10 (193.16)	v	38.851				
	h	1146.4				
	s	1.7926				
15 (212.99)	v		29.906	37.986	57.931	
	h		1192.7	1287.3	1534.7	
	s		1.8137	1.9243	2.1312	
100 (327.82)	v			5.5875	8.6576	
	h			1279.3	1532.3	
	s			1.7089	1.9209	
200 (381.81)	v			2.7246	4.3098	
	h			1269.1	1529.5	
	s			1.6243	1.8430	
360 (434.43)	v			1.4460	2.3774	3.2291
	h			1250.6	1525.0	1799.8
	s			1.5446	1.7758	1.9375

So, using the Steam Tables, the amount of heat that must be removed from the system in order to lower the temperature from 450°C to 400°C, at 1 Atm, would be:

$$\Delta h = h_i - h_f = 3382.8 \text{ kJ/kg} - 3278.5 \text{ kJ/kg} = 104.3 \text{ kJ/kg}$$

Therefore, for most practical purposes, the Mollier Diagram and the Steam Tables are equivalent insofar as thermodynamic system analyses are concerned. The 4.3% difference between the two approaches is due mainly to the small amount of inaccuracy in reading of the scale of the Mollier Diagram.

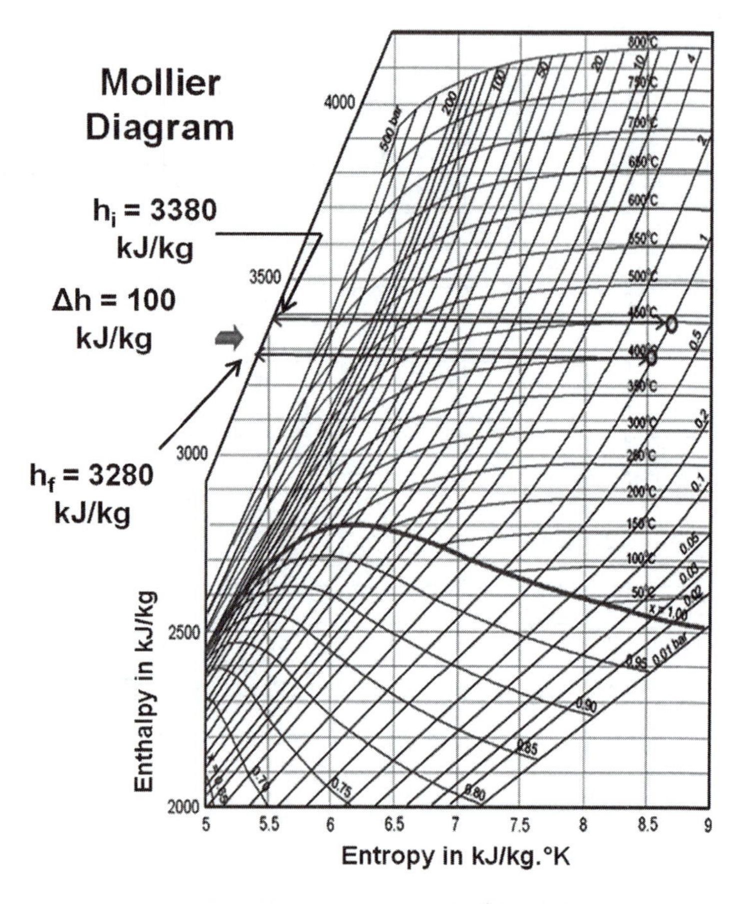

Figure 5-1. Mollier Diagram, SI/Metric Units

Chapter 5—Self Assessment Problems & Questions

1. Using the saturated liquid enthalpy value for $\mathbf{h_L}$ and the saturated vapor enthalpy value for $\mathbf{h_V}$, at 0.2 MPa and 120.2°C, as listed in the saturated steam tables in Appendix B, calculated the value for $\mathbf{h_{fg}}$.

2. Calculate the enthalpy of 450 psia and 970°F superheated steam.

3. Determine the enthalpy of saturated water at 50°C and 1 Bar.

4. Determine the enthalpy and specific volume for 14 psia steam with a quality of 65%.

Chapter 6

Phases of Water and Associated Thermodynamics

PHASES OF SUBSTANCE

Thermodynamic properties and phases of a substance are defined and determined by two important properties; namely, temperature and pressure.

Three most common phases of a substance are as follows:

1) Solid
2) Liquid
3) Gaseous

These three phases, subcategories within these phases, and other pertinent thermodynamic terms are defined below.

Solid

The shape and volume of a substance, in solid phase, is non-volatile. A substance in its solid phase does not adapt itself to the shape or volume of its container. The temperature—under atmospheric pressure conditions—at which a substance attains the solid phase, is called the ***freezing point***.

Liquid

The shape of a substance, in liquid phase, is volatile. A substance in its liquid phase adapts itself to the shape or of its container. The temperature—under atmospheric pressure conditions—at which a substance attains the liquid phase is called the ***melting point***.

Gas

The shape and volume of a substance, in gaseous phase, is volatile. Gas is a state of matter consisting of a collection of molecular or atomic particles that lacks definite shape or volume. The temperature—under atmospheric pressure conditions—at which a substance attains the gaseous phase, is called the ***boiling point***.

Sensible Heat

Sensible heat is the heat required, or absorbed, in raising the temperature of a substance, without a change in phase. **Example:** Heat required to raise the temperature of water from 60°F to 80°F, at sea level, or a pressure of 1 atm. Calculation of sensible heat is demonstrated in other chapters.

Latent Heat

Latent heat is the heat required or absorbed in changing the phase of a substance. Latent heats for fusion, sublimation and vaporization of water are listed—in SI/Metric Units and US/Imperial Units—in Table 6-1. Consider the following examples as illustration of how to use this table:

Example A: Latent Heat of Fusion, $\mathbf{h_{SL}}$, which is the heat required to fully melt ice to liquid water, is **334 kJ/kg**.

Example B: Latent Heat of Sublimation, $\mathbf{h_{SV}}$, which is the heat required to fully evaporate ice to saturated vapor phase, is **2838 kJ/kg**.

Example C: Latent Heat of Vaporization, $\mathbf{h_{fg}}$, which is the heat required to fully evaporate water, at sea level—or a pressure of 1 atm—and 100°C, is **2260 kJ/kg**.

Table 6-1. Latent Heats for Water Phase Transformation, SI Units

	Latent Heat Fusion kJ/kg	**Latent Heat Sublimation kJ/kg**	**Latent Heat Vaporization kJ/kg**
kJ/kg	334	2838	2260 at 100°C
BTU/lbm	143.4	1220	970.3
kcal/kg	79.7	677.8	539.1

Saturation Temperature

Saturation temperature, at a given pressure, is the temperature below which a gas or vapor would condense to liquid phase. For example, the saturation temperature at standard atmospheric pressure of 101 kPa (0.1014MPa) is 100°C. See the circled pressure and temperature values in Table 6-2. Now, if the pressure is reduced to 12.4 kPa (0.0124 MPa), the saturation temperature would drop to 50°C. In other words, if the pressure is reduced to 12.4 kPa (0.0124 MPa), the water's boiling point would be reduced to 50°C.

Saturation Pressure

Saturation pressure, at a given temperature, is the pressure above which a gas or vapor would condense to liquid phase. For example, the saturation pressure at a temperature of 100°C is 101 kPa (0.1014MPa). See the circled pressure and temperature values in Table 6-2. Now, if the pressure is increased to (1.5547 MPa), the saturation temperature rises to 200°C. In other words, if the pressure is increased to 1,554.7 kPa, the water's boiling point would double, from 100°C to 200°C. This also means that if the initial temperature and pressure conditions are 100°C is 101 kPa, and the pressure is escalated by almost 15 fold, up to 1,5547 kPa, the water would no longer be in saturated water phase; it will instead fall back into subcooled liquid phase—see the discussion on subcooled liquids in the next section.

Same example in the US unit realm would be that of saturation pressure at a given temperature of say 209°F. The saturation pressure at the given temperature of 209°F, as circled in Table 6-3, would be 14 psia. If, however, the pressure is raised, for instance, to 100 psia, the saturation temperature would rise to 328°F. In other words, if the initial conditions are changed such that while the temperature remains the same, i.e. 209°F, the pressure is increased from 14 psia to 100 psia, the water would fall into the subcooled liquid state.

Note that increasing the boiling point of the water by raising the pressure on the surface of the water is the same principle that is employed in pressure cookers. By raising the pressure and the temperature, water's enthalpy or heat content is raised, thus accelerating the cooking or decomposition of the contents.

Subcooled Liquid

When the temperature of a liquid is less than its boiling point, at a given pressure, it is referred to be in a subcooled state or phase. **Example:** Water at room temperature (77°F, 25°C), at sea level (1-Atm or 1-Bar), is considered to be subcooled, in that, addition of a small amount of heat will not cause the water to boil.

- If, at a certain pressure, temperature is a determining variable, a substance is said to be in a subcooled liquid phase when its temperature is ***below*** the saturation temperature value corresponding to its pressure.

- Conversely, if at a certain temperature, pressure is a determining variable, a substance is said to be in a subcooled liquid phase when its pressure is ***greater*** than the saturation pressure value corresponding to that temperature.

Saturated Liquid

When the temperature of a liquid is almost at its boiling point, such that addition of a small amount of heat energy would cause the liquid to boil, it is said to be saturated. In other words, it is saturated with heat and cannot accept additional heat without evaporating into vapor phase. **Example:** Water at 212°F, or 100°C, at sea level.

Saturated Vapor

Vapor that has cooled off to the extent that it is almost at the boiling point, or saturation point, and on the verge of condensing, is called a saturated vapor.

Liquid-vapor Phase

A substance is said to be in a liquid-vapor phase when its temperature is equal to or slightly greater than the saturated temperature value corresponding to its pressure. When water is in the liquid-vapor phase, in most cases, a portion of the total volume of water has evaporated; the remaining portion is in saturated water state.

Superheated Vapor

Superheated vapor is vapor that has absorbed heat beyond its boiling point. Loss of a small amount of heat would not cause superheated vapor to condense.

- If, at a certain pressure, temperature is the determining variable, then a vapor is said to be in a superheated vapor state when its temperature ***exceeds*** the saturation temperature corresponding.

- Conversely, if at a certain temperature, pressure is the determining variable, a vapor is said to be in a superheated vapor state when its pressure is ***less*** than the saturation pressure corresponding.

EXAMPLE PROBLEM 1

Answer the following questions for water at a temperature of 153°F and pressure of 4 psia:

a) Heat content for saturated water.

b) Specific heat (Btu/lbm) required to evaporate the water.

c) If the water were evaporated, what would the saturated vapor heat content be?

d) What state or phase would the water be in at the stated temperature and pressure?

e) What would be the entropy of the water while it is in saturated liquid phase?

f) What would be the specific volume of the water while it is in saturated vapor phase?

Solution/Answers

a) At 153°F and pressure of 4 psia, the water is in saturated liquid form. According to saturated steam table excerpt in Table 6-3, the saturated water enthalpy is 120.8 Btu/lbm. This value is listed under column labeled h_L, in Table 6-3, in the row representing temperature of 153°F and pressure of 4 psia.

Answer: The enthalpy or heat content for saturated water, at the given temperature and pressure, is **120.8 Btu/lbm.**

b) The specific heat, in Btu/lbm, required to evaporate the water from saturated liquid phase to saturated vapor phase, is represented by the term $\mathbf{h_{fg}}$. The value of h_{fg}, for saturated water at 153°F and a pressure of 4 psia, as read from Table 6-3 is 1006 Btu/lbm. See circled values in Table 6-3.

Answer: $\mathbf{h_{fg\ at153°F\ and\ 4\ psia}}$ = **1006 Btu/lbm**

c) Saturated vapor heat content, if the water were evaporated, would be the value for $\mathbf{h_{v\ at153°F\ and\ 4\ psia}}$, and from Table 6-3 this value is 1127 Btu/lbm.

Answer: $\mathbf{h_{v\ at153°F\ and\ 4\ psia}}$ = **1127 Btu/lbm**

d) The water would be in **saturated liquid phase** at the stated temperature and pressure. All stated saturation temperatures and pressures, in the saturated steam tables, represent the current state of water in saturated liquid phase.

Answer: Saturated liquid phase

e) The entropy of water at 153°F and a pressure of 4 psia, in saturated liquid phase, as read from Table 6-3, would be s_L= 0.22 Btu/(lbm.°R). Note that s_L value is retrieved form the table and not the s_V value. This is because the problem statement specifies the ***liquid phase.***

Answer: s_L= 0.22 Btu/(lbm.°R)

f) The specific volume of water at 153°F and a pressure of 4 psia, in saturated vapor phase, as read from Table 6-3 would be ν_V = 90 **ft³/lbm**. Note that ν_V value is retrieved form the table and not the ν_L value. This is because the problem statement specifies the *vapor phase.*

Answer: ν_V = 90 **ft³/lbm**

Table 6-2. Properties of Saturated Steam by Temperature, SI Units

Properties of Saturated Steam By Temperature									
Metric/SI Units									
	Abs.	Specific Volume		Enthalpy			Entropy		
Temp.	Press.	m^3/kg		kJ/kg			kJ/kg.°K		Temp.
°C	MPa	Sat. Liquid	Sat. Vapor	Sat. Liquid	Evap.	Sat. Vapor	Sat. Liquid	Sat. Vapor	°C
		v_L	v_V	h_L	h_{fg}	h_V	s_L	s_V	
20	0.0023	0.001	57.76	83.9	2454	2537	0.30	8.67	20
50	0.0124	0.001	12.02	209	2382	2591	0.70	8.07	50
100	0.1014	0.001	1.672	419	2257.	2675	1.31	7.35	100
200	1.5547	0.001	0.127	852	1940	2792	2.33	6.43	200

Table 6-3. Properties of Saturated Steam By Pressure, US Units

Properties of Saturated Steam By Pressure									
US/Imperial Units									
Abs. Press. psia	Temp in °F	Specific Volume ft^3/lbm		Enthalpy BTU/lbm			Entropy BTU/(lbm.°R)		Abs. Press. psia
		Sat. Liquid v_L	Sat. Vapor v_V	Sat. Liquid h_L	Evap. h_{fg}	Sat. Vapor h_V	Sat. Liquid s_L	Sat. Vapor s_V	
1.0	102	0.0161	333.	69.73	1036	1105	0.133	1.978	1.0
4.0	153	0.0164	90.0	120.8	1006	1127	0.220	1.862	4.0
10.0	193	0.0166	38.42	161.2	982	1143	0.2836	1.788	10.0
14.0	209	0.0167	28.0	177.6	972	1149	0.308	1.761	14.0
100	328	0.0177	4.4	298.5	889	1188	0.474	1.603	100

PHASE TRANSFORMATION OF WATER AT CONSTANT PRESSURE

The process of phase transformation of water, under constant pressure, is illustrated in **Figure 6-1**.

In segment A, to the extreme left in **Figure 6-1**, water is depicted in subcooled liquid phase, at a certain temperature T°C, well below the saturation temperature at the given pressure. As heat is added to the water, as shown in segment B, the temperature of the water rises to T°C + ΔT°C, thus causing the water to expand and achieve the ***saturated liquid state***. This heat, that simply increases the temperature of the water without causing evaporation, is ***sensible heat***.

As more heat is added into the system, some of the saturated water transforms into saturated vapor, resulting in a mixture of saturated water and saturated vapor phases, as shown in segment C. The heat added to transform the saturated water into saturated vapor is ***latent heat***. During this phase transformation, the water temperature stays constant, at $T_{Saturation}$. Further addition of heat leads to the transformation of all of the saturated water into saturated vapor, as shown in segment D.

The last stage of this process is represented in segment E of Figure 6-1. This stage shows the transformation of water from saturated vapor state to ***superheated vapor state***, through introduction of more heat. Since the water maintains the vapor state in this stage, the heat added is sensible heat.

The phase transformation process described above can also be followed in graphical form as shown in Figure 6-2. For instance, the state of water represented by segment A, in Figure 6-1, lies in the subcooled region, to the left of the graph in Figure 6-2. Water represented by segment B lies directly on the saturated water line in Figure 6-2. Water represented by segment C lies within the "bell curve," shown as the shaded region on the graph in Figure 6-2. Water represented by segment D lies directly on the saturated vapor line. As more heat is added and water transitions into state represented by segment E, it

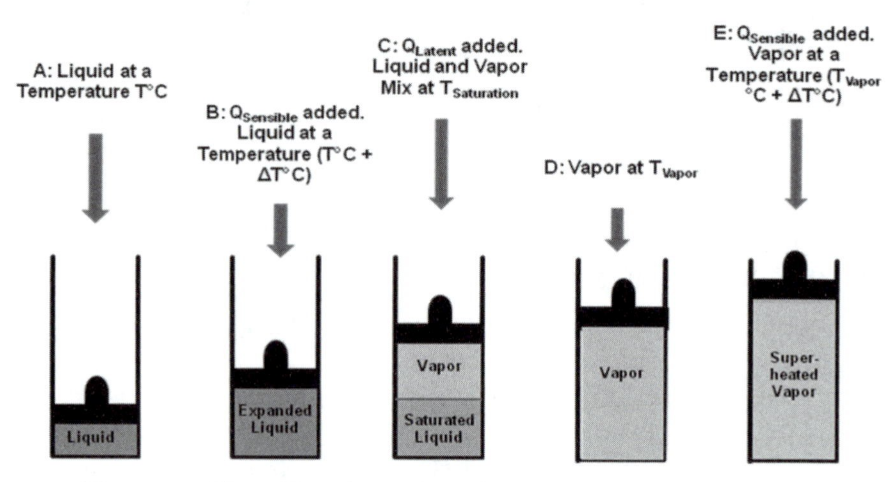

Figure 6-1. Phase Transformation of Water at Constant Pressure

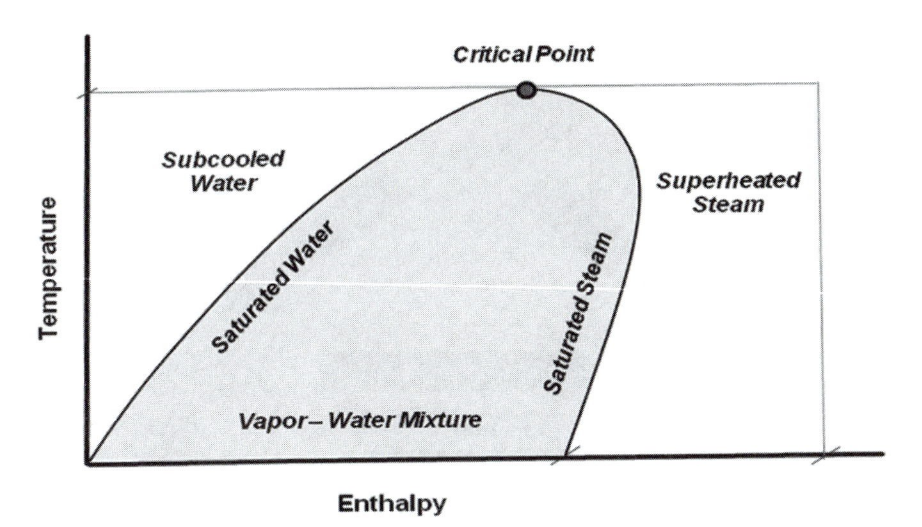

Figure 6-2. Phase Transformation of Water at Constant Pressure, Depicted in Graphical Form

enters the superheated vapor region, shown to the right of the saturated vapor curve on Figure 6-2.

Phase and state transformation of water can also be viewed, from physical perspective, as shown in Figure 6-3. This diagram depicts the transformation of water from solid phase to vapor phase as addition of heat drives the phase transformation process, counterclockwise, from ice to vapor state. Once in vapor phase, removal of heat would drive the phase transformation cycle, clockwise, from vapor to ice phase—unless the transformation constitutes sublimation.

Phase transformation from ice directly into vapor, as shown in the left half of the diagram in Figure 6-3, represents the sublimation process. Of course, sublimation can be achieved in reverse through removal of heat, resulting in direct phase transformation from vapor to solid state, or ice.

TYPES OF PHASE TRANSFORMATION

Three common phase transformations, and associated latent heat values, have been discussed in earlier sections of this chapter. In addition to providing tabular cross-referencing between various phases of water, Table 6-4 expands on the topic of phase transformation and

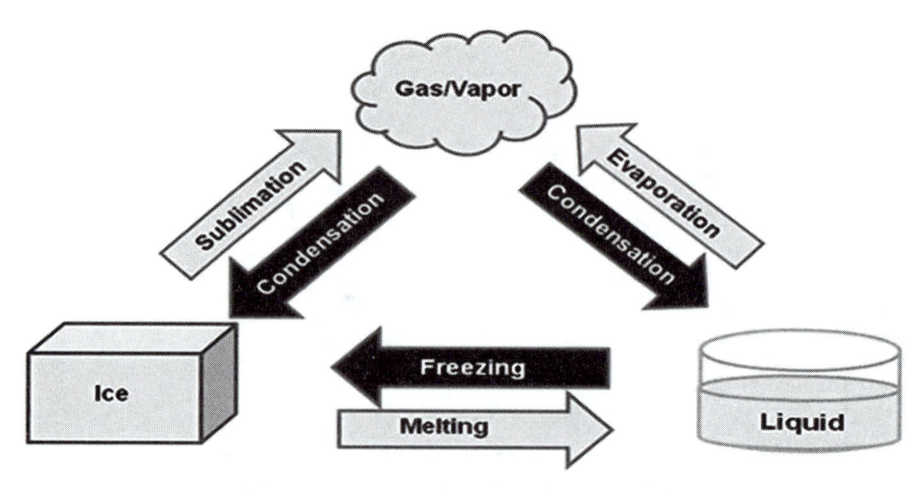

Figure 6-3. Phase Transformation of Water

Table 6-4. Types of Phase Transformation

	To			
From	**SOLID**	**LIQUID**	**GAS**	**PLASMA**
SOLID	Solid-Solid Transfor mation	MELTING or FUSION	SUBLIMATION	N/A
LIQUID	FREEZ-ING	N/A	BOILING or EVAPORATION	N/A
GAS	DEPOSI-TION	CONDENSA-TION	N/A	IONIZA-TION
PLASMA	N/A	N/A	RECOMBINATION or DEIONIZATION	N/A

lists other types of transformation and phases that don't commonly involve water. Plasma is stated as one of the phases in Table 6-4. Plasma is, essentially, ionized gas. Plasma, or ionized gas, is used in various processes, including, vaporized metal deposition on substances such as semiconductor substrates—for fabrication of integrated circuits or "chips"—and flat glass panels, for addition of energy conservation characteristics.

Ideal Gas

Any gas that behaves in accordance with the ideal gas laws is said to be an ideal gas. Highly superheated vapor behaves like an ideal gas and is treated as such. This is demonstrated in later chapters of this text.

Real Gas

Any gas that does not behave according to the ideal gas laws is said to be a real gas. Saturated vapor is considered to be a real gas.

Critical Point

If ***temperature*** and ***pressure*** of a liquid are increased, eventually, a state is reached where ***liquid and vapor phases coexist,*** and are indistinguishable. This point is referred to as a "critical point." See point D in Figure 6-4.

Critical Properties

Properties, such as temperature, specific volume, density and pressure of a substance at the critical point are referred to as critical properties. See Table 6-5.

Triple Point

Triple point of a substance is a state in which ***solid, liquid and gaseous phases coexist.*** (See Table 6-6.)

Comparison—Triple Point vs. Critical Point:

Both triple point and critical point have been explained earlier sections of this chapter. In this section, we will compare these two points for clarification and distinction. Figure 6-4, is used to illustrate the differences between these two points. The graph depicted in Figure 6-4 plots pressure as a function of temperature. Note that water—or any other substance in question—can exist in liquid or gaseous phases, simultaneously, to the right of the line labeled A-B. And, since the critical point "D" lies in the region to the right of line A-B, critical point can only involve two phases, namely, the gaseous phase and the liquid phase. Also, note that T_{Cr}, the temperature at the critical

Table 6-5. Critical Properties of Select Substances

Substance	Critical Temperature		Critical Pressure	
	°C	°K	atm	kPa
Argon	−122.4	151	48.1	4,870
Ammonia	132.4	405.5	111.3	11,280
Bromine	310.8	584.0	102	10,300
Cesium	1,664.85	1,938.00	94	9,500
Chlorine	143.8	417.0	76.0	7,700
Fluorine	−128.85	144	51.5	5,220
Helium	−267.96	5.19	2.24	227
Hydrogen	−239.95	33.2	12.8	1,300
Krypton	−63.8	209	54.3	5,500
Neon	−228.75	44.4	27.2	2,760
Nitrogen	−146.9	126	33.5	3,390
Oxygen	−118.6	155	49.8	5,050
CO_2	31.04	304.19	72.8	7,380
H_2SO_4	654	927	45.4	4,600
Xenon	16.6	289.8	57.6	5,840
Lithium	2,950	3,220	652	66,100
Mercury	1,476.9	1,750.1	1,720	174,000
Sulfur	1,040.85	1,314.00	207	21,000
Iron	8,227	8,500	N/A	N/A
Gold	6,977	7,250	5,000	510,000
Aluminum	7,577	7,850	N/A	N/A
Water	373.946	647.096	217.7	22,060

Table 6-6. Triple Point Properties of Select Substances

Substance	Triple Point Temperature, T (K)	Triple Point Pressure, (kPa)
Ammonia	195.4	6.076
Argon	83.81	68.9
Butane	134.6	7×10^{-4}
Ethane	89.89	8×10^{-4}
Ethanol	150	4.3×10^{-7}
Ethylene	104	0.12
Hydrogen	13.84	7.04
Hydrogen Chloride	158.96	13.9
Methane	90.68	11.7
Nitric oxide	109.5	21.92
Nitrogen	63.18	12.6
Nitrous Oxide	182.34	87.85
Oxygen	54.36	0.152
Water	273.16	0.6117
Zinc	692.65	0.065

point—or critical temperature—is substantially higher than the, the triple point temperature. At this higher temperature T_{Cr}, solid phase cannot exist, for most practical purposes. On the other hand, at the triple point, substance exists in all three phases: solid, liquid and gaseous—simultaneously. As shown in Figure 6-4, triple point C lies in the region where solid, liquid and gaseous phases coexist. The solid phase is to the left of the dashed line E-C, liquid phase in the B-C-D region and vapor phase lies to the right of the line formed by points A, C and D.

Another key distinction between triple point and critical point, from pressure point of view, is that critical point pressure, P_{Cr}, is significantly higher than the critical pressure, P_t.

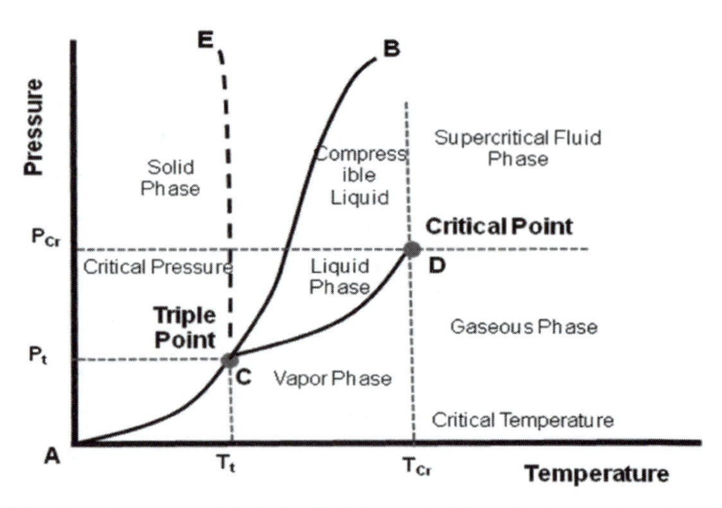

Figure 6-4. Comparison of Triple Point vs. Critical Point, a Graphical Perspective

Chapter 6—Self Assessment Problems and Questions

1. A boiler is relocated from sea level to a location that is at an elevation of 10,000 ft MSL. Using Table 6A-1 and Table 6-3, determine the temperature at which the water will boil if the boiler is assumed to be open to atmosphere.

2. In problem (1), if the objective is just to heat the water close to the boiling point, will the boiler consume more or less fuel than it did when it was located at the sea level?

3. Answer the following questions for water at a temperature of 193°F and pressure of 10 psia:

   a) Heat content for saturated water.

   b) Specific heat (Btu/lbm) required to evaporate the water.

   c) If the water were evaporated, what would the saturated vapor heat content be?

   d) What state or phase would the water be in at the stated temperature and pressure?

e) What would the entropy of the water be while it is in saturated liquid phase?

f) What would the specific volume of the water be while it is in saturated vapor phase?

g) What would the phase of the water if the pressure is increased to 20 psia while keeping the temperature constant at 193°F?

Table 6A-1.

Altitude With Mean Sea Level as Ref.		Absolute Pressure in Hg Column		Absolute Atmospheric Pressure		
Feet	Meters	Inches Hg Column	mm Hg Column	psia	kg/cm^2	kPa
0	0	29.9	765	14.7	1.03	101
500	152	29.4	751	14.4	1.01	99.5
1000	305	28.9	738	14.2	0.997	97.7
1500	457	28.3	724	13.9	0.979	96
2000	610	27.8	711	13.7	0.961	94.2
2500	762	27.3	698	13.4	0.943	92.5
3000	914	26.8	686	13.2	0.926	90.8
3500	1067	26.3	673	12.9	0.909	89.1
4000	1219	25.8	661	12.7	0.893	87.5
4500	1372	25.4	649	12.5	0.876	85.9
5000	1524	24.9	637	12.2	0.86	84.3
6000	1829	24	613	11.8	0.828	81.2
7000	2134	23.1	590	11.3	0.797	78.2
8000	2438	22.2	568	10.9	0.768	75.3
9000	2743	21.4	547	10.5	0.739	72.4
10000	3048	20.6	526	10.1	0.711	69.7
15000	4572	16.9	432	8.29	0.583	57.2
20000	6096	13.8	352	6.75	0.475	46.6

Chapter 7

Laws of Thermodynamics

INTRODUCTION

This chapter explores major categories of thermodynamic systems based on their interaction with the surroundings or environment. The three major categories of thermodynamic systems are introduced and key differences between them are explained. Differentiation between the three types of thermodynamic systems is reinforced through tabular cross-referencing of characteristic properties.

Since open thermodynamic systems are somewhat more common than other type of thermodynamic systems, a detailed case study—involving a fossil fuel powered steam power generation system—is undertaken and explained in this chapter. This case study highlights the practical significance and application of steam based thermodynamic systems. The detailed discussion and step by step analysis of each thermodynamic process involved provides the energy engineers an opportunity to understand and hone skills associated with thermodynamic analysis of steam based power harnessing and power generating systems.

The material in this chapter and the case study build upon and utilize the thermodynamic concepts, principles, laws and computational methods covered in foregoing chapters. For the reader's convenience—and to make the analysis versatile—the case study in this chapter is presented in SI/metric units as well as US/imperial units.

MAJOR CATEGORIES OF THERMODYNAMIC SYSTEMS

Thermodynamic systems can be categorized in myriad ways. However, in this chapter we will focus on categorization of thermodynamic systems based on their interaction with the surroundings or environment. From thermodynamic system and environment interface perspective, thermodynamic systems can be categorized as follows:

I. Open Thermodynamic Systems
II. Closed Thermodynamic Systems
III. Isolated Thermodynamic Systems

Open Thermodynamic Systems

Open thermodynamic systems are systems in which, in addition to the exchange of heat energy with the surroundings, mass or matter are free to cross the system boundary. Also, in open thermodynamic systems, work is performed on or by the system. The type of open thermodynamic systems where entering mass flow rate is the same as the exiting mass flow rate is referred to as a ***Steady Flow Open System***. Examples of Steady Flow Open Systems include pumps, compressors, turbines and heat exchangers.

Closed Thermodynamic Systems

Closed thermodynamic systems are systems in which no mass crosses the system boundary. Energy, however, can cross through the system boundary in form of heat or work. Examples of closed thermodynamic systems include: sealed pneumatic pistons and refrigerant in a refrigeration system.

Isolated Thermodynamic Systems

Isolated thermodynamic systems are systems in which ***no*** work is performed by or on the system; no heat is added or extracted from the system and no matter flows in or out of the system. Imagine a rigid sealed steel cylinder containing liquid nitrogen. This steel cylinder is heavily insulated and is placed inside another sealed steel container such that cylinder's walls do not come in contact with the outside steel container. If vacuum is now created between the outer container and the inner gas cylinder, you would have a thermodynamic system that is "isolated" for most practical purposes. In that, there would be negligible, in any, heat transfer between the liquid nitrogen and the environment or surroundings outside the outer sealed container. The liquid nitrogen is contained in a sealed steel container, with fixed volume; therefore, no work can be performed by the nitrogen (the system) or the environment (e.g. the air) outside the outer container. In addition, because of the containment or isolation attained through the steel cylinder and the outer sealed container, there would be no transfer of mass or matter.

Table 7-1 facilitates comparison and cross-referencing of the three categories of thermodynamic systems. In addition, this table permits an examination of the three types of thermodynamic systems on the basis of three important thermodynamic process attributes, namely: (1) Mass Flow Across the Boundary, (2) Work Flow Across the Boundary, and (3) Heat Exchange with the Surroundings.

Table 7-1. Thermodynamic System Definition, Categorization and Cross-referencing

Transition of Heat, Mass and Work in Thermodynamic Systems			
Type of Thermodynamic System	**Open**	**Closed**	**Isolated**
Mass Flow Across Boundary	**Yes**	**No**	**No**
Work Performed On or By the System	**Yes**	**Yes**	**No**
Heat Exchange with the Surroundings	**Yes**	**Yes**	**No**

LAWS OF THERMODYNAMICS:

Engineering discipline and study of science, in general, utilize principles and laws for developing or deriving equations. These equations are mathematical representation of the engineering or scientific principles and laws. Of course, the key purpose for deriving or developing equations is to be able to define or determine the value of unknown entities or unknown variables.

In our continued effort to expand our list of tools for thermodynamic system analyses, at this juncture, we will explore the first law of thermodynamics.

First Law of Thermodynamics:

The first law of thermodynamics is a statement of the law of conservation of energy in the thermodynamics realm. In other words, the net energy entering a thermodynamic system is equal to the net change in the internal energy of the system plus the work performed by the system. All energy and work is accounted for at all points in the thermodynamic system. Of course, energy status at one point is compared with another point in the system as the laws of thermodynamics are applied to derive equations for system or process analysis.

As we transform the first law of conservation of energy into an equation with practical application and significance, it is important to reiterate that energy, heat and work are mathematically equivalent. Energy, heat and work can, therefore, be added or subtracted linearly in an equation.

Mathematical Statement of the **First Law of Thermodynamics** in a ***Closed*** Thermodynamic System is as follows:

$$Q = \Delta U + \Delta KE + \Delta PE + W \qquad \textbf{Eq. 7-1}$$

Where,

- **ΔU** = Change in Internal Energy
- **ΔKE** = Change in Kinetic Energy
- **ΔPE** = Change in Potential Energy
- **Q** = Heat energy entering (+), or leaving (-) the system
- **W** = Work performed by the system on the surroundings is positive and work performed by the surroundings on the system is negative.

Mathematical Statement of the **First Law of Thermodynamics** in an ***Open*** Thermodynamic System, also referred to as SFEE, or Steady Flow Energy Equation, would be as follows:

$$Q = \Delta U + \Delta FE + \Delta KE + \Delta PE + W \qquad \textbf{Eq. 7-2}$$

Where,

- **ΔU** = Change in Internal Energy
- **ΔFE** = Change in Flow Energy = **Δ(PV)**
- **ΔKE** = Change in Kinetic Energy
- **ΔPE** = Change in Potential Energy
- **Q** = Heat energy entering (+), or leaving (-) the system
- **W** = Work performed by the system on the surroundings is positive and work performed by the surroundings on the system is negative.

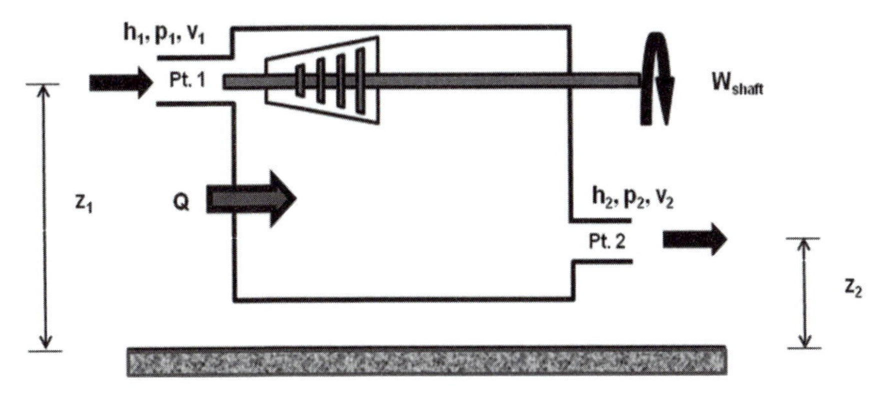

Figure 7-1. Open thermodynamic system with a turbine.

The Steady Flow Energy Equation, or the SFEE, representing the application of the First Law of Thermodynamics in an Open Thermodynamic System can be written in the specific, or per unit mass, form as:

$$q = \Delta u + \Delta FE_{specific} + \Delta KE_{specific} + \Delta PE_{specific} + w_{shaft} \qquad \text{Eq. 7-3}$$

Since,

$$\Delta h = \Delta u + \Delta FE_{specific} \qquad \text{Eq. 7-4}$$

The SFEE can be written, in a more practical form, for a turbine open system as:

$$q = \Delta h + \Delta KE_{specific} + \Delta PE_{specific} + w_{shaft} \qquad \text{Eq. 7-5}$$

Or,

$$q = (h_2 - h_1) + 1/2 \,.\, (v_2^2 - v_1^2) + g.(z_2 - z_1) + w_{shaft} \qquad \text{Eq. 7-6}$$

This Steady Flow Energy Equation is stated in the SI or metric realm.

Where,

h_1 = Enthalpy of the steam entering the turbine, in kJ/kg.
h_2 = Enthalpy of the steam exiting the turbine, in kJ/kg.
v_1 = Velocity of the steam entering the turbine, in m/s.
v_2 = Velocity of the steam exiting the turbine, in m/s.
z_1 = Elevation of the steam entering the turbine, in meters.
z_2 = Elevation of the steam exiting the turbine, in meters.

q = Specific heat added or removed from the turbine system, in kJ/kg.

w_{shaft} = Specific work or work per unit mass; measured in kJ/kg.

W_{shaft} = Work performed by the turbine shaft, measured in kJ.

g = Acceleration due to gravity, 9.81 m/s²

Since,

$$P_{shaft} = w_{shaft} \cdot \dot{m} \qquad \text{Eq. 7-7}$$

Where,

$\dot{m}$ = Mass flow rate of the system, in kg/sec or lbm/sec.

This SI version of the SFEE can be written, in a more useful form, for a power calculation in turbine type open system as:

$$\dot{m}\,.(q) = \dot{m}\,[\,(h_2 - h_1) + 1/2\,.\,(v_2^2 - v_1^2) + g.(z_2 - z_1)] + \dot{m}\,.\,w_{shaft} \qquad \text{Eq. 7-8}$$

Since,

$$\dot{Q} = \dot{m}\,.\,(q) \qquad \text{Eq. 7-9}$$

Where,

$\dot{Q}$ = Flow rate of heat added or removed from the turbine system, in kJ/second.

Therefore, the SI version of the SFEE can be written as:

$$\dot{Q} = \dot{m}\,[(h_2 - h_1) + 1/2\,.\,(v_2^2 - v_1^2) + g.(z_2 - z_1)] + P_{shaft} \qquad \text{Eq. 7-10}$$

The Open System Steady Flow Energy Equation can be rewritten, for application in the US or Imperial unit realm, as:

$$\dot{m}\,.\,(q) = \dot{m}\,[\,(h_2 - h_1) + 1/2\,.\,(v_2^2 - v_1^2)/(g_c\,.\,J) + g\,.\,(z_2 - z_1)/(g_c\,.\,J)\,] + \dot{m}\,.\,w_{shaft} \qquad \text{Eq. 7-11}$$

Or,

$$\dot{Q} = \dot{m}\,[\,(h_2 - h_1) + 1/2\,.\,(v_2^2 - v_1^2)/(g_c\,.\,J) + g\,.\,(z_2 - z_1)/(g_c.J)\,] + P_{shaft} \qquad \text{Eq. 7-12}$$

Where,

- $\mathbf{h_1}$ = Enthalpy of the steam entering the turbine, in Btu/lbm.
- $\mathbf{h_2}$ = Enthalpy of the steam exiting the turbine, in Btu/lbm.
- $\mathbf{v_1}$ = Velocity of the steam entering the turbine, in ft/s.
- $\mathbf{v_2}$ = Velocity of the steam exiting the turbine, in ft/s.
- $\mathbf{z_1}$ = Elevation of the steam entering the turbine, in ft.
- $\mathbf{z_2}$ = Elevation of the steam exiting the turbine, in ft.
- $\mathbf{q}$ = Specific heat added or removed from the turbine system, in Btu/lbm
- $\dot{\mathbf{Q}}$ = Flow rate of heat added or removed from the turbine system, in Btus/sec.
- $\mathbf{w_{shaft}}$ = Specific work or work per unit mass; measured in Btu/lbm.
- $\mathbf{W_{shaft}}$ = Work performed by the turbine shaft, measured in Btus.
- $\dot{\mathbf{m}}$ = Mass flow rate of the system, in lbm/sec.
- $\mathbf{g}$ = Acceleration due to gravity, 32.2 ft/s^2
- $\mathbf{g_c}$ = Gravitational constant, 32.2 lbm-ft/lbf-sec^2
- $\mathbf{J}$ = 778 ft-lbf/Btu = Joules constant

Second Law of Thermodynamics

The Second Law of Thermodynamics can be stated in multiple ways. Some of the more common and practical statements of the Second Law of Thermodynamics are, briefly, discussed in this section

The Second Law of Thermodynamics is also known as the law of increasing entropy. While quantity of total energy remains constant in the universe as stipulated by the First Law of Thermodynamics, the Second Law of Thermodynamics states that the amount of usable, work producing, energy in the universe continues to decline; irretrievably lost in form of unusable energy.

Since entropy is defined as unusable energy, the Second Law of Thermodynamics can be interpreted to state that in a closed system, such as the universe, entropy continues to increase. In other words, the second law of thermodynamics states that the net entropy must always increase in practical, irreversible cyclical processes.

The second law of thermodynamics can also be stated mathematically in form of Eq. 7-13.

$$\Delta S \geq \int \frac{\delta Q}{T} \qquad \textbf{Eq. 7-13}$$

Equation 7-13 stipulates that the increase in entropy of a

thermodynamic system must be greater than or equal to the integral of the incremental heat absorbed, divided by the temperature during each incremental heat absorption.

Kevin-Planck statement of the second law of thermodynamics, effectively, implies that it is impossible to build a cyclical engine that has an efficiency of 100%.

CASE STUDY 7-1—SI UNITS

Technical feasibility of a toping cycle cogenerating power plant is being studied at Station "Zebra." This facility is to be stationed in a remote Arctic region. The objective of this plant is to produce steam and generate electricity for an Arctic Environmental Monitoring and Deep Sea Mining Facility. Due to saltwater corrosion risk, it has been established that local glacier ice will be harvested and utilized for steam production purposes. A natural gas boiler is to be used to generate steam. The average temperature of the glacier ice hovers at -10°C, through the year. The glacier ice is to be melted and converted to 500°C, 2.5 MPa, steam; the steam enters the turbine at this temperature and pressure. The steam is discharged by the turbine at 150°C and 50 kPa. This discharged steam is used for the mining process and to heat the station. The condensate is used as potable, utility and process water. Assume that the turbine represents an open, steady flow, thermodynamic system. In other words, the SFEE, Steady Flow Energy Equation applies. Also assume the potable water demand and flow, along the thermodynamic process stages to be relatively negligible. Station Zebra would operate on a 24/7 schedule. For simplicity, the thermodynamic process flow for this system is laid out in Figure 7-2. **Note:** In an actual project setting, developing such a process flow diagram would constitute the first order of business as an energy engineer begins analyzing this process.

a) Estimate the mass flow rate $\dot{\mathbf{m}}$ for generation of 10 MW electrical power if the rate of turbine casing heat loss, **Q**, is 30 kJ/s (or 0.03 MW), exit velocity of steam, $\mathbf{v_2}$, is 35 m/s, entrance velocity of steam, $\mathbf{v_1}$, is 15 m/s, steam exit elevation is $\mathbf{z_2}$ = 1m, steam intake elevation is $\mathbf{z_1}$ = 0.5m. Assume electric power generator efficiency of 90%. Extrapolate the answer into approximate truck loads per hour. Assume truck capacity of approximately 10 cubic meters.

b) Assume that the power station is generating 10 MW of electric power. Calculate the amount of total heat energy needed, in Btu's/hr, to convert -10°C harvested ice to 500°C, 2.5 MPa steam per hour.

c) Calculate the volume, in cu-ft, of natural gas required to power up the station, each day. Assume 98% burner efficiency.

d) If the natural gas transportation cost is **$4.85/DT** in addition to the well head or commodity cost stated in Table 7-2, what would be the annual fuel cost of operating this station?

e) What is the overall energy efficiency of the power station?

f) If heat is added to the steam turbines, would the steady flow energy process in the turbine system constitute an adiabatic process or a non-adiabatic process?

g) What is the change in entropy, **Δs**, in the turbine system?

Case Study Solution Strategy

Before embarking on the analyses and solution for this case study, let's highlight some important facts from the case study statement.

1) As apparent from the case study statement, the working fluid or system consists of water, in various phases.

2) Water is introduced into the overall thermodynamic system in form of -10°C ice and is then heated during various sensible and latent stages. It is, finally, fed into the turbine as 500°C, 2.5 MPa superheated steam.

3) Unlike a typical Rankin Cycle Heat Engine, the steam exhausted from the turbine is not condensed, pumped and recycled through the boilers to repeat the heat cycle. Instead, fresh ice is harvested, melted and introduced as working fluid.

4) Even though Figure 7-2 shows potable water being removed from the system at Stages 2 and 3, it is assumed to be negligible. Thus supporting the assumption that mass flow rate of the working fluid stays constant through the system.

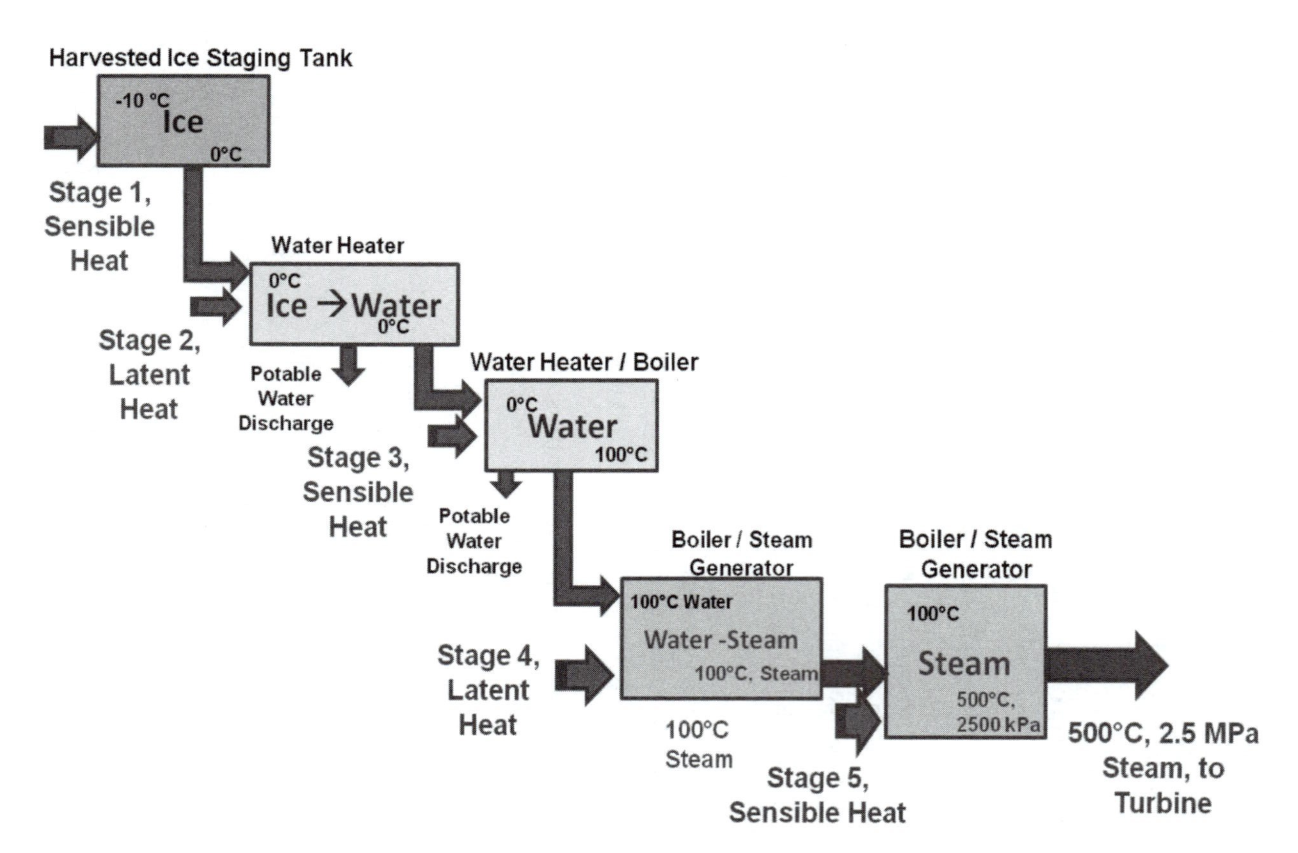

Figure 7-2. Thermodynamic Process Flow Diagram, Case Study 7-1

Table 7-2. Energy Content and Cost Comparison of Common Fuels

Energy Content and Cost Comparison										
Fuel Type or Energy Source	Heat Content						Approx. Cost		US Sept. 2010 Average	
	DT/gal	BTU/gal	MMBTU/Barrel	BTU/Cu-ft.	DT/MCF	Price Per Gallon*	$/DT*	BTU/kWh	$/kWh	$/MMBTU
Propane	0.092	91,600	3.35	2,488		$1.42	$15.38			27.29
Diesel/No. 2 Fuel Oil	0.138	140,000	5.6			$2.65	$19.17			23.429
No. 6 Fuel Oil	0.144	143,888	6.8							
Natural Gas				1,034	1.034	$1.00	$4.15			4.15
Electricity							$30.01	3,412	0.1024	30.01

* Note: These costs represent a January 3, 2011 snapshot of wholesale or industrial market costs.

5) Fuel heat content and fuel cost information is available through Table 7-2.

6) Specific heat and latent heat data for water and ice are provided through Tables 7-3 and 7-4, for sensible heat and latent heat calculations, respectively.

7) Most of the data pertinent to the application of SFEE equation, at the turbine, are given, including the velocities, temperatures, pressures and elevations.

8) Final output of the power generating station is given in terms of the 10 MW rating of the generator and its stated efficiency of 90%.

The overall thermodynamic process flow can be tiered into stages that involve either sensible heating or latent heating. All heating stages of this comprehensive process are depicted in Figure 7-2. Each stage of the overall process is labeled with pertinent entry and exit temperature and pressure, as applicable. Each stage is named as either Sensible or Latent Stage. Furthermore, each stage states the phase of water at point of entry and exit.

a) Estimate the mass flow rate $\dot{m}$ for generation of 10 MW electrical power output if the rate of turbine casing heat loss, Q, is 30 kJ/s (or 0.03 MW), exit velocity of steam, v_2, is 35 m/s, entrance velocity of steam, v_1, is 15 m/s, steam exit elevation is $z_2 = 1$m, steam intake elevation is $z_1 = 0.5$m. Assume electric power generator efficiency of 90%. Extrapolate the answer into approximate truck loads per hour. Assume truck capacity of approximately 10 cubic meters.

Solution

This part of the case study can be analyzed and solved by simply focusing on the very last stage and applying the SFEE in form of Eq. 7-10. The computation of mass flow rate does not require assessment of the heat required at the various stages of the overall thermodynamic process, in this case study, because of the following key assumption included in the problem statement:

> "Also assume the potable water demand and flow, along the thermodynamic process stages to be relatively negligible. "

Table 7-3. Specific Heat, c_p, for Select Liquids and Solids

Approximate Specific Heat, c_p, for Selected Liquids and Solids			
Substance	**c_p in J/gm K**	**c_p in cal/gm K or BTU/lb F**	**Molar C_p J/mol K**
Aluminum	0.9	0.215	24.3
Bismuth	0.123	0.0294	25.7
Copper	0.386	0.0923	24.5
Brass	0.38	0.092	...
Gold	0.126	0.0301	25.6
Lead	0.128	0.0305	26.4
Silver	0.233	0.0558	24.9
Tungsten	0.134	0.0321	24.8
Zinc	0.387	0.0925	25.2
Mercury	0.14	0.033	28.3
Alcohol(ethyl)	2.4	0.58	111
Water	**4.186**	**1**	**75.2**
Ice (-10 C)	**2.05**	**0.49**	**36.9**
Granite	0.79	0.19	...
Glass	0.84	0.2	

In other words the mass flow rate is assumed to be constant through out the process, and any discharge of water during individual stages of the overall process is negligible.

The turbine segment of the overall power generating system is illustrated in Figure 7-3. The enthalpy values are obtained from the superheated steam table excerpt in Table 7-5.

Apply SFEE, in form of **Eq. 7-10**.

$$\dot{Q} = \dot{m}\,[(h_2 - h_1) + 1/2\,.\,(v_2^2 - v_1^2) + g.(z_2 - z_1)] + P_{shaft} \qquad \text{Eq. 7-10}$$

Table 7-4. Latent Heat for Phase Transformation of Water

	Latent Heat Fusion $\mathbf{h_{sl}}$ **(ice/water)**	**Latent Heat Sublimation** $\mathbf{h_{ig}}$	**Latent Heat Vaporization** $\mathbf{h_{fg}}$
kJ/kg	333.5	2838	2257
BTU/lbm	143.4	1220	970.3
kcal/kg	79.7	677.8	539.1

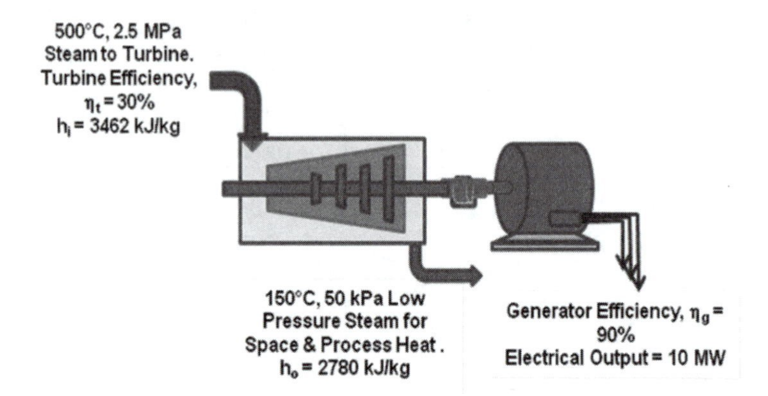

Figure 7-3. Case Study 7-1, Mass Flow Rate Analysis

Rearrangement of Eq. 7-10 yields:

$$\mathbf{m = \dot{Q} - P_{shaft)}/[(h_2 - h_1) + 1/2\ .\ (v_2^2 - v_1^2) + g.(z_2 - z_1)]}$$

Given:

$\dot{Q}$ = - 0.03 MW = - 0.03x10^6 W = - 0.03x10^6 J/s

$\mathbf{P_{shaft}}$ = **(10 MW)/η_g**

= (10 MW)/(0.9)

= 11.11x10^6 W

= 11.11x10^6 J/s

$\mathbf{h_2 = h_o} = 2780\ kJ/kg = 2780x10^3\ J/kg$ {See Table 7-5}
$\mathbf{h_1 = h_i} = 3462\ kJ/kg = 3462x10^3\ J/kg$ {See Table 7-5}
$\mathbf{v_2} = 35\ m/s$
$\mathbf{v_1} = 15\ m/s$
$\mathbf{z_2} = 1m$
$\mathbf{z_1} = 0.5m$

Apply Eq. 7-10, in its rearranged form as follows:

$$\dot{m} = (\dot{Q} - P_{shaft})/[(h_2 - h_1) + 1/2 \,.\, (v_2^2 - v_1^2) + g.(z_2 - z_1)]$$

$$\dot{m} = (-0.03x10^6\ J/s - 11.11x10^6\ J/s)/[(2780x10^3\ J/kg - 3462x10^3\ J/kg) + 1/2\ \{(35m/s)^2 - (15m/s)^2\} + 9.81\ m/s^2.(1m - 0.5m)]$$

$$\dot{m} = (-\ 0.03x10^6\ J/s - 11.11x10^6\ J/s)/(-\ 682{,}000\ J/kg + 500\ J/kg + 4.9J/kg)$$

$\dot{m}$ **= 16.35 kg/sec,**

Or,

$\dot{m}$ **= {(16.35 kg/907.2kg/ton)/sec }.(3600 sec/hour)**
$\dot{m}$ **= 65 tons/hour, or**

Since the density of ice = 916.8 kg/m^3,
The volumetric flow rate, $\dot{V}$, would be:

$\dot{V}$ **= (16.35 kg/sec).(3600 sec/hr)/(916.8 kg/m^3)**
$\dot{V}$ **= 64.19 cu-meters/hr**

At an estimated 10 cubic meters per truck load, this volumetric mass flow rate would amount to:

$\dot{V}$ **= (64.19 cu-meters/hr)/10**

Or,

$\dot{V}$ **= 6.4 truck loads per hour**

b) Assume that the power station is generating 10 MW of electric power. Calculate the amount of heat needed, in Btu's/hr, to convert -10°C harvested ice to 500°C, 2.5 MPa steam, per hour.

Table 7-5. Excerpt, Superheated Steam Table, SI Units.

Properties of Superheated Steam Metric/SI Units						
Abs. Press. in MPa		**Temp.** in °C	"**v**" in m^3/kg	"**h**" in kJ/kg	"**s**" in kJ/kg.°K	
(Sat. T, °C)		**150**	**300**	**500**	**650**	**800**
0.05	v	**3.889**	**5.284**	**7.134**		
(81.33)	*h*	**2780.1**	**3075.5**	**3488.7**		
	s	**7.9401**	**8.5373**	**9.1546**		
0.1	v	**1.9367**	**2.6398**	**3.5656**		
(99.61)	*h*	**2776.6**	**3074.5**	**3488.1**		
	s	**7.6147**	**8.2171**	**8.8361**		
1.0	v		**0.2580**	**0.3541**	**0.4245**	**0.4944**
(179.89)	*h*		**3051.7**	**3479.0**	**3810.5**	**4156.2**
	s		**7.1247**	**7.7640**	**8.1557**	**8.5024**
2.5	v		**0.0989**	**0.13998**	**0.1623**	**0.1896**
(223.99)	*h*		**3008.8**	**3462.1**	**3799.7**	**4148.9**
	s		**6.6438**	**7.3234**	**7.7056**	**8.0559**
3.0	v		**0.0812**	**0.1162**	**0.1405**	**0.1642**
(233.86)	*h*		**2994.3**	**3457.0**	**3797.0**	**4147.0**
	s		**6.5412**	**7.2356**	**7.6373**	**7.9885**
4.0	v		**0.0589**	**0.0864**	**0.1049**	**0.1229**
(250.36)	*h*		**2961.7**	**3445.8**	**3790.2**	**4142.5**
	s		**6.3638**	**7.0919**	**7.4989**	**7.8523**

Solution

Part (b) of this case study does require accounting for heat added during each of the five (5) stages of the overall process. Therefore, this part is divided into five sub-parts, each involving either sensible or latent heat calculation, based on the entry and exit temperature and phase status.

Table 7-3 lists specific heat for water and ice. These heat values will be used in the sensible heat calculations associated with Part (b). Table 7-4 lists latent heat values for water. These values will be used to compute the latent heats associated with stages that involve phase transformation.

(i) **Calculate the heat required to heat the ice from -10°C to 0°C:**
Since there is no change in phase involved, the entire heat absorbed by the ice (working substance) in this stage would be sensible heat.

The first stage of the overall power generating system is illustrated in Figure 7-4.

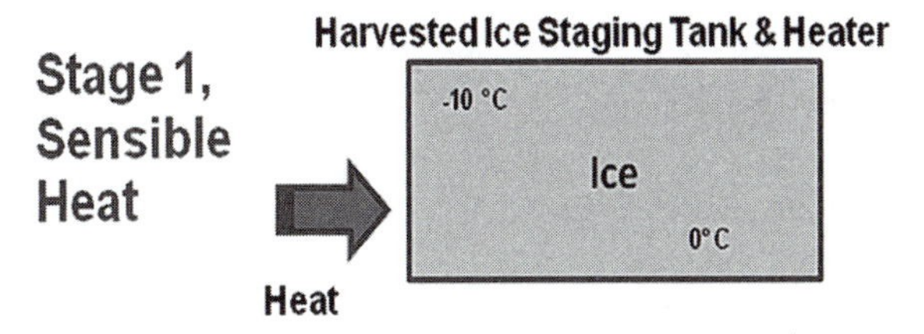

Figure 7-4. Case Study 7-1 Stage 1 Sensible Heat Calculation

Given:

$T_i = -10°C$
$T_f = 0°C$
$c_{ice} = 2.05$ kJ/kg. °K {Table 7-3}

Utilizing the given information:

$$\Delta T = T_f - T_i$$
$$\therefore \Delta T = 0 - (-10°C)$$
$$= +10°C$$

Since **ΔT** represents the ***change*** in temperature and not a specific absolute temperature,

$$\therefore \Delta T = +10°C = +10°K$$

Mathematical relationship between sensible heat, mass of the working substance, specific heat of the working substance and change in temperature can be stated as:

$$Q_{s(heat\ ice)} = m \cdot c_{ice} \cdot \Delta T \qquad \textbf{Eq. 7-14}$$

And,

$$\dot{Q}_{s(heat\ ice)} = \dot{m} \cdot c_{ice} \cdot \Delta T \qquad \textbf{Eq. 7-15}$$

Where,

$Q_{s(heat\ ice)}$ = Sensible heat required to heat the ice over ΔT
$\dot{Q}_{s(heat\ ice)}$ = Sensible heat flow rate required to heat the ice over ΔT
m = Mass of ice being heated
c_{ice} = Specific heat of ice = **2.05 kJ/kg. °K**
ΔT = Change in temperature, in **°C or °K**
$\dot{m}$ = Mass flow rate of water/ice
= **16.35 kg/sec**

Or,

$\dot{m}$ = (16.35 kg/sec) . (3600 sec/hr)
= **58,860 kg/hr**

Then, by application of **Eq. 7-15**:

$$\dot{Q}_{s(heat\ ice)} = (58{,}860\ kg/hr) \cdot (2.05\ kJ/kg.\ °K) \cdot (10\ °K)$$

Or,

$$\dot{Q}_{s(heat\ ice)} = \mathbf{1{,}206{,}630\ kJ/hr}$$

Since there are 1.055 kJ per Btu,

$$\dot{Q}_{s(heat\ ice)} = \mathbf{(1{,}206{,}630\ kJ/hr)/(1.055\ kJ/Btu)}$$

Or,

$$\dot{Q}_{s(heat\ ice)} = \mathbf{1{,}143{,}725\ Btu/hr}$$

(ii) Calculate the heat required to melt the ice at 0°C.
Since change in phase ***is*** involved in this case, the heat absorbed by the ice (working substance) in this stage would be latent heat.

The 2nd stage of the overall power generating system is illustrated in Figure 7-5.

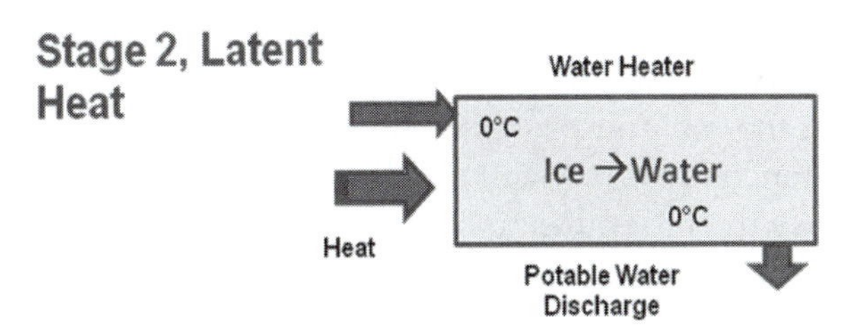

Figure 7-5. Case Study 7-1 Stage 2 Latent Heat Calculation

Mathematical relationship between latent heat, mass of the working substance, and the heat of fusion of ice can be stated as:

$$Q_{l(latent\ ice)} = h_{sl\ (ice)} \cdot m \qquad \textbf{Eq. 7-16}$$

And,

$$\dot{Q}_{l(latent\ ice)} = h_{sl\ (ice)} \cdot \dot{m} \qquad \textbf{Eq. 7-17}$$

Where,

$Q_{l(latent\ ice)}$ = Latent heat required to melt a specific mass of ice, isothermally

$\dot{Q}_{l(latent\ ice)}$ = Latent heat flow rate required to melt a specific mass of ice, isothermally, over a period of time

m = Mass of ice being melted

$\dot{m}$ = Mass flow rate of water/ice
= 60 tons/hr
= (60 tons/hr) . (907.2 kg/ton)
= **58,860 kg/hr,** same as part (a) (i)

$h_{sl\ (ice)}$ = Heat of fusion for ice or water
= **333.5 kJ/kg** {Table 7-4}

Then, by application of **Eq. 7-17**:

$$\dot{Q}_{l(latent\ ice)} = h_{sl\ (ice)} \cdot \dot{m}$$

$$\dot{Q}_{l(latent\ ice)} = (333.5\ kJ/kg).\ (58{,}860\ kg/hr)$$

$$\dot{Q}_{l(latent\ ice)} = \mathbf{19{,}629{,}810\ kJ/hr}$$

Since there are 1.055 kJ per Btu,

$$\dot{Q}_{l(latent\ ice)} = \mathbf{(19{,}629{,}810\ kJ/hr)/(1.055\ kJ/Btu)}$$

Or,

$$\dot{Q}_{l(latent\ ice)} = \mathbf{18{,}606{,}455\ Btu/hr}$$

Note that the specific heat required to melt ice is called heat of fusion because of the fact that the water molecules come closer together as heat is added in the melting process. The water molecules are held apart at specific distances in the crystallographic structure of solid ice. The heat of fusion allows the molecules to overcome the crystallographic forces and "**fuse**" to form liquid water. This also explains why the density of water is higher than the density of ice.

(iii) Calculate the heat reqd. to heat the water from 0°C to 100°C:
The 3rd stage of the overall power generating system is illustrated in Figure 7-6. Since no phase change is involved in this stage, the heat absorbed by the water in this stage would be sensible heat.

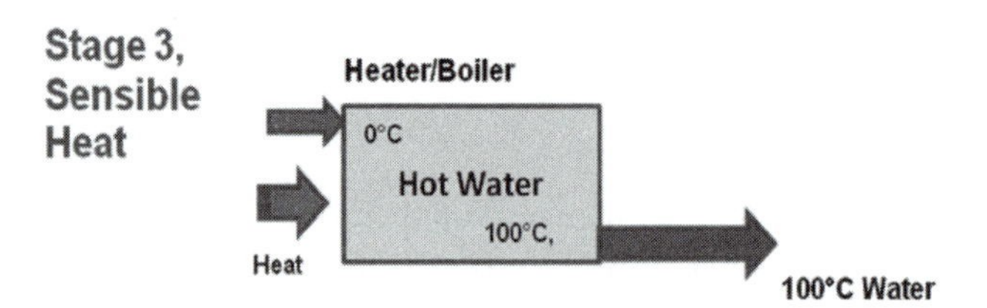

Figure 7-6. Case Study 7-1 Stage 3 Sensible Heat Calculation

Given:

$T_i = 0°C$

$T_f = 100°C$

$c_{p\text{-water}} = 4.19$ kJ/kg. °K {Table 7-3}

Utilizing the given information:

$$\Delta T = T_f - T_i$$

$$\therefore \Delta T = 100°C - 0°C$$

$$= 100°C$$

Since **ΔT** represents the ***change*** in temperature and not a specific absolute temperature,

$$\therefore \Delta T = 100°K$$

Mathematical relationship between sensible heat, mass of the working substance, specific heat of water (the working substance), and change in temperature can be stated as:

$$Q_{s(water)} = m \cdot c_{p\text{-}water} \cdot \Delta T \qquad \textbf{Eq. 7-18}$$

And,

$$\dot{Q}_{s(water)} = \dot{m} \cdot c_{p\text{-}water} \cdot \Delta T \qquad \textbf{Eq. 7-19}$$

Where,

$Q_{s(water)}$ = Sensible heat required to heat the water over ΔT

$\dot{Q}_{s(water)}$ = Sensible heat ***flow rate*** required to heat the water over ΔT

m = Mass of water being heated

$c_{p\text{-}water}$ = Specific heat of water = **4.19 kJ/kg. °K**

$\dot{m}$ = Mass flow rate of water = **58,860 kg/hr**, as calculated in part (a)

ΔT = Change in temperature, in **°C or °K**

Then, by applying **Eq. 7-19**:

$$\dot{Q}_{s(water)} = \dot{m} \cdot c_{p\text{-}water} \cdot \Delta T$$

$$\dot{Q}_{s(water)} = (58{,}860 \text{ kg/hr}) \cdot (4.19 \text{ kJ/kg. °K}) \cdot (100 \text{ °K})$$

$$\dot{Q}_{s(water)} = \mathbf{24{,}662{,}340 \text{ kJ/hr}}$$

Since there are 1.055 kJ per Btu,

$$\dot{Q}_{s(water)} = \mathbf{(24{,}662{,}340 \text{ kJ/hr})/(1.055 \text{ kJ/Btu})}$$

Or,

$$\dot{Q}_{s(water)} = \mathbf{23{,}376{,}626 \text{ Btu/hr}}$$

(iv) Calculate the heat required to convert 100°C water to 100°C steam:
The 4th stage of the overall power generating system is illustrated in Figure 7-7. Since change in phase ***is*** involved in this case, the heat absorbed by the water in this stage would be latent heat.

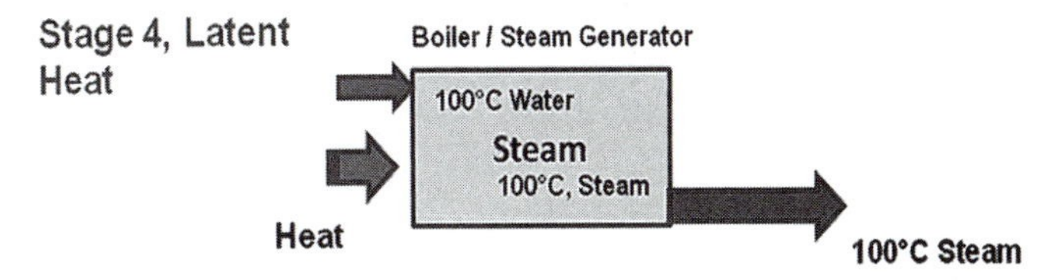

Figure 7-7. Case Study 7-1 Stage 4 Latent Heat Calculation

Mathematical relationship between latent heat of vaporization for water, $\mathbf{h_{fg(water)}}$, mass of the water, and the total heat of vaporization of water, $\mathbf{Q_{l(latent\ water)}}$, can be stated as:

$$Q_{l(latent\ water)} = h_{fg\ (water)} \cdot m \qquad \textbf{Eq. 7-20}$$

And,

$$\dot{Q}_{l(latent\ water)} = h_{fg\ (water)} \cdot \dot{m} \qquad \textbf{Eq. 7-21}$$

Where,

$\mathbf{Q_{l(latent\ water)}}$ = Latent heat of vaporization of water required to evaporate a specific mass of water, isothermally

$\mathbf{\dot{Q}_{l(latent\ water)}}$ = Latent heat of vaporization *flow rate* required to evaporate a specific mass of water, isothermally, over a given period of time

$\mathbf{m}$ = Mass of water being evaporated

$\mathbf{\dot{m}}$ = Mass flow rate of water
= 60 tons/hr
= (60 tons/hr) . (907.2 kg/ton)
= **58,860 kg/hr,** same as part (a) (i)

$\mathbf{h_{fg\ (water)}}$ = latent heat of vaporization for water
= **2257 kJ/kg** {From the steam tables and Table 7-4}

Then, by application of **Eq. 7-21**:

$$\dot{Q}_{l(latent\ water)} = h_{fg\ (water)} \cdot \dot{m}$$

$$\dot{Q}_{l(latent\ water)} = (2257\ kJ/kg).\ (58{,}860\ kg/hr)$$

$$\dot{Q}_{l(latent\ water)} = \mathbf{132{,}487{,}020\ kJ/hr}$$

Since there are 1.055 kJ per Btu,

$$\dot{Q}_{l(latent\ water)} = \mathbf{(132{,}487{,}020\ kJ/hr)/(1.055\ kJ/Btu)}$$

Or,

$$\dot{Q}_{l(latent\ water)} = \mathbf{125{,}921{,}346\ Btu/hr}$$

(v) Calculate the heat reqd. to heat the steam from 100°C, 1-atm (102 KPa, or 1-bar) to 500°C, 2.5 MPa superheated steam:

The 5th stage of the overall power generating system is illustrated in Figure 7-8. Since this stage involves no phase change, the heat absorbed by the steam is sensible heat.

In superheated steam phase, the heat required to raise the temperature and pressure of the steam can be determined using the enthalpy difference between the initial and final conditions.

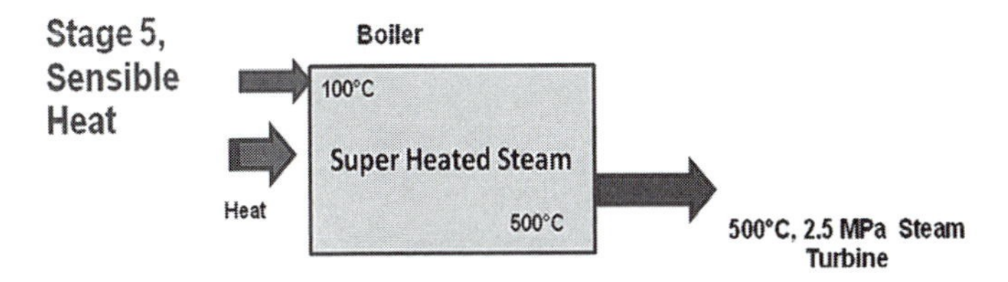

Figure 7-8. Case Study 7-1 Stage 5 Sensible Heat Calculation

Given:

T_i = 100°C

P_i = 1- Atm. Note: At 100°C, the saturation pressure is 1- Atm, 1-Bar, or 102 kPa

T_f = 500°C

P_f = 2.5 MPa

For the initial and final temperature and pressure conditions stated above, the enthalpy values, as read from saturated steam table excerpt in Table 7-5a and the superheated steam table excerpt in Table 7-5, are as follows:

Table 7-5a. Excerpt, Saturated Steam Table, SI.

Properties of Saturated Steam By Temperature, Metric/SI Units									
Temp. °C	Abs. Press. MPa	Specific Volume m³/kg		Enthalpy kJ/kg			Entropy kJ/kg		Temp. °C
		Sat. Liquid v_L	Sat. Vapor v_V	Sat. Liquid h_L	Evap. h_{fg}	Sat. Vapor h_V	Sat. Liquid s_L	Sat. Vapor s_V	
20	0.002339	0.0010018	57.7610	83.920	2454.1	2537.5	0.2965	8.6661	20
50	0.012351	0.0010121	12.0280	209.34	2382.7	2591.3	0.7038	8.0749	50
100	0.101420	0.0010435	1.6719	419.10	2257.0	2675.6	1.3070	7.3541	100
200	1.554700	0.0011565	0.1272	852.39	1940.7	2792.1	2.3308	6.4303	200

$\mathbf{h_i = 2676\ kJ/kg\ at\ 100°C,\ 1\text{-}Atm}$
$\mathbf{h_f = 3462\ kJ/kg\ at\ 500°C,\ 2.5\ MPa}$

Equations for determining the heat required to boost the steam from **100°C, 1-Atm** to **500°C, 2.5 MPa** are as follows:

$$\Delta Q_{steam} = (h_f - h_i) \cdot m \qquad \textbf{Eq. 7-22}$$

$$\dot{Q}_{steam} = (h_f - h_i) \cdot \dot{m} \qquad \textbf{Eq. 7-23}$$

Where,

ΔQ_{steam} = Addition of heat required for a specific change in enthalpy
$\dot{Q}_{steam}$ = Rate of addition of heat for a specific change in enthalpy
h_i = Initial enthalpy
h_f = Final enthalpy
m = Mass of steam being heated
$\dot{m}$ = Mass flow rate of steam as calculated in part (a) of this case study
= **58,860 kg/hr** {From Part (a)}

Then, by applying **Eq. 7-23**:

$$\dot{Q}_{steam} = (h_f - h_i) \cdot \dot{m}$$

$$\dot{Q}_{steam} = (3462\ kJ/kg - 2676\ kJ/kg).\ (58{,}860\ kg/hr)$$

$$\dot{Q}_{steam} = 46{,}263{,}960\ kJ/hr$$

Since there are 1.055 kJ per Btu,

$$\dot{Q}_{s(water)} = (46{,}263{,}960 kJ/hr)/(1.055\ kJ/Btu)$$

Or,

$$\dot{Q}_{s(water)} = 43{,}852{,}095\ Btu/hr$$

After assessing the heat added, per hour, during each of the five (5) stages of the steam generation process, add all of the heat addition rates to compile the total heat addition rate for the power generating station. The tallying of total heat is performed in Btu's/hr, as well as in kJ/hr.

Total Heat Addition Rate in kJ/hr

Total Heat Required to Generate 500°C, 2.5 MPa steam from -10°C Ice, at 58,860 kg/hr

= 1,206,630 kJ/hr + 19,629.810 kJ/hr + 24,662,340 kJ/hr + 132,487,020 kJ/hr + 46,263,960 kJ/hr

= **224,609,760 kJ/hr**

Total Heat Addition Rate in Btu's/hr:

Total Heat Required to Generate 500°C, 2.5 MPa steam from -10°C Ice, at 58,860 kg/hr

= 1,143,725 Btu/hr + 18,606,455 Btu/hr + 23,376,626 Btu/hr + 125,580,114 Btu/hr + 43,852,095 Btu/hr

= **212,559,014 Btu/h**

c) Calculate the volume, in cu-ft, of natural gas required to power up the station, each day. Assume 98% burner efficiency.

Solution

This part of Case Study 7-1 requires computation of the amount (volume) of natural gas required to power up the station each day. This calculation is straight forward after the derivation of the total energy required, per hour, in part (b). However, the hourly energy requirement must be "scaled up" to account for the 98% efficiency of the boiler burner.

The hourly energy requirement, in kJ or Btu's, can be extended into daily usage. The daily energy usage can then be converted into the volume of natural gas required, based on natural gas energy content listed in Table 7-2.

Total Energy Required Per Day = (212,559,014 Btu/hr)/(0.98) x 24 hr

= **5,205,526, 873 Btu's**

Since Natural Gas Energy Content = 1034 Btu/cu-ft,

Total volume of natural gas required per day

= (5,205,526, 873 Btu's)/1034 Btu/cu-ft

= **5,034,358 cu-ft**

d) If the natural gas transportation cost is $ 4.85/DT in addition to the well head or commodity cost stated in Table 7-2, what would be the annual fuel cost of operating this station?

Solution

Part (d) of the case study relates to the computation of total annual cost of fuel for the power generation station. This requires the extrapolation of daily energy consumption into ***annual*** energy consumption. The annual energy consumption is then multiplied with the ***total,*** delivered, cost rate in **$/DT** to obtain the annual cost in dollars.

From part (b), the heat energy required for operating the power station = **212,559,014** Btu/hr

Then, based on 24/7 operating schedule assumption, the total energy required per year would be:

= (212,559,014 Btu/hr)/(0.98) x 8760 hrs/year
= 1.900 E+12 Btu's

Since there are 1,000,000 Btu's, or 1 MMBtu, per DT, the Total Annual Energy Consumption in DT would be:

= **1.900 E+12 Btu's**/(1,000,000 Btu's/DT)
= **1,900,017 DT's**

The total cost of natural gas, per DT =

Cost at the Source/Well + Transportation Cost
= $4.15/DT + $4.85/DT
= **$9.00/DT**

Then, the Total Annual Cost for Producing 10MW of Power with Natural Gas

= (1,900,017 DT) . ($9.00/DT)
= $ 17,100,156

e) What is the overall energy efficiency of the power station ?

Solution/Answer:

Part (e) entails determination of the overall efficiency of the power generating station. Efficiency calculation, in this case study, requires knowledge of the total electrical energy (or power) produced and the total energy (or power) consumed through the combustion of natural gas. The output of the overall power generating system is ostensible from the given system output rating of 10 MW. The total heat consumption by the system is derived in parts (b), (c) and (d). The overall efficiency of the system can then be assessed by dividing the output power (or energy) by the input power (or energy).

From Part (c):

Total Energy Required Per Day, in Btu's = 5,205,526, 873 Btu's

Since there are 1.055kJ per Btu,

The Total Energy Input, Each Day, in kJ
= 5,205,526, 873 Btu's x 1.055 kJ/Btu
= 5,491,830,851 kJ

This fuel energy usage can be converted into Watts of Mega Watts (MW) by dividing the energy usage by the total number of seconds in a day . This is because the **5,491,830,851 kJ** of energy is used over a period of a day.

∴ **The System Power Input**
= (5,491,830,851 kJ)/(24 hr x 3600 sec/hr)
= 63,563 kJ/sec or 63,563 kW

Since there are 1000 kW per MW,

The System Power Input, in MW
= (63,563 kW)/(1,000 kW/MW)
= 63.563 MW

Since Power Output = 10.00 MW, { Given }
Total Station Energy Efficiency, in Percent
= (Power Output/Power Input) x 100
= 10 MW/63.563 MW x 100
= 16%

f) If heat is added to the steam turbines, would the steady flow energy process in the turbine system constitute an adiabatic process or a non-adiabatic process?

Solution/Answer:

The answer to Part (f) of the case study lies simply in the definition for adiabatic process.

Adiabatic process is a thermodynamic process in which no heat either enters or leaves the thermodynamic system boundary. Therefore, this process is a **Non-Adiabatic Process.**

g) What is the change in entropy, Δs, in the turbine system?

Solution/Answer:

Part (g) involves computation of the change in entropy in the turbine segment of the overall thermodynamic process. As characteristic of the turbine stage of a typical heat engine cycle, there is a small change in entropy as the working fluid travels from the intake to the exit point of the turbine. This is evident as the entropy is assessed, through the superheated steam tables, for the incoming steam at 500°C and 2.5MPa, and the outgoing steam at 150°C and 50 kPa.

To determine the change in entropy, Δs, in the turbine system we need to identify $\mathbf{s}_i$, the entropy of 500°C, 2.5 MPa steam entering the steam turbine and $\mathbf{s}_f$, the entropy of 150°C, 50 kPa steam exiting the turbine.

$$\Delta s = s_i - s_f$$

$\mathbf{s}_i$ **= 7.3234 kJ/kg. °K**{From superheated steam tables, See Table 7-5}

$\mathbf{s}_f$**= 7.9401 kJ/kg. °K**{From superheated steam tables, See Table 7-5}

$$\begin{aligned} \therefore \Delta s &= s_i - s_f \\ &= 7.3234 \text{ kJ/kg. °K} - 7.9401 \text{ kJ/kg. °K} \\ &= -\,0.6167 \text{ kJ/kg. °K} \end{aligned}$$

Note: The negative sign signifies the loss of heat through the turbine casing.

CASE STUDY 7-1. US/IMPERIAL UNITS, WITH ILLUSTRATION OF INTERPOLATION METHOD

Technical feasibility of a toping cycle cogenerating power plant is being studied at Station "Zebra." This facility is to be stationed in a remote Arctic region. The objective of this plant is to produce steam and generate electricity for an Arctic Environmental Monitoring and Deep Sea Mining Facility. Due to saltwater corrosion risk, it has been established that local glacier ice will be harvested and utilized for steam production purposes. A natural gas boiler is to be used to generate steam. The average temperature of the glacier ice hovers at 14°F, through the year. The glacier ice is to be melted and converted to 932°F, 362.6 psia, steam; the steam enters the turbine at this temperature and pressure. The steam is discharged by the turbines at 302°F and 7.252 psia. This discharged steam is used for the mining process and to heat the station. The condensate is used as potable, utility and process water. Assume that the turbine represents an open, steady flow, thermodynamic system. In other words, the SFEE, Steady Flow Energy Equation applies. Also assume the potable water demand and flow, along the thermodynamic process stages to be relatively negligible. Station Zebra would operate on a 24/7 schedule. For simplicity, the thermodynamic process flow for this system is laid out in Figure 7-9. **Note:** In an actual project setting, developing such a process flow diagram would constitute the first order of business as an energy engineer begins analyzing this process.

a) Estimate the mass flow rate $\dot{\mathbf{m}}$ for generation of 10 MW electrical power if the rate of turbine casing heat loss, $\dot{\mathbf{Q}}$, is 28.435 Btu/s (or 0.03 MW), exit velocity of steam, $\mathbf{v_2}$, is 114.83 ft/sec, entrance velocity of steam, $\mathbf{v_1}$, is 49.21 ft/sec, steam exit elevation is $\mathbf{z_2} = 3.28$ ft, steam intake elevation is $\mathbf{z_1} = 1.64$ ft. Assume electric power generator efficiency of 90%. Extrapolate the answer into approximate truck loads per hour. Assume truck capacity of approximately 353.15 ft^3.

b) Assume that the power station is generating 10 MW of electric power. Calculate the amount of total heat energy needed, in Btu's/hr, to convert 14°F harvested ice to 932°F, 362.6 psia steam per hour.

c) Calculate the volume, in cu-ft, of natural gas required to power up the station, each day. Assume 98% burner efficiency.

d) If the natural gas transportation cost is $4.85/DT in addition to the well head or commodity cost stated in Table 7-6, what would be the annual fuel cost of operating this station?

e) What is the overall energy efficiency of the power station?

f) If heat is added to the steam turbines, would the steady flow energy process in the turbine system constitute an adiabatic process or a non-adiabatic process?

g) What is the change in entropy, **Δs**, in the turbine system?

CASE STUDY SOLUTION STRATEGY

Before embarking on the analyses and solution for this case study, let's highlight some important facts from the case study statement.

1) As apparent from the case study statement, the working fluid or system consists of water, in various phases.

2) Water is introduced into the overall thermodynamic system in form of 14°F ice and is then heated during various sensible and latent stages. It is, finally, introduced into the turbine as 932°F, 362.6 psia superheated steam.

3) Unlike a typical Rankine Cycle Heat Engine, the steam exhausted from the turbine is not condensed, pumped and recycled through the boilers to repeat the heat cycle. Instead, fresh ice is harvested, melted and introduced as working fluid.

4) Even though Figure 7-9 shows potable water being removed from the system at Stages 2 and 3, it is assumed to be negligible. Thus supporting the assumption that mass flow rate of the working fluid stays constant through the system.

5) Fuel heat content and fuel cost information is available through Table 7-6.

6) Specific heat and latent heat data for water and ice are provided through Tables 7-7 and 7-8, for sensible heat and latent heat calculations, respectively.

7) Most of the data pertinent to the application of SFEE equation, at the turbine, is given, including the velocities, temperatures, pressures and elevations.

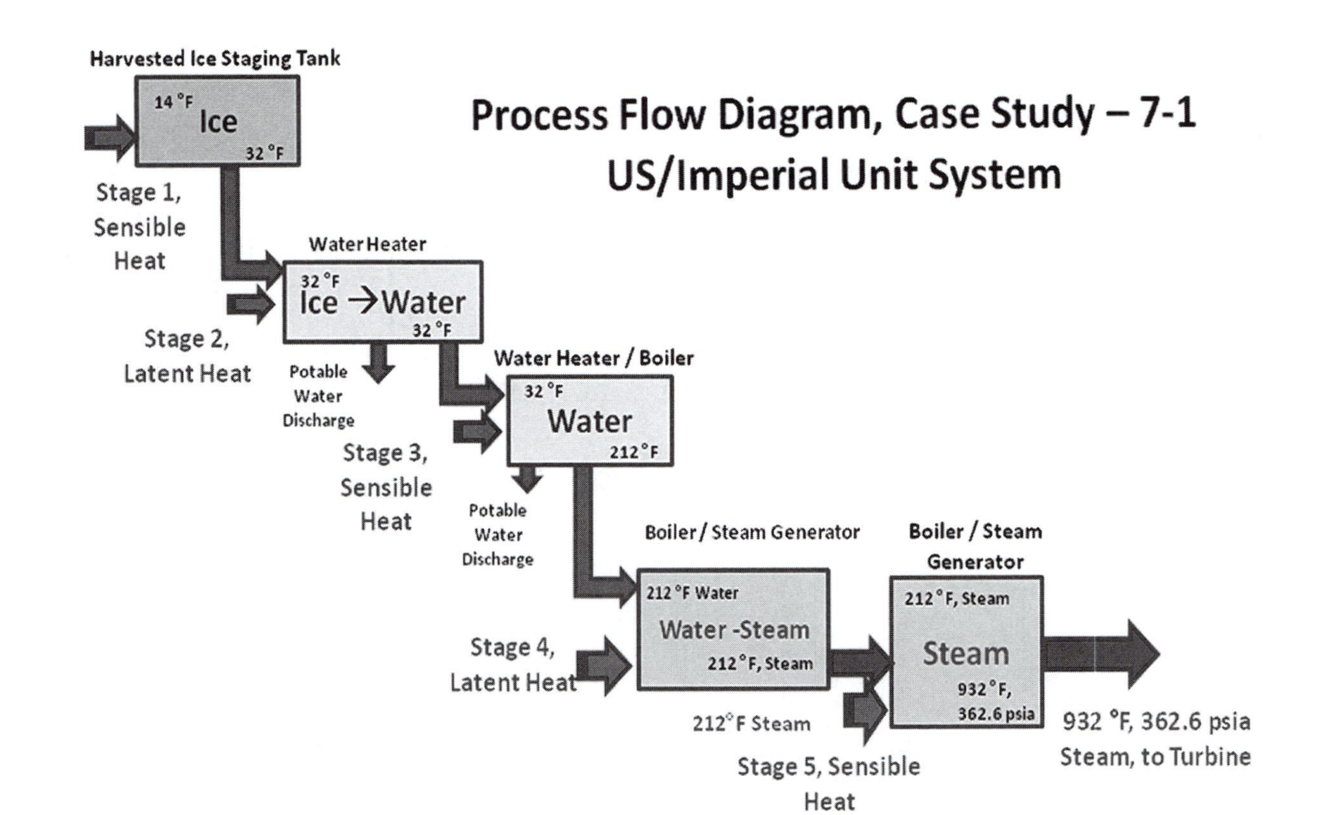

Figure 7-9. Thermodynamic Process Flow Diagram, Case Study 7-1

Table 7-6. Energy Content and Cost Comparison of Common Fuels

Energy Content and Cost Comparison										
Fuel Type or Energy Source	Heat Content						Approx. Cost		US Sept. 2010 Average	
	DT/gal	BTU/gal	MMBTU/Barrel	BTU/Cu-ft.	DT/MCF	Price Per Gallon*	$/DT*	BTU/kWh	$/kWh	$/MMBTU
Propane	0.092	91,600	3.35	2,488		$1.42	$15.38			27.29
Diesel/No. 2 Fuel Oil	0.138	140,000	5.6			$2.65	$19.17			23.429
No. 6 Fuel Oil	0.144	143,888	6.8							
Natural Gas				1,034	1.034	$1.00	$4.15			4.15
Electricity							$30.01	3,412	0.1024	30.01

* Note: These costs represent a January 3, 2011 snapshot of wholesale or industrial market costs.

8) Final output of the power generating station is given in terms of the 10 MW rating of the generator and its stated efficiency of 90%.

The overall thermodynamic process flow can be tiered into stages that involve either sensible heating or latent heating. All heating stages of this comprehensive process are depicted in Figure 7-9. Each stage of the overall process is labeled with pertinent entry and exit temperature and pressure, as applicable. Each stage is named as either Sensible or Latent Stage. Furthermore, each stage shows the phase of water at point of entry and exit.

a) Estimate the mass flow rate $\dot{m}$ for generation of 10 MW electrical power output if the rate of turbine casing heat loss, $\dot{Q}$, is 28.435 Btu/s (or

Table 7-7. Specific Heat, $c_{p,}$ for Selected Liquids and Solids

Approximate Specific Heat, $c_{p,}$ for Selected Liquids and Solids			
Substance	c_p in J/gm K	c_p in cal/gm K or	Molar C_p
		BTU/lb F	J/mol K
Aluminum	0.9	0.215	24.3
Bismuth	0.123	0.0294	25.7
Copper	0.386	0.0923	24.5
Brass	0.38	0.092	...
Gold	0.126	0.0301	25.6
Lead	0.128	0.0305	26.4
Silver	0.233	0.0558	24.9
Tungsten	0.134	0.0321	24.8
Zinc	0.387	0.0925	25.2
Mercury	0.14	0.033	28.3
Alcohol(ethyl)	2.4	0.58	111
Water	4.186	1	75.2
Ice (-10 C)	2.05	0.49	36.9
Granite	0.79	0.19	...
Glass	0.84	0.2	

Table 7-8. Latent Heat for Phase Transformation of Water

	Latent Heat Fusion h_{sl}	**Latent Heat Sublimation** h_{ig}	**Latent Heat Vaporization** h_{fg}
kJ/kg	333.5	2838	2257
BTU/lbm	143.4	1220	970.3
kcal/kg	79.7	677.8	539.1

0.03 MW), exit velocity of steam, $\mathbf{v_2}$, is 114.83 ft/sec, entrance velocity of steam, $\mathbf{v_1}$, is 49.21 ft/sec, steam exit elevation is $\mathbf{z_2} = 3.28$ ft, steam intake elevation is $\mathbf{z_1} = 1.64$ ft. Assume electric power generator efficiency of 90%. Extrapolate the answer into approximate truck loads per hour. Assume truck capacity of approximately 353.15 ft^3.

Solution:

This part of the case study can be analyzed and solved by simply focusing on the very last stage and applying the SFEE in form of Eq. 7-12. The computation of mass flow rate does not require assessment of the heat required at the various stages of the overall thermodynamic process, in this case study, because of the following key assumption included in the problem statement:

"Also assume the potable water demand and flow, along the thermodynamic process stages to be relatively negligible. "

In other words the mass flow rate is assumed to be constant through out the process, and any discharge of water during individual stages of the overall process is negligible.

The turbine segment of the overall power generating system is illustrated in Figure 7-10. The enthalpy values are obtained from the superheated steam table excerpts in and Table 7-10.

Figure 7-10. Case Study 7-1, Mass Flow Rate Analysis, US Units

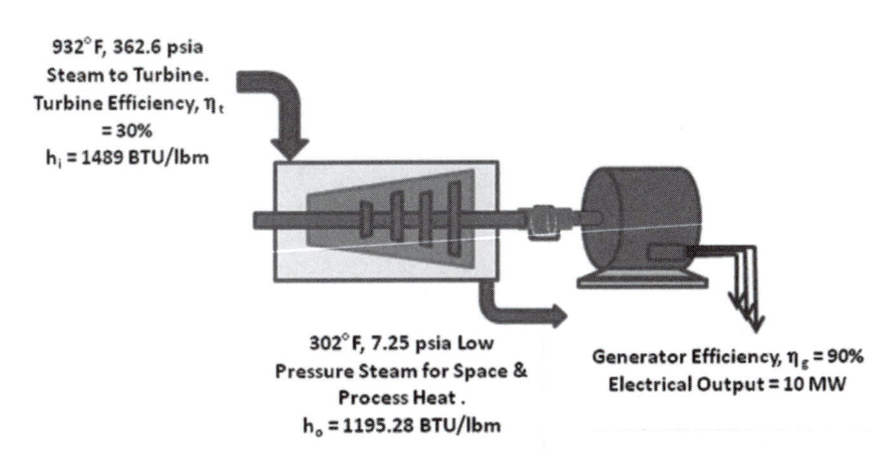

Single and Double Interpolation of Steam Table Data for Enthalpy Determination

Interpolation is often required when the retrieving data from tables such as the Saturated Steam Tables or the Superheated Steam Tables. Interpolation, specifically in steam tables, is needed when the given pressure, temperature or both don't coincide with the standard pressure and temperature values on the given tables.

This case study, in the US unit realm, offers an opportunity to study the interpolation method. Even though interpolation method is being illustrated on the basis of steam tables here, this technique can be employed for interpolation of other types of tabular data, as well. In this case study, enthalpy values need to be retrieved from the Superheated Steam Tables in Appendix B. The pressure and temperature for the initial (turbine entry) enthalpy, h_i, are 362.6 psia and 932°F, respectively. The pressure and temperature for the final (turbine exit) enthalpy, h_f, are 7.25 psia and 302°F, respectively. However, as you examine the superheated steam tables for these parameters, exact match for this data is not available in the tables. See Tables 7-9 and 7-10, for excerpts from the superheated steam tables in Appendix B.

As far as the initial point is concerned, h_i can be obtained through only one interpolation, or single interpolation; the interpolation associated with the temperature of 932°F, by rounding of the pressure to 360 psia. The magnitude of error in the enthalpy value, due to the 2.6 psia difference between 362.6 psia and 360 psia, is insignificant. This single interpolation approach was adopted in the derivation of the value of $h_i = 1488.76$ Btu/lbm.

The formula for single interpolation, applied between the stated or available enthalpy values for 900°F and 1000°F, at 360 psia, is as follows:

$$\mathbf{h_{i\ at\ 932°F,\ 362.6\ psia} = h_{i\ at\ 932°F,\ 360\ psia}}$$
$$\mathbf{= ((h_{i\ at\ 1000°F,\ 360\ psia} - h_{i\ at\ 900°F,\ 360\ psia})/(1000°F - 900°F)).(932-900) + h_{i\ at\ 900°F,\ 360\ psia}}$$

Substituting enthalpy values and other given data from superheated steam table excerpt, shown in Table 7-9:

$$\mathbf{h_{i\ at\ 932°F,\ 362.6\ psia} = h_{i\ at\ 932°F,\ 360\ psia}}$$
$$\mathbf{= ((1525\ Btu/lbm - 1471.7\ Btu/lbm)/(1000°F - 900°F)).(932°F - 900°F) + 1471.7\ Btu/lbm}$$
$$\mathbf{= 1488.76\ Btu/lbm}$$

Note: The available enthalpy values are circled in Table 7-9.

Double interpolation method is employed in deriving $\mathbf{h_f}$, the enthalpy value at 7.25 psia and 302°F. As apparent from the superheated steam tables in Appendix B, this value is not readily available and, therefore, double interpolation must be conducted between the enthalpy values given for 5 psi, 300°F, and 10 psi, 350°F, to derive $\mathbf{h_{f\ at\ 7.25\ psia\ and\ 302°F}}$. Where, $\mathbf{h_{f\ at\ 7.25\ psia\ and\ 302°F}}$ is the final enthalpy—enthalpy at the turbine exit point—at 7.25 psia and 302°F. The double interpolation approach, as applied here, will entail three steps.

First step involves determination of $\mathbf{h_{f,\ at\ 7.25\ psia\ and\ 300°F}}$, the enthalpy value at 7.25 psia and 300°F. The enthalpy values available and used in this first interpolation step are circled in Table 7-10. The following formula sums up the mathematical approach to this first step:

$$\mathbf{h_{f\ at\ 300°F,\ 7.25\ psia}}$$
$$\mathbf{= ((h_{f\ at\ 300°F,\ 5\ psia} - h_{i\ at\ 300°F,\ 10\ psia})/(10\ psia - 5\ psia)).(10\ psia - 7.25\ psia)}$$
$$\mathbf{+ h_{f\ at\ 300°F,\ 10\ psia}}$$

Substituting enthalpy values and other given data from superheated steam table excerpt, shown in Table 7-10:

Table 7-9. Superheated Steam Table Excerpt, US/Imperial Units

Properties of Superheated Steam US/Imperial Units					
Abs. Press. psia		**Temp. °F**	**Note: v is in ft^3/lbm, *h* is in BTU/lbm and *s* is in BTU/(lbm-°R)**		
(Sat. Temp. °F)		500	700	900	1000
260 (404.45)	*v*	2.062	2.5818	3.0683	3.3065
	h	1262.5	1370.8	1475.2	1527.8
	s	1.5901	1.6928	1.7758	1.8132
360 (434.43)	*v*	1.446	1.8429	2.2028	2.3774
	h	1250.6	1365.2	1471.7	1525
	s	1.5446	1.6533	1.7381	1.7758
450 (456.32)	*v*	1.1232	1.4584	1.7526	1.8942
	h	1238.9	1360	1468.6	1522.4
	s	1.5103	1.6253	1.7117	1.7499
600 (486.25)	*v*			1.3023	1.411
	h			1463.2	1518
	s			1.577	1.7159

$\mathbf{h_{f\ at\ 300°F,\ 7.25\ psia}}$

= ((1194.8 Btu/lbm – 1193.8)/(10 psia -5 psia)).(10 psia-7.25 psia) + 1193.8 Btu/lbm

= **1194.35 Btu/lbm**

Second interpolation step involves determination of $\mathbf{h_{f,\ at\ 7.25\ psia\ and\ 350°F}}$, the enthalpy value at 7.25 psia and 350°F. The enthalpy values available and used in this interpolation step are circled in Table 7-10. The following formula sums up the mathematical approach associated with this interpolation step:

$\mathbf{h_{f\ at\ 350°F,\ 7.25\ psia}}$
$= \mathbf{((h_{f\ at\ 350°F,\ 5\ psia} - h_{i\ at\ 350°F,\ 10\ psia})/(10\ psia - 5\ psia)).(10\ psia - 7.25\ psia) + h_{f\ at\ 350°F,\ 10\ psia}}$

Substituting enthalpy values and other given data from superheated steam table excerpt, shown in Table 7-10:

$\mathbf{h_{f\ at\ 350°F,\ 7.25\ psia}}$
$= \mathbf{((1218\ Btu/lbm - 1217.2_{)}/(10\ psia - 5\ psia)).(10\ psia - 7.25\ psia) + 1217.2\ Btu/lbm}$
$= \mathbf{1217.64\ Btu/lbm}$

The final step in the double interpolation process, as applied in this case, involves interpolating between $\mathbf{h_{f\ at\ 300°F,\ 7.25\ psia}}$ and $\mathbf{h_{f\ at\ 350°F,\ 7.25\ psia}}$ the enthalpy values derived in the first two steps above, to obtain the desired final enthalpy $\mathbf{h_{f\ at\ 302°F,\ 7.25\ psia.}}$

The formula for this final step is as follows:

$\mathbf{h_{f\ at\ 302°F,\ 7.25\ psia}}$
$= \mathbf{((h_{f\ at\ 350°F,\ 7.25\ psia} - h_{f\ at\ 300°F,\ 7.25\ psia})/(350°F - 300°F)).(302°F - 300°F) + h_{f\ at\ 300°F,\ 7.25\ psia}}$

Substituting enthalpy values derived in the first two steps above:

$h_{f\ at\ 302°F,\ 7.25\ psia}$
$= ((1217.64\ Btu/lbm - 1194.35\ Btu/lbm)/(350°F - 300°F)).(302°F - 300°F) + 1194.35\ Btu/lbm$

$h_{f\ at\ 302°F,\ 7.25\ psia} = \mathbf{1195.28\ Btu/lbm}$

With the key enthalpy values for turbine entry and exit points identified, we can proceed with the computation of the mass flow rate.

The **Open System Steady Flow Energy Equation** for power computation in the US or Imperial unit realm is:

Table 7-10. Superheated Steam Table Excerpt, US/Imperial Units

Properties of Superheated Steam US/Imperial Units					
Abs. Press. psia (Sat. T, °F)		**Temp. °F**	Note: v is in ft³/lbm, *h* is in BTU/lbm and *s* is in BTU/(lbm-°R)		
		200	300	350	500
5 (162.18)	*v*		90.248	96.254	
	h		1194.8	1218	
	s		1.937	1.9665	
10 (193.16)	*v*	38.851	44.993	48.022	
	h	1146.4	1193.8	1217.2	
	s	1.7926	1.8595	1.8893	
15 (212.99)	*v*		29.906		37.986
	h		1192.7		1287.3
	s		1.8137		1.9243
100 (327.82)	*v*				5.5875
	h				1279.3
	s				1.7089

$$\dot{Q} = \dot{m}\,[\,(h_2 - h_1) + 1/2\,.\,(v_2^2 - v_1^2)/(g_c.J) + g.(z_2 - z_1)/(g_c.J)\,] + P_{shaft} \qquad \text{Eq. 7-12}$$

Rearrangement of Eq. 7-12 yields:

$$\dot{m} = (\dot{Q} - P_{shaft})/[\,(h_2 - h_1) + 1/2\,.\,(v_2^2 - v_1^2)/(g_c.J) + g.(z_2 - z_1)/(g_c.J)\,] \qquad \text{Eq. 7-24}$$

Where,

$\mathbf{h_1}$ = Enthalpy of the steam entering the turbine, in Btu/lbm.
$\mathbf{h_2}$ = Enthalpy of the steam exiting the turbine, in Btu/lbm.
$\mathbf{v_1}$ = Velocity of the steam entering the turbine, in ft/s.
$\mathbf{v_2}$ = Velocity of the steam exiting the turbine, in ft/s.
$\mathbf{z_1}$ = Elevation of the steam entering the turbine, in ft.
$\mathbf{z_2}$ = Elevation of the steam exiting the turbine, in ft.
$\dot{\mathbf{Q}}$ = Flow rate of heat added or removed from the turbine system, in Btus/sec.
$\dot{\mathbf{m}}$ = Mass flow rate of the system, in lbm/sec.
$\mathbf{g}$ = Acceleration due to gravity, 32.2 ft/s²
$\mathbf{g_c}$ = Gravitational constant = 32.2 lbm-ft/lbf-sec²
$\mathbf{J}$ = 778 ft-lbf/Btu = Joules constant

Given:

$\dot{Q}$ = - 0.03 MW = - 0.03x10⁶ W = - 0.03x10⁶ J/s

Since there are **1055 Joules per Btu**,

$\dot{Q}$ = (- 0.03x10⁶ J/s)/(1055 J/Btu)
= **- 28.4 Btu/s**

$\mathbf{P_{shaft}}$ = **(10 MW)/η_g**
= (10 MW)/(0.9)
= 11.11x10⁶ W
= 11.11x10⁶ J/s

Since there are 1055 Joules per Btu,

$\mathbf{P_{shaft}}$ = (11.11x10⁶ J/s)/(1055 J/Btu)
= **10,530 Btu/s**

$\mathbf{h_2}$ = **1195 Btu/lbm**, from Appendix B and interpolation stated above.
$\mathbf{h_1}$ = **1489 Btu/lbm** from Appendix B and interpolation stated above.
$\mathbf{v_2}$ = **114.83 ft/sec**
$\mathbf{v_1}$ = **49.21 ft/sec**
$\mathbf{z_2}$ = **3.28ft**
$\mathbf{z_1}$ = **1.64 ft**

By applying Eq. 7-24:

$$\dot{m} = (\dot{Q} - P_{shaft})/[(h_2 - h_1) + 1/2 \cdot (v_2^2 - v_1^2)/(g_c \cdot J) + g \cdot (z_2 - z_1)/(g_c \cdot J)] \qquad \text{Eq. 7-24}$$

ṁ = (- 28.4 Btu/s - 10,530 Btu/s)/[(1195 Btu/lbm – 1489 Btu/lbm) + 1/2 . ((114.83 ft/sec)^2 – (49.21 ft/sec)^2)/(32.2 lbm-ft/lbf-sec^2. 778 ft-lbf/Btu) + (32.2 ft/s^2).(3.28ft – 1.64 ft)/(32.2 lbm-ft/lbf-sec^2 .778 ft-lbf/Btu)]

For clarity, the **ṁ** computation equation, Eq. 7-24, with known values substituted, can be stated alternatively as:

$$\dot{m} = \frac{(-28.4 Btu/s - 10{,}530 Btu/s)}{\left[\left(1195 Btu/lbm - 1489 Btu/lbm + 1/2 \frac{(114.83 ft/s)^2}{(32.2 lbm-ft/lbf-s^2 \cdot (778 ft-lbf/Btu)} + (32.2 ft/s^2) \cdot \frac{3.28 ft - 1.64 ft)}{(32.2 lbm-ft/lbf-s^2 \cdot (778 ft-lbf/Btu)}\right)\right]}$$

ṁ = 35.94 lbm/sec,

Or,

ṁ = (35.94lbm / sec) . (3600 sec / hour)

ṁ = 129,382 lbm/hr or 58,687 kg/hr

Or,

ṁ = {(35.94lbm / 2000 lbm / ton) / sec }.(3600 sec / hour)

ṁ = 65 tons/hour

The density data for of ice and water is as follows:

Density of water = 1000 kg / m^3 or **62.4 lb/cu.-ft at 4°C.**
Density of ice = 917 kg / m^3 or **57.26 lbm/cu-ft**

Since our objective is to determine the volumetric flow rate of ice, we will introduce the density of ice in determination of the volumetric flow rate. Therefore, the volumetric flow rate, $\dot{V}$, would be:

$\dot{V}$ = (35.94 lbm/sec).(3600 sec/hr)/(57.26 lbm/ft^3)
$\dot{V}$ = 2260 cu-ft/hr

At an estimated **353.15 ft^3 per truck load**, this volumetric mass flow rate would amount to:

$$\dot{V} = \textbf{(2260 cu-ft/hr)/353.15 ft}^3\textbf{ per truck load}$$

Or,

$$\dot{V} = \textbf{6.4 truck loads per hour}$$

b) Assume that the power station is generating 10 MW of electric power. Calculate the amount of heat needed, in Btu's/hr, to convert 14°F harvested ice to 932°F, 362.6 psia steam, per hour.

Solution

Similar to the metric unit version, part (b) of the US unit version of Case Study 7-1 requires accounting for heat added during each of the five (5) stages of the overall process. Therefore, this part is divided into five sub-parts, each involving either sensible or latent heat calculation, based on the entry and exit temperature and phase status.

Table 7-7 lists specific heat for water and ice. These heat values will be used in the sensible heat calculations associated with Part (b). Table 7-8 lists latent heat values for water. These values will be used to compute the latent heats associated with stages that involve phase transformation.

(i) Calculate the heat required to raise the temperature of the ice from 14°F to 32°F.

Since there is no change in phase involved, the entire heat absorbed by the ice (working substance) in this stage would be sensible heat.

First stage of the overall power generating system is illustrated in Figure 7-11.

Given:

$T_i = 14°F$

$T_f = 32°F$

$c_{ice} = 0.49$ Btu/lbm.°F {Table 7-7}

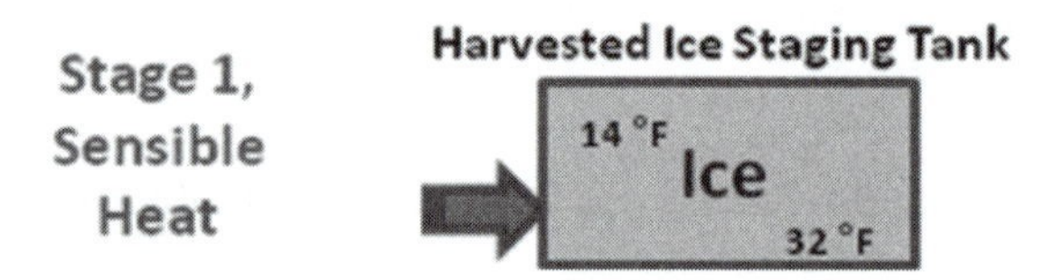

Figure 7-11. Case Study 7-1—US Units, Stage 1 Sensible Heat Calculation

Utilizing the given information:

$$\Delta T = T_f - T_i$$
$$\therefore \Delta T = 32 - 14°F$$
$$= 18°F$$

Since **ΔT** represents the ***change*** in temperature and not a specific absolute temperature,

$$\Delta T = 18°F = 18°R$$

Mathematical relationship between sensible heat, mass of the working substance, specific heat of the working substance and change in temperature can be stated as:

$$Q_{s(heat\ ice)} = m \cdot c_{ice} \cdot \Delta T \qquad \textbf{Eq. 7-14}$$

And,

$$\dot{Q}_{s(heat\ ice)} = \dot{m} \cdot c_{ice} \cdot \Delta T \qquad \textbf{Eq. 7-15}$$

Where,

$Q_{s(heat\ ice)}$ = Sensible heat required to heat the ice over ΔT
$\dot{Q}_{s(heat\ ice)}$ = Sensible heat flow rate required to heat the ice over ΔT
m = Mass of ice being heated
c_{ice} = Specific heat of ice = **0.49 Btu/lbm.°F= 0.49 Btu/lbm.°R** Since **Δ°F = Δ°R**
ΔT = Change in temperature, in **°F or °R = 18°F= 18°R**
$\dot{m}$ = Mass flow rate of water / ice = **35.94 lbm/sec**, or, **129,382 lbm/hr** as computed in part (a)

Then, by applying **Eq. 7-15**:

$$\dot{Q}_{s(heat\ ice)} = (129{,}382\ lbm/hr) \cdot (0.49\ Btu/lbm.°F) \cdot (18°F)$$

Or,

$$\dot{Q}_{s(heat\ ice)} = 1{,}141{,}149\ Btu/hr$$

(ii) Calculate the heat required to melt the ice at 32°F.

Since change in phase ***is*** involved in this case, the heat absorbed by the ice (working substance) in this stage would be latent heat.

The 2nd stage of the overall power generating system is illustrated in Figure 7-12.

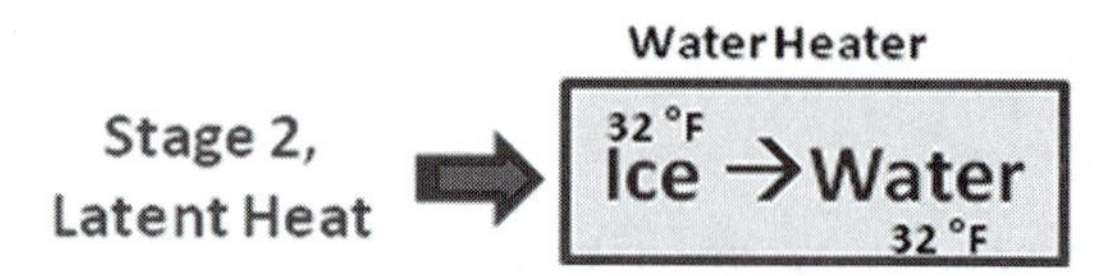

Figure 7-12. Case Study 7-1—US Units, Stage 2 Latent Heat Calculation

Mathematical relationship between latent heat, mass of the working substance, and the heat of fusion of ice can be stated as:

$$\mathbf{Q_{l(latent\ ice)} = h_{sl\ (ice)} \cdot m} \qquad \textbf{Eq. 7-16}$$

And,

$$\mathbf{\dot{Q}_{l(latent\ ice)} = h_{sl\ (ice)} \cdot \dot{m}} \qquad \textbf{Eq. 7-17}$$

Where,

$\mathbf{Q_{l(latent\ ice)}}$ = Latent heat required to melt a specific mass of ice, isothermally

$\mathbf{Q_{l(latent\ ice)}}$ = Latent heat flow rate required to melt a specific mass of ice, isothermally, over a period of time

$\mathbf{m}$ = Mass of ice being melted

$\mathbf{\dot{m}}$ = Mass flow rate of water/ice = **129,382 lbm/hr** as computed in part (a)

$\mathbf{h_{sl\ (ice)}}$ = Heat of fusion for ice or water = **143.4 Btu/lbm** {Table 7-8}

Then, by application of **Eq. 7-17**:

$$\mathbf{\dot{Q}_{l(latent\ ice)} = h_{sl\ (ice)} \cdot \dot{m}}$$

$$\mathbf{\dot{Q}_{l(latent\ ice)} = (143.4\ Btu/lbm)\ .\ (129{,}382\ lbm/hr)}$$

$$\mathbf{\dot{Q}_{l(latent\ ice)} = 18{,}553{,}380\ Btu/hr}$$

Note that the specific heat required to melt ice is called heat of fusion because of the fact that the water molecules come closer together as heat is added in the melting process. The water molecules are held apart at specific distances in the crystallographic structure of solid ice. The heat of fusion allows the molecules to overcome the crystallographic forces and **"fuse"** to form liquid water. This also explains why the density of water is higher than the density of ice.

(iii) Calculate the heat reqd. to heat the water from 32°F to 212°F:

The 3rd stage of the overall power generating system is illustrated in Figure 7-13. Since no phase change is involved in this stage, the heat absorbed by the water in this stage would be sensible heat.

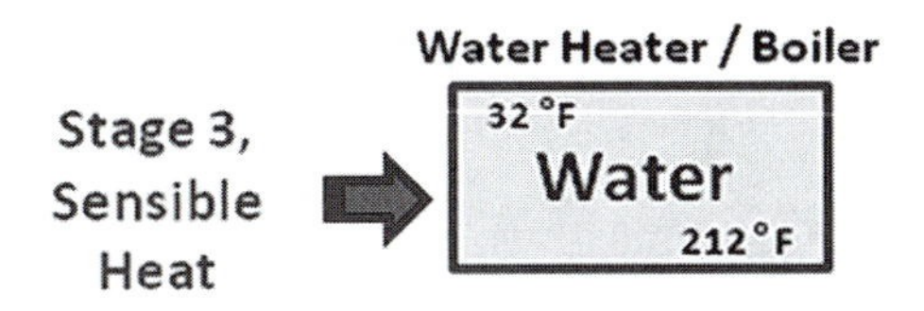

Figure 7-13. Case Study 7-1, US Units, Stage 3 Sensible Heat Calculation

Given:

$\mathbf{T_i = 32°F}$

$\mathbf{T_f = 212°F}$

c_{water} = Specific heat of water = **1.0 Btu/lbm.°F**

= **1.0 Btu/lbm.°R** {Since Δ°F= Δ°R} {Table 7-7}

Utilizing the given information:

$$\Delta T = T_f - T_i$$
$$\therefore \Delta T = 212°F - 32°F$$
$$= 180°F$$

Since **ΔT** represents the ***change*** in temperature and not a specific absolute temperature,

$$\therefore \Delta T = 180°R = 180°F$$

Mathematical relationship between sensible heat, mass of the working substance, specific heat of water (the working substance), and change in temperature can be stated as:

$$Q_{s(water)} = m \cdot c_{p\text{-}water} \cdot \Delta T \qquad \text{Eq. 7-18}$$

And,

$$\dot{Q}_{s(water)} = \dot{m} \cdot c_{p\text{-}water} \cdot \Delta T \qquad \text{Eq. 7-19}$$

Where,

$\mathbf{Q_{s(water)}}$ = Sensible heat required to heat the water over ΔT

$\mathbf{\dot{Q}_{s(water)}}$ = Sensible heat ***flow rate*** required to heat the water over ΔT

$\mathbf{m}$ = Mass of water being heated

$c_{p\text{-water}}$ = Specific heat of water = **1 Btu/lbm. °R**
= **1 Btu/lbm.°F = 1 Btu/lbm. °R** {Table 7-7}

$\dot{m}$ = Mass flow rate of water = **129,382 lbm/hr** as computed in part (a)

ΔT = Change in temperature, in **°F or °R = 180°F**

Then, by applying **Eq. 7-19**:

$$\dot{Q}_{s(water)} = \dot{m} \cdot c_{p\text{-water}} \cdot \Delta T$$

$$\dot{Q}_{s(water)} = (129{,}382 \text{ lbm/hr}) \cdot (1 \text{ Btu/lbm.°F}) \cdot (180°\text{F})$$

$$\dot{Q}_{s(water)} = 23{,}288{,}760 \text{ Btus/hr}$$

(iv) Calculate the heat required to convert 212°F water to 212°F steam:

The 4th stage of the overall power generating system is illustrated in Figure 7-14. Since change in phase *is* involved in this case, the heat absorbed by the water in this stage would be latent heat.

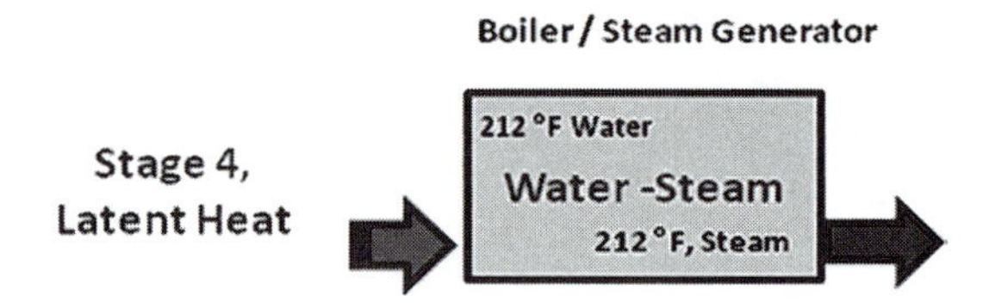

Figure 7-14. Case Study 7-1 Stage 4 Latent Heat Calculation

Mathematical relationship between latent heat of vaporization for water, $h_{fg(water)}$, mass of the water, and the total heat of vaporization of water, $Q_{l(latent\ water)}$, can be stated as:

$$Q_{l(latent\ water)} = h_{fg\ (water)} \cdot m \qquad \textbf{Eq. 7-20}$$

And,

$$\dot{Q}_{l(latent\ water)} = h_{fg\ (water)} \cdot \dot{m} \qquad \textbf{Eq. 7-21}$$

Where,

$Q_{l(latent\ water)}$ = Latent heat of vaporization of water required to evaporate a specific mass of water, isothermally

$\dot{Q}_{l(latent\ water)}$ = Latent heat of vaporization ***flow rate*** required to evaporate a specific mass of water, isothermally, over a given period of time

m = Mass of water being evaporated

$\dot{m}$ = Mass flow rate of water = **129,382 lbm/hr,** same as part (a) (i)

$h_{fg\ (water)}$ = latent heat of vaporization for water
= **970.3 Btu/lbm** {From the steam tables and Table 7-8}

Then, by application of **Eq. 7-21**:

$$\dot{Q}_{l(latent\ water)} = h_{fg\ (water)} \cdot \dot{m}$$

$$\dot{Q}_{l(latent\ water)} = (970.3\ Btu/lbm).\ (129{,}382\ lbm/hr)$$

$$\dot{Q}_{l(latent\ water)} = 125{,}539{,}355\ Btus/hr$$

(v) Calculate the heat reqd. to heat the steam from 212°F, 1-atm (102 KPa, or 1-bar) to 932°F, 362.6 psia superheated steam

The 5th stage of the overall power generating system is illustrated in Figure 7-15. Since this stage involves no phase change, the heat absorbed by the steam is sensible heat.

In superheated steam phase, the heat required to raise the temperature and pressure of the steam can be determined using the enthalpy difference between the initial and final conditions.

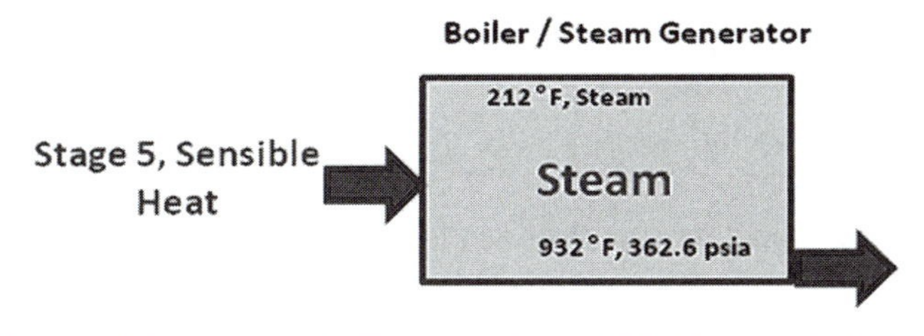

Figure 7-15. Case Study 7-1 Stage 5 Sensible Heat Calculation

Given:

T_i **= 212°F**

P_i **= 1- Atm.** Note: At 212°F, the saturation pressure is 1- Atm, 1-Bar, or 102 kPa

T_f **= 932°F**

$\mathbf{P_f = 362.6\ psia}$

For the initial and final temperature and pressure conditions stated above, the enthalpy values—as read from superheated steam table excerpt in Table 7-9, as interpolated in part (a) of this case study, and from Table 7-11—are as follows:

$\mathbf{h_{i\ at\ 212°F,\ 1\text{-}Atm} = 1149.4\ Btu/lbm}$

$\mathbf{h_{f\ at\ 932°F,\ 362.6\ psia} = h_{f\ at\ 932°F,\ 360\ psia} = 1488.76\ Btu/lbm}$

Equations for determining the heat required to boost the steam from **212°F, 1-Atm** to **932°F, 362.6 psia** are as follows:

$$\Delta Q_{steam} = (h_f - h_i) \cdot m \qquad \textbf{Eq. 7-22}$$

$$\dot{Q}_{steam} = (h_f - h_i) \cdot \dot{m} \qquad \textbf{Eq. 7-23}$$

Where,

ΔQ_{steam} = Addition of heat required for a specific change in enthalpy
$\dot{Q}_{steam}$ = Rate of addition of heat for a specific change in enthalpy
h_i = Initial enthalpy
h_f = Final enthalpy
m = Mass of steam being heated
$\dot{m}$ = **129,382 lbm/hr,** same as part (a)

Then, by applying **Eq. 7-23**:

$$\dot{Q}_{steam} = (h_f - h_i) \cdot \dot{m}$$

$$\dot{Q}_{steam} = (1488.76\ Btu/lbm - 1149.4\ Btu/lbm) \cdot (129{,}382\ lbm/hr)$$

$$\dot{Q}_{steam} = 43{,}907{,}076\ Btu/hr$$

After assessing the heat added, per hour, during each of the five (5) stages of the steam generation process, add all of the heat addition rates to compile the total heat addition rate for the power generating station. The tallying of total heat is performed in Btu's/hr, as well as in kJ/hr.

Table 7-11. Saturated Steam Table Excerpt, US/Imperial Units

Properties of Saturated Steam By Pressure
US/Imperial Units

Abs. Press. psia	Temp. °F	Specific Volume ft^3/lbm		Enthalpy Btu/lbm			Entropy Btu/(lbm.°R)		Abs. Press. psia
		Sat. Liquid v_L	Sat. Vapor v_V	Sat. Liquid h_L	Evap. h_{fg}	Sat. Vapor h_V	Sat. Liquid s_L	Sat. Vapor s_V	
1.0	101.69	0.016137	333.51	69.728	1036	1105.4	0.1326	1.9776	1.0
4.0	152.91	0.016356	90.628	120.89	1006.4	1126.9	0.2198	1.8621	4.0
14.0	209.52	0.016697	28.048	177.68	972.0	1149.4	0.3084	1.7605	14.0
100	327.82	0.017736	4.4324	298.57	889.2	1187.5	0.4744	1.6032	100

Total Heat Addition Rate in Btu's/hr:

Total Heat Required to Generate 932°F, 362.6 psia steam from 1°F Ice, at 129,382 lbm/hr

= 1,141,149 Btu/hr + 18,553,380 Btu/hr + 23,288,760 Btus/hr + 125,539,355 Btus/hr + 43,907,076 Btu/hr

= **212,429,719 Btu/hr**

c) **Calculate the volume, in cu-ft, of natural gas required to power up the station, each day. Assume 98% burner efficiency.**

Solution:

This part of Case Study 7-1 requires computation of the amount (volume) of natural gas required to power up the station each day. This calculation is straight forward after the derivation of the total energy required, per hour, in part (b). However, the hourly energy requirement must be "scaled up" to account for the 98% efficiency of the boiler burner.

The hourly energy requirement, in kJ or Btu's, can be extended into daily usage. The daily energy usage can then be converted into the volume of natural gas required, based on natural gas energy content listed in Table 7-6.

Total Energy Required Per Day = (212,429,719 Btu/hr)/(0.98) x 24 hr

= **5,098,313,255 Btus**

Since Natural Gas Energy Content = **1034 Btu/cu-ft,**

Total volume of natural gas required per day

= **(5,098,313,255 Btus)/1034 Btu/cu-ft**

= **4,930,670 cu-ft**

d) **If the natural gas transportation cost is $ 4.85/DT in addition to the well head or commodity cost stated in Table 7-6, what would be the annual fuel cost of operating this station?**

Solution:

Part (d) of the case study relates to the computation of total annual cost of fuel for the power generation station. This requires the extrapolation of daily energy consumption into ***annual*** energy consumption. The annual energy consumption is then multiplied with the ***total***, delivered, cost rate in $/DT to obtain the annual cost in dollars.

From part (b), the heat energy required for operating the power station = 196,909,695 Btu/hr

Then, based on 24/7 operating schedule assumption, the total energy required per year would be:

= (212,429,719 Btu/hr)/(0.98) x 8760 hrs/year
= 1.8989 E+12 Btu's

Since there are 1,000,000 Btu's per DT, the Total Annual Energy Consumption in DT would be:

= 1.8989 E+12 Btu's/(1,000,000 Btu's/DT)
= 1,898,862 DT's

The total cost of natural gas, per DT =
Cost at the Source/Well + Transportation Cost
= $4.15/DT + $4.85/DT
= $9.00/DT

Then, the Total Annual Cost for Producing 10MW of Power with Natural Gas
= $ 17,089,754

e) What is the overall energy efficiency of the power station ?

Solution/Answer:

Part (e) entails determination of the overall efficiency of the power generating station. Efficiency calculation, in this case study, requires knowledge of the total electrical energy (or power) produced and the total energy (or power) consumed through the combustion of natural gas. The output of the overall power generating system is ostensible from the given system output rating of 10 MW. The total heat consumption by the system is derived in parts (b), (c) and (d). The overall efficiency of the system can then be assessed by dividing the output power (or energy) by the input power (or energy).

From Part (c):

Total Energy Required Per Day, in Btu's = 5,098,313,255 Btus
Since there are 1.055kJ per Btu,

The Total Energy Input, Each Day, in kJ
= 5,098,313,255 Btus x 1.055 kJ/Btu
= 5,378,720,484 kJ

This fuel energy usage can be converted into Watts of Mega Watts (MW) by dividing the energy usage by the total number of seconds in a day. This is because the **5,378,720,484 kJ** of energy is used over a period of a day.

∴ **The System Power Input**
= (5,378,720,484 kJ)/(24 hr x 3600 sec/hr)
= 62,254 kJ/sec or 62,254 kW

Since there are 1000 kW per MW,
The System Power Input, in MW
= (62,254 kW)/(1,000 kW/MW)
= 62.254 MW

Since Power Output = 10.00 MW { Given }

Total Station Energy Efficiency, in Percent
= Power Output/Power Input x 100
= 10 MW/62.254MW x 100
= **16%**

f) If heat is added to the steam turbines, would the steady flow energy process in the turbine system constitute an adiabatic process or a non-adiabatic process?

Solution/Answer:

The answer to Part (f) of the case study lies simply in the definition for adiabatic process.

Adiabatic process is a thermodynamic process in which no heat either enters or leaves the thermodynamic system boundary. Therefore, this process is a **Non-Adiabatic Process.**

g) What is the change in entropy, Δs, in the turbine system?

Solution/Answer:

Part (g) involves computation of the change in entropy in the turbine segment of the overall thermodynamic process. As characteristic of the turbine stage of a typical heat engine cycle, there is a small change in entropy as the working fluid travels from the intake to the exit point of the turbine. This is evident as the entropy is assessed, through the superheated steam tables, for the incoming steam at 932°F and 362.6 psia, and the outgoing steam at 302°Fand 7.252 psia.

To determine the change in entropy, Δs, in the turbine system we need to identify $\mathbf{s}_i$, the entropy of 932°F, 362.6 psia steam entering the steam turbine and $\mathbf{s}_f$, the entropy of 302°F, 7.252 psia steam exiting the turbine. As evident from superheated steam table excerpts in Tables 7-9 and 7-10, the desired entropy values are not readily available. Interpolation is required to obtain the entropy values at the entry and exit points of the turbine.

The formula for single interpolation, applied between the stated or available entropy values for 900°F and 1000°F, at 360 psia, is as follows:

$$s_{i \text{ at } 932°F,\ 362.6 \text{ psia}} = s_{i \text{ at } 932°F,\ 360 \text{ psia}}$$

$$= ((s_{i \text{ at } 1000°F,\ 360 \text{ psia}} - s_{i \text{ at } 900°F,\ 360 \text{ psia}})/(1000°F - 900°F)).(932-900) + s_{i \text{ at } 900°F,\ 360 \text{ psia}}$$

Substituting entropy values and other given data from superheated steam table excerpt, shown in Table 7-9:

$$s_{i \text{ at } 932°F,\ 362.6 \text{ psia}} = s_{i \text{ at } 932°F,\ 360 \text{ psia}}$$

$$= ((1.7758 \text{ Btu/lbm-°F} - 1.7381 \text{ Btu/lbm-°F})/(1000°F - 900°F)).(932°F - 900°F) + 1.7381 \text{ Btu/lbm-°F}$$

$$\therefore s_{i \text{ at } 932°F,\ 362.6 \text{ psia}} = 1.7502 \text{ Btu/lbm-°F}$$

Note: The available entropy values are circled in Table 7-12.

Table 7-12. Superheated Steam Table Excerpt, US/Imperial Units

Properties of Superheated Steam **US/Imperial Units**					
Abs. Press. psia (Sat. T, °F)		**Temp. °F**	Note: v is in ft^3/lbm, h is in BTU/lbm and s is in BTU/(lbm-°R)		
		500	700	900	1000
260 (**404.45**)	*v*	2.062	2.5818	3.0683	3.3065
	h	1262.5	1370.8	1475.2	1527.8
	s	1.5901	1.6928	1.7758	1.8132
360 (**434.43**)	*v*	1.446	1.8429	2.2028	2.3774
	h	1250.6	1365.2	1471.7	1525
	s	1.5446	1.6533	1.7381	1.7758
450 (**456.32**)	*v*	1.1232	1.4584	1.7526	1.8942
	h	1238.9	1360	1468.6	1522.4
	s	1.5103	1.6253	1.7117	1.7499
600 (**486.25**)	*v*			1.3023	1.411
	h			1463.2	1518
	s			1.577	1.7159

Double interpolation method is employed in deriving $\mathbf{s_f}$, the entropy value at 7.25 psia and 302°F. As apparent from the superheated steam tables in Appendix B, this value is not readily available and, therefore, double interpolation must be conducted between the entropy values given for 5 psi, 300°F, and 10 psi, 350°F, to derive $\mathbf{s_{f\ at\ 7.25\ psia\ and\ 302°F}}$. Where, $\mathbf{s_{f\ at\ 7.25\ psia\ and\ 302°F}}$ is the final entropy—entropy at the turbine exit point—at 7.25 psia and 302°F. The double interpolation approach, as applied here, will entail three steps.

First step involves determination of $\mathbf{s_{f,\ at\ 7.25\ psia\ and\ 300°F}}$, the entropy value at 7.25 psia and 300°F. The entropy values available and used in this first interpolation step are circled in Table 7-13. The following formula sums up the mathematical approach to this first step:

$$\mathbf{s_{f\ at\ 300°F,\ 7.25\ psia}} = \mathbf{((s_{f\ at\ 300°F,\ 5\ psia} - s_{i\ at\ 300°F,\ 10\ psia})/(10\ psia - 5\ psia)).(10\ psia - 7.25\ psia) + s_{f\ at\ 300°F,\ 10\ psia}}$$

Substituting entropy values and other given data from superheated steam table excerpt, shown in Table 7-13:

$$\mathbf{s_{f\ at\ 300°F,\ 7.25\ psia}} = \mathbf{((1.937\ Btu/lbm\text{-}°F - 1.8595\ Btu/lbm\text{-}°F)/(10\ psia - 5\ psia)).(10\ psia - 7.25\ psia) + 1.8595\ Btu/lbm\text{-}°F}$$

$$= \mathbf{1.902125\ Btu/lbm\text{-}°F}$$

Second interpolation step involves determination of $\mathbf{s_{f,\ at\ 7.25\ psia\ and\ 350°F}}$, the entropy value at 7.25 psia and 350°F. The entropy values available and used in this interpolation step are circled in Table 7-13. The following formula sums up the mathematical approach associated with this interpolation step:

$$\mathbf{s_{f\ at\ 350°F,\ 7.25\ psia}} = \mathbf{((s_{f\ at\ 350°F,\ 5\ psia} - s_{i\ at\ 350°F,\ 10\ psia})/(10\ psia - 5\ psia)).(10\ psia - 7.25\ psia) + s_{f\ at\ 350°F,\ 10\ psia}}$$

Substituting entropy values and other given data from superheated steam table excerpt, shown in Table 7-13.

Table 7-13. Superheated Steam Table Excerpt, US/Imperial Units

Properties of Superheated Steam US/Imperial Units					
Abs. Press. psia (Sat. T, °F)		**Temp. °F**	Note: v is in ft³/lbm, *h* is in BTU/lbm and ***s*** is in BTU/(lbm-°R)		
		200	300	350	500
5 (162.18)	*v*		90.248	96.254	
	h		1194.8	1218	
	s		1.937	1.9665	
10 (193.16)	*v*	38.851	44.993	48.022	
	h	1146.4	1193.8	1217.2	
	s	1.7926	1.8595	1.8893	
15 (212.99)	*v*		29.906		37.986
	h		1192.7		1287.3
	s		1.8137		1.9243
100 (327.82)	*v*				5.5875
	h				1279.3
	s				1.7089

$s_{f \text{ at } 350°F,\ 7.25 \text{ psia}}$

= **((1.9665 Btu/lbm-**°F – **1.8893 Btu/lbm-**°F)**/(10 psia -5 psia)).(10 psia-7.25 psia) + 1.8893 Btu/lbm-**°F

= **1.93176 Btu/lbm-**°F

The final step in the double interpolation process, as applied in this case, involves interpolating between $s_{f \text{ at } 300°F,\ 7.25 \text{ psia}}$ and $s_{f \text{ at } 350°F,\ 7.25 \text{ psia}}$ the entropy values derived in the first two steps above, to obtain the desired final entropy $s_{f \text{ at } 302°F,\ 7.25 \text{ psia}}$.

The formula for this final step is as follows:

$s_{f \text{ at } 302°F, 7.25 \text{ psia}}$
$= ((s_{f \text{ at } 350°F, 7.25 \text{ psia}} - s_{f \text{ at } 300°F, 7.25 \text{ psia}})/(350°F - 300°F)).(302°F - 300°F) + s_{f \text{ at } 300°F, 7.25 \text{ psia}}$

Substituting entropy values derived in the first two steps above:

$s_{f \text{ at } 302°F, 7.25 \text{ psia}}$
$= ((1.932 \text{ Btu/lbm-°F} - 1.9021 \text{ Btu/lbm-°F})/(350°F - 300°F)).(302°F - 300°F) + 1.902125 \text{ Btu/lbm-°F}$

$s_{f \text{ at } 302°F, 7.25 \text{ psia}} = 1.9033 \text{Btu/lbm-°F}$

With the key entropy values for turbine entry and exit points identified, we can proceed with the computation of **Δs:**

$s_{f \text{ at } 302°F, 7.25 \text{ psia}} = 1.9033 \text{Btu/lbm-°F}$
$s_{i \text{ at } 932°F, 362.6 \text{ psia}} = 1.7502 \text{ Btu/lbm-°F}$

Then,

$$\begin{aligned}\Delta s &= s_{f \text{ at } 302°F, 7.25 \text{ psia}} - s_{i \text{ at } 932°F, 362.6 \text{ psia}} \\ &= 1.7502 \text{ Btu/lbm-°F} - 1.9033 \text{Btu/lbm-°F} \\ &= -0.1531 \text{ Btu/lbm-°F}\end{aligned}$$

Note: The negative sign signifies the loss of heat through the turbine casing.

Chapter 7 —Self Assessment Problems & Questions

1. Why is the efficiency of this power plant in Case Study 7-1 rather low (17%)?

2. Using the steam tables in Appendix B and the Double Interpolation Method described in Case Study 7-1, US Unit Version, determine the exact enthalpy of a superheated steam at a pressure of 400 psia and temperature of 950°F.

3. In Case Study 7-1, as an energy engineer you have been retained by Station Zebra to explore or develop an alternative integrated steam turbine and electric power generating system that is capable of generating 10 MW of power with only 60 truck loads, or 54,432 kg, of ice per hour. With all other parameters the same as in the original Case Study 7-1 scenario, determine the total heat flow rate needed, in kJ/hr, to produce 10 MW of electrical power.

4. If all of the working fluid, or steam, discharged from the turbine in Case Study 7-1 is reclaimed, reheated and returned to the turbine, what would be the overall system efficiency?

Chapter 8

Thermodynamic Processes

INTRODUCTION

This chapter explores some of the mainstream thermodynamic processes, heat engines and heat engine cycles. Fundamentals of thermodynamic processes, heat engines, heat engine cycles and associated systems are explained and illustrated through process flow diagrams, graphs, tables and pictures. Practical significance, application, analytical methods and computational techniques associated with heat engine cycles and thermodynamic processes are demonstrated through case study, examples and self-assessment problems.

THERMODYNAMIC PROCESSES

Thermodynamic processes are processes that entail heat, internal energy, enthalpy, entropy, work, pressure, temperature and volume. In this section, we will explore the following thermodynamic processes and illustrate these processes with practical examples:

1. Adiabatic Process
2. Isenthalpic Process
3. Isochoric Process
4. Isothermal Process
5. Isobaric Process
6. Isentropic Process

Adiabatic Process

Adiabatic process is a thermodynamic process in which no heat either enters or leaves the thermodynamic system boundary. An adiabatic process can also be explained through the following mathematical statements or equations:

$$\Delta U = - W \qquad \textbf{Eq. 8-1}$$

$$\Delta Q = 0 \qquad \textbf{Eq. 8-2}$$

Equations 8-1 and 8-2 essentially state that in an adiabatic process, wherein no heat is gained or lost, any work performed on the system or by the system is transformed into a net change in the internal energy of the system. As specifically stated above, Eq. 8-1 represents a scenario where negative work is involved. In other words, work is being performed by the surroundings onto the system. And, since no heat is transferred to or from the environment in an adiabatic process, the work performed by the surroundings onto the system, in this case, is converted into an equivalent amount of increase in the internal energy of the system. This explanation of the adiabatic process is validated by the fact that the units for work "W" and internal energy "U" are the same; Joules in the SI/Metric realm and Btu's in the US/Imperial realm.

Equation 8-1 can be restated as:

$$- \Delta U = W \qquad \textbf{Eq. 8-1a}$$

It is important to note that while this restatement of the Eq. 8-1 keeps the equation mathematically equivalent to the original version, the physical significance changes. Equation 8-1a represents a scenario where work is positive and is performed by the system onto the environment or surroundings. Since this is an adiabatic process, there is no transfer or exchange of heat. Therefore, in this case, work is performed by the system, onto the surroundings, at the expense of the internal energy of the system. The negative ΔU signifies a reduction in the internal energy of the system.

Work performed in adiabatic processes, such as that performed by the compressor on the refrigerant, would be represented by Eq. 8-3.

$$W = \int_{V_1}^{V_2} p.dV \qquad \textbf{Eq. 8-3}$$

In a reversible adiabatic process, such as the compression stroke in an internal combustion gasoline engine, the product of pressure and volume is represented as shown in **Eq. 8-4.**

$$pV^{\gamma} = \text{Constant} \qquad \textbf{Eq. 8-4}$$

Where,

γ = Degrees of freedom of the gas molecules; i.e. 7/5 for nitrogen and oxygen.
p = Pressure in SI or US (Imperial) units
V = Volume in SI or US (Imperial) units

Figure 8-1 represents a "**real gas**" scenario, where temperature does change to some degree. At an ***inversion point***, however, the temperature ***does not*** change during a throttling process. A real gas tends to behave like a ideal gas as temperature approaches the inversion point. Adiabatic process in an "**ideal gas**" scenario is depicted in Figure 8-2.

ADIABATIC PROCESS EXAMPLE I—
THROTTLING PROCESS IN A REFRIGERATION SYSTEM

Throttling process in a refrigeration system is an example of an adiabatic process that occurs in the expansion valve; where a high pressure liquid system (refrigerant) is allowed to expand to a low pressure liquid, without absorption or release of heat energy.

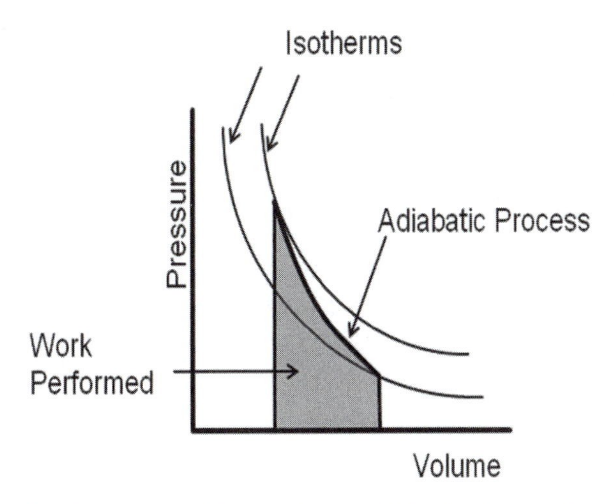

Figure 8-1. Work performed in a real gas, adiabatic, thermodynamic process

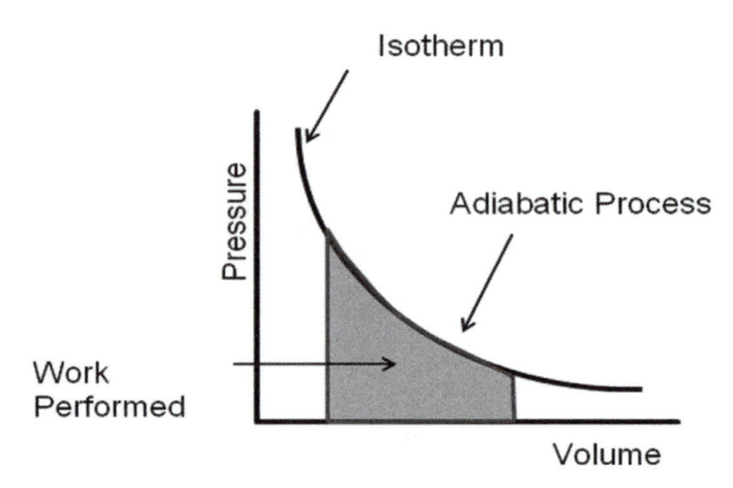

Figure 8-2. Work performed in an ideal adiabatic thermodynamic process

ADIABATIC PROCESS EXAMPLE II—
COMPRESSOR SEGMENT OF A REFRIGERATION SYSTEM

The compression segment of the refrigeration cycle is an adiabatic process. During compression the refrigerant, or the system, is compressed from low pressure vapor phase to high pressure vapor phase. No heat is exchanged with the environment during this compression process. The work performed on the refrigerant is negative work, or, "**– W.**" Since no heat is released by the system, this negative work, in accordance with Eq. 8-1 and the law of conservation of energy, is transformed into internal energy of the refrigerant. Compression of vapor, therefore, is an adiabatic process. As discussed later in this text, the compression segment of the refrigeration cycle is not just adiabatic but also an ***isentropic*** process.

Isenthalpic or **Isoenthalpic Process**

An **isenthalpic,** or **isoenthalpic,** process is a thermodynamic process in which no change in enthalpy occurs, or $\Delta h = 0$, or $h_1 = h_2$.
A steady-state, steady-flow process, would be isenthalpic if the following conditions are met:

1) The thermodynamic process is adiabatic—meaning, no heat is exchanged with the environment.
2) Work is neither performed by the system onto the surroundings

nor is it performed by the surroundings onto the system.

3) There is no change in the kinetic energy of the system or fluid.

Isenthalpic Process Example—
Throttling Process in a Refrigeration System

Refrigeration system throttling process is an example of an isenthalpic process. See Figure 8-3. Throttling of a high pressure liquid refrigerant to a low pressure liquid phase is an adiabatic process; i.e. no heat is exchanged with the environment. Moreover, no work is done on or by the surroundings, and there is no change in the kinetic energy of the fluid. Note that during the throttling process shown in Figure 8-3, the process adopts a vertical downward path, dropping the pressure precipitously while the enthalpy stays unchanged, thus, rendering the process isenthalpic. In other words, all three requirements or conditions, stated above, for an isenthalpic process are met during the throttling segment of the refrigeration cycle.

Other examples of practical isoenthalpic processes include lifting of a relief valve or safety valve on a pressurized vessel. The specific enthalpy of the fluid inside the pressure vessel is the same as the specific

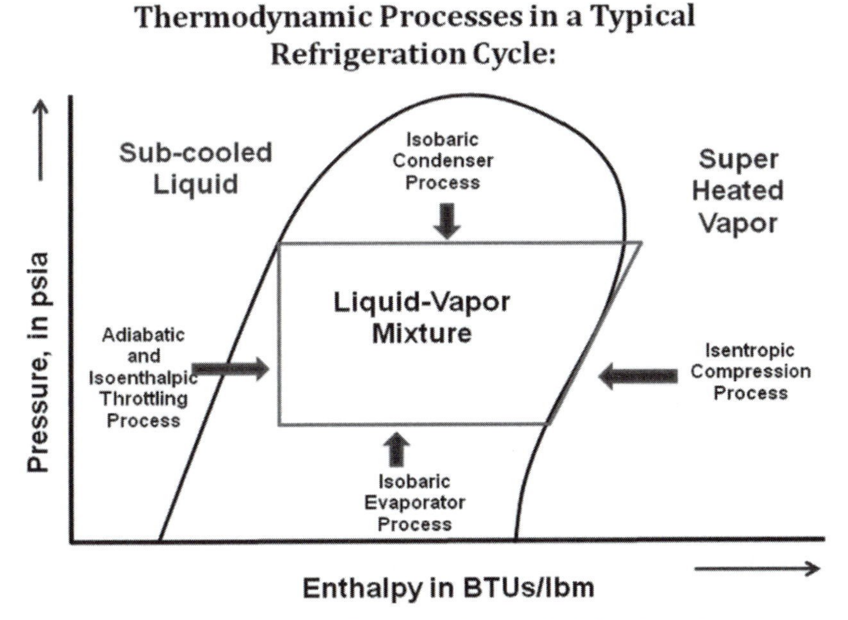

Figure 8-3. Thermodynamic Processes in a Typical Refrigeration Cycle

enthalpy of the same fluid immediately after it escapes the vessel. In such a scenario, the temperature and velocity of the escaping fluid can be calculated if the enthalpy is known.

In Figures 8-1 and 8-2, an isenthalpic process follows the isotherm line at a specific temperature, and along the isotherm the following relationship between enthalpy, temperature and specific heat holds true:

$$dh = c_p dT = 0$$

Additional examples of isenthalpic process are referenced later in this chapter, under the heat engine cycle discussion.

Constant Pressure or Isobaric Process

An isobaric process is a thermodynamic process in which the pressure remains constant. See Figure 8-4, where the curve represents an isobar. Even though the temperature varies as a function of the entropy in this graph, the pressure stays constant.

Isobaric Process Example I:
Evaporation Stage of a Refrigeration Cycle

Evaporation stage of a refrigeration cycle represents an isobaric process in that the pressure remains constant as the low pressure liquid system evaporates or changes phases from liquid to gaseous by ***absorbing the heat*** energy from the air passing through the heat exchanger. This absorption of heat by the system—refrigerant or the working fluid—from the surroundings (ambient air) is shown in Figure 8-4 as the shaded area under the isobar, between entropies $\mathbf{s_1}$ and $\mathbf{s_2}$.
In an isobaric process:

$$\Delta p = 0 \text{ and,}$$
$$Q = \Delta H$$

The latter mathematical statement, $Q = \Delta H$, implies that in this isobaric process, the heat absorbed by the refrigerant, during the evaporation phase, results in a net increase in the enthalpy of the refrigerant.

Some of the equations with practical applications in closed system isobaric processes are listed below:

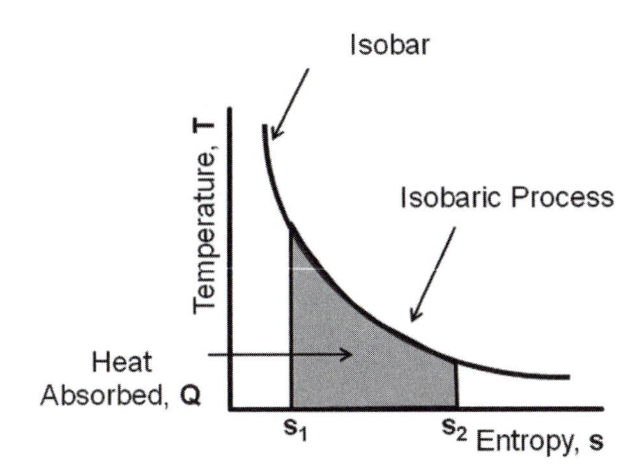

Figure 8-4. Heat absorbed in an isobaric thermodynamic process

$$T_2 = T_1 \left(\frac{\upsilon_2}{\upsilon_1}\right) \qquad \text{Eq. 8-5}$$

$$\upsilon_2 = \upsilon_1 \left(\frac{T_2}{T_1}\right) \qquad \text{Eq. 8-6}$$

$$q = h_2 - h_1 \qquad \text{Eq. 8-7}$$

$$q = c_p\,(T_2 - T_1) \qquad \text{Eq. 8-8}$$

$$q = c_v\,(T_2 - T_1) + p(v_2 - v_1) \qquad \text{Eq. 8-9}$$

$$W = p(v_2 - v_1) \qquad \text{Eq. 8-10}$$

$$W = R(T_2 - T_1) \qquad \text{Eq. 8-11}$$

Where,

q = Heat per unit mass or, Q/m
= Total Heat/Unit Mass, in Btu/lbm or kJ/kg
p = Pressure in lbf/ft^2 or Pa
υ_1 = Initial specific volume in ft^3/lbm or m^3/kg
υ_2 = Final specific volume in ft^3/lbm or m^3/kg
h_1 = Initial enthalpy in Btu/lbm or kJ/kg
h_2 = Final enthalpy in Btu/lbm or kJ/kg
T_1 = Initial Temperature, in °F, °C, °K, R

T_2 = Final Temperature, in °F, °C, °K, R
R = Specific gas constant, in ft-lbf / lbm-°R or kJ / kg.°K
W = Specific work in ft-lbf / lbm or kJ / kg. W also represents total work in ft-lbf or kJ

Note that Eq. 8-5 and Eq. 8-6 are derived from the ideal gas law, as stated in form of Eq. 8-12, with pressure held constant in an isobaric process.

$$\frac{p_1 v_1}{T_1} = \frac{p_2 v_2}{T_2} \qquad \textbf{Eq. 8-12}$$

Isobaric Process Example II:
Isobaric Segments of an Ideal Cycle Heat Engine

As shown in Figure 8-5, ideal heat engine cycle segments represented by processes A-B and C-D are isobaric processes because during these two processes within the heat engine cycle, the pressure remains constant as the volume expands and contracts, respectively. The working fluid performs work on the surroundings as it expands, from point A to B. This work is considered ***positive***. The surroundings perform work on the system, or working fluid, from point C to D; resulting in ***negative*** work. As explained in greater detail under the heat cycle section, this negative work is in form of the condensate pump performing work on the condensed vapor as it is pressurized to saturated liquid phase.

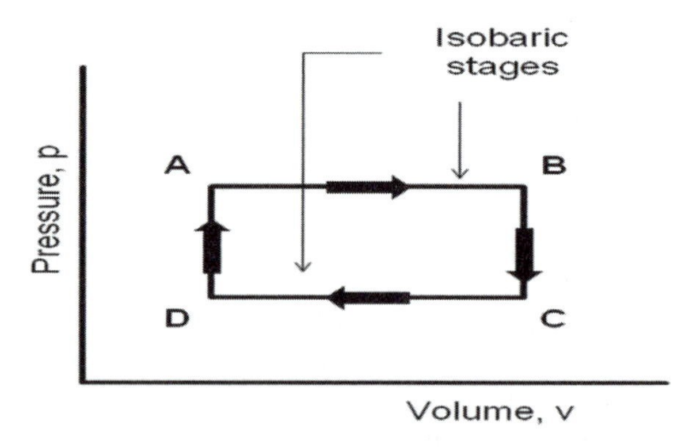

Figure 8-5. Isobaric Process in an Ideal Heat Engine Cycle

Constant Temperature or Isothermal Process

An isothermal process is a thermodynamic process in which the temperature stays constant. In isothermal processes, there is no change in internal energy because internal energy is directly related to temperature. This is validated by Eq. 8-17. Furthermore, as stipulated by Eq. 8-18, there is no change in enthalpy.

Some of the equations with practical applications in closed system isothermal processes are listed below:

$$T_2 - T_1 \text{ or, } \Delta T = 0 \qquad \text{Eq. 8-13}$$

$$P_2 = P_1\left(\frac{\upsilon_1}{\upsilon_2}\right) \qquad \text{Eq. 8-14}$$

$$\upsilon_2 = \upsilon_1\left(\frac{P_1}{P_2}\right) \qquad \text{Eq. 8-15}$$

$$q = w \qquad \text{Eq. 8-16}$$

$$u_2 - u_1 = 0 \qquad \text{Eq. 8-17}$$

$$h_2 - h_1) = 0 \qquad \text{Eq. 8-18}$$

$$Q = nR^*T \ln\left(\frac{P_1}{P_2}\right) \qquad \text{Eq. 8-19}$$

Where,

q = Heat per unit mass or, Q/m = Total Heat/Unit Mass, in Btu/lbm or kJ/kg

Q = Total heat in Btu or kJ. Q also denotes molar heat in Btu/lbmol or kJ/kmol

p_1 = Initial Pressure in lbf/ft^2 or Pa

p_2 = Final Pressure in lbf/ft^2 or Pa

υ_1 = Initial specific volume in ft^3/lbm or m^3/kg

υ_2 = Final specific volume in ft^3/lbm or m^3/kg

h_1 = Initial enthalpy in Btu/lbm or kJ/kg

h_2 = Final enthalpy in Btu/lbm or kJ/kg

T_1 = Initial Temperature, in °F, °C, °K, R

T_2 = Final Temperature, in °F, °C, °K, R

n = Number of moles, in lbmole or kmol

R = Specific gas constant, in ft-lbf/lbm-°R or kJ/kg.°K

R^* = Universal gas constant, in ft-lbf/lbmole-°R or kJ/kmol-°K

w = Specific work in ft-lbf/lbm or kJ/kg. W also represents total work in ft-lbf or kJ

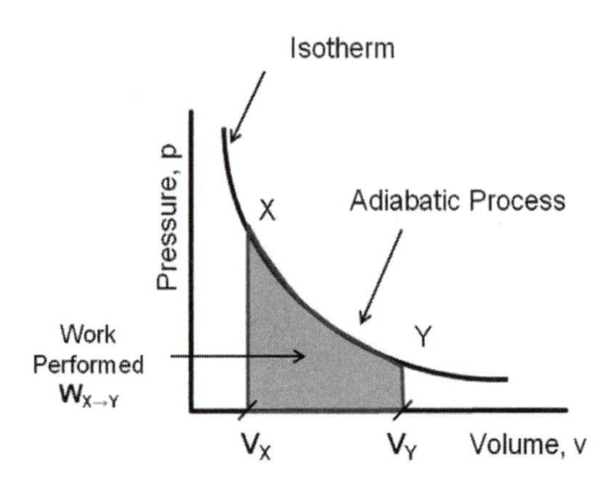

Figure 8-6. Isothermal Process in an Heat Engine Heat Cycle

Isothermal Process Example I: Steam Generation Process

The, latent, water evaporation stage in the steam generation process is an isothermal process because the temperature of the water and saturated vapor remains constant until all evaporation is concluded.

The work "**W**" performed by, or on, an isothermal system, is shown in Figure 8-6. Where work $\mathbf{W} = \mathbf{W}_{X \to Y}$.

And,

$$W_{X \to Y} = -\int_{V_X}^{V_Y} p.dV \qquad \textbf{Eq. 8-20}$$

Example Problem 8-1:

Consider the collision of the cast iron block with the compressed air filled shock absorbing system described in Case Study 2, part (e), Chapter 1. A cooling jacket is installed on the cylinder to maintain the temperature constant, at 20°C. For simplicity, assume that 1.3 kg of air is present in the shock absorbing system cylinder. Calculate the amount of heat, in Btu's, the cooling jacket would need to remove each time a block is stopped.

Solution:

According to Eq. 8-19, the heat removed to maintain constant temperature in an isothermal compression process would be:

$$Q = nR^*T \ln\left(\frac{P_1}{P_2}\right)$$

Where,

Q	=	Total heat removed
n	=	Number of kmols of gas involved =?
R^*	=	Universal gas constant = 8.314 kJ/kmol.°K or, 1545 ft-lbf/lbmole.°R. See Table 8-1.
p_1	=	Initial pressure = 101.3 kPa = 1 Bar
p_2	=	Final pressure = 202.6 kPa = 2 Bar
T	=	20 °C + 273 = 293°K

Since the number of kmols is not given, it needs to be derived using given mass of air, i.e. 1.3 kg and the molecular weight listed in Table 8-1.

The molecular weight of air, from Table 8-1 is 28.97 kg/kmol.

Table 8-1. Molecular Weights of Common Gases Associated with Combustion Reactions and Byproducts

Gas	Molecular Weight	Gas	Molecular Weight
Acetylene, C_2H_2	26.04	Natural Gas	19
Air	28.966	Nitric Oxide, NO_2	30.006
Ammonia (R-717)	17.02	Nitrogen, N_2	28.0134
Argon, Ar	39.948	Nitrous Oxide	44.012
N-Butane, C_4H_{10}	58.12	Oxygen, O_2	31.9988
Iso-Butane	58.12	Ozone	47.998
Carbon Dioxide, CO_2	44.01	Propane, C_3H_8	44.097
Carbon Disulphide	76.13	Propylene	42.08
Carbon Monoxide, CO	28.011	R-11	137.37
Ethane, C_2H_6	30.07	R-12	120.92
Ethyl Alcohol	46.07	R-22	86.48
Ethylene, C_2H_4	28.054	R-114	170.93
Helium, He	4.02	R-123	152.93
Hydrogen, H_2	2.016	R-134a	102.03
Hydrogen Chloride	36.461	R-611	60.05
Hydrogen Sulfide	34.076	Sulfur	32.02
Methane, CH_4	16.044	Sulfur Dioxide	64.06
Methyl Alcohol	32.04	Sulfuric Oxide	48.1
Methyl Butane	72.15	Water Vapor/Steam,	18.02

Therefore,

n = Number of kmols of air = 1.3 kg/28.97 kg/kmol
= 0.0449 kmols

Then, from Eq. 8-19:

$$Q = (0.0449 kmol)(8.314 kJ/kmol°K)(293°K)\ \ln\left(\frac{1}{2}\right)$$

= **-76 kJ of heat removed.**

Since 1kJ = 0.95 Btu,

$\mathbf{Q_{removed,\ in\ Btu's}}$ **= (-76 kJ).(0.95 Btu/kJ)**
= -71.2 Btus

Constant Volume Process

A constant volume process is also referred to as an isometric process or an iso-volumetric process. In a constant volume thermodynamic process, the volume of a closed system remains constant while other parameters, i.e. pressure, internal energy and temperature vary.

Pressure energy, work, volume and pressure are related by the following equation:

$\Delta W = P\Delta V$, where P is pressure

Since the volume is constant in an isometric process,

$$\Delta V = 0$$

And,

$$\Delta W = 0$$

Therefore, in an isometric or constant volume process, by application of First Law of Thermodynamics:

$$Q = \Delta U$$

In other words, in an isometric or isochoric process, the heat added to the system is transformed into the higher level of the system's internal energy "U."

Constant Volume Process Example I:
Superheated steam generation in a "rigid" constant volume boiler.

Most boilers consist of rigid vessels, tanks, channels or tubes. Since these systems are rigid, as the water is boiled, evaporated into saturated steam, and heated further into superheated steam, the temperature and pressure of the water or steam increase but the volume remains constant. There are two basic approaches to boiler design: (1) Water Tube Boiler and (2) Fire Tube Boiler. The water tube boilers are somewhat more common. Figure 8-7 illustrates the fundamental design concept for Water Tube Boilers.

Constant Volume Process Example II: Ideal Heat Engine

Processes or paths D-A and B-C in an ideal heat engine cycle, as shown in Figure 8-5 Pressure–Volume diagram, provide additional examples of constant volume or isometric thermodynamic processes. In paths D-A and B-C, while the pressure increases and decreases, respectively, the volume stays constant.

Isentropic or Constant Entropy Process:

Isentropic process, in a thermodynamic system, is a process in which the entropy of the system stays constant. Any reversible adiabatic process is an isentropic process.

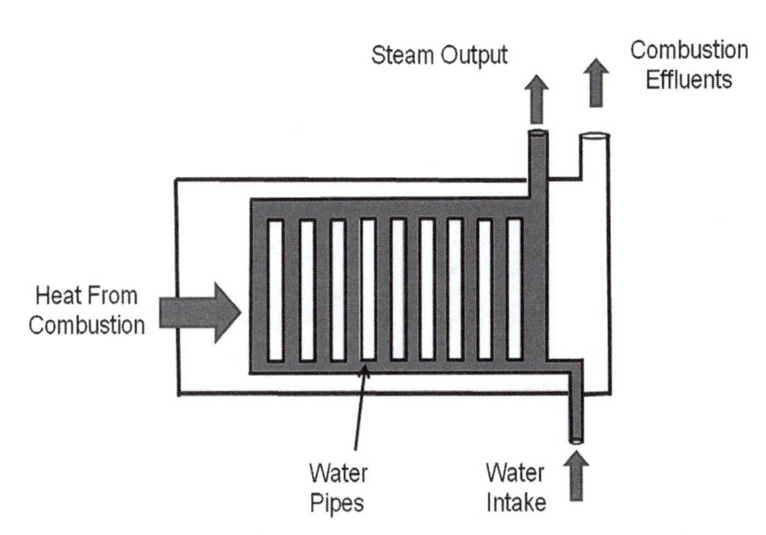

Figure 8-7. Fire-tube Boiler Design and Operation Concept Diagram

Isentropic Process Example I: Ideal Heat Engine—Carnot Cycle

Process paths D′-A′ and B′-C′, as shown in the Temperature-Entropy diagram for a Carnot Cycle Heat Engine in Figure 8-8, are examples of isentropic process. Note: The Carnot Cycle is explained later in this chapter. In paths D′-A′ and B′-C′, while the temperature rises and drops, respectively, the entropy "**s**" stays constant.

Throttling Process and Inversion Point:

Throttling process in a thermodynamic system is an adiabatic process which consists of a significant pressure drop but no change in the system enthalpy. Furthermore, in a throttling process, no heat is exchanged with the surrounding and no work is performed on or by the system. Since the enthalpy in a throttling process stays constant, a throttling process is also an isenthalpic (constant enthalpy) process.

In a throttling process:

$$p_2 << p_1$$

$$\Delta H = 0, \text{ and } \Delta h = 0$$

Where,

ΔH = Change in "absolute" enthalpy, measured in Btus, Joules (J) or kilo Joules (kJ)

Δh = Change in specific enthalpy, measured in Btus/lbm,

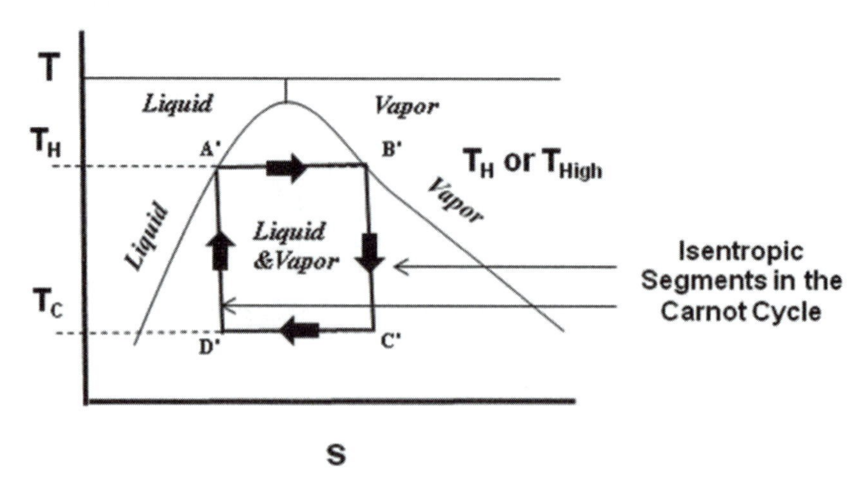

Figure 8-8. Isentropic Processes in a Carnot Cycle, an Ideal Cycle Heat Engine

Joules (J)/gram or kilo Joules (kJ)/kilogram (kg)

p_2 = Final, lower, pressure

p_1 = Initial, higher, pressure

In ideal gas systems, throttling processes are constant temperature processes. In real gas scenarios, temperature change does occur when the gas is throttled. However, for every real gas, under a given set of conditions, there is a temperature point at which no temperature change occurs when the gas is throttled. This temperature is called an ***inversion point.*** For air, maximum inversion temperature is 603°K.

Thermodynamic Equilibrium

A thermodynamic system is said to be in equilibrium when it is in a thermal, chemical, mechanical, convectional and radiative state of balance. A thermodynamic system in equilibrium experiences no thermal, chemical, mechanical, radiative and convectional changes when isolated or insulated from the surroundings.

Quasistatic or Quasiequilibrium Process

Some thermodynamic systems are in equilibrium at the beginning of a process and are in equilibrium toward the end of a process. However, these systems may deviate from equilibrium at interim points of the process. Such processes are referred to as ***quasistatic* or *quasiequilibrium processes.*** Such processes are said to constitute infinitesimal steps. Since the property changes in each of these steps are small, for all intensive purposes, these steps are assumed to represent short equilibrium phases.

Polytropic Process

Polytropic processes pertain to gases and are processes that function in accordance with the polytropic equation of state:

$$(v_1)^n p_1 = p_2 (v_2)^n \qquad \textbf{Eq. 8-21}$$

Where, "n" is a polytropic exponent and is an intrinsic property of the equipment and not the gaseous system. For instance, the polytropic exponent "n" for air compressors ranges from 1.25 to 1.3.

Note the difference between the polytropic process equation **Eq. 8-21** and ideal gas law equation **Eq. 8-15a.**

$$\upsilon_2 = \upsilon_1 \left(\frac{P_1}{P_2} \right) \qquad \textbf{Eq. 8-15}$$

Equation 8-15, which is also a mathematical representation of an ideal gas law referred to as the Boyles Law, can be rearranged and written as:

$$v_1 p_1 = p_2 v_2 \qquad \textbf{Eq. 8-15a}$$

A comparison between **Eq. 8-15a** and **Eq. 8-21** reveals that a salient difference between the behavior of an ideal gas, under ideal conditions, and ideal gas (air) operating under specific equipment specification is that under specific equipment constraints the exponent "n" of volumes υ_1 and υ_2 is not equal to 1. Hence, "**n**" is used as an exponent for volume in **Eq. 8-21**.

Reversible Process

A thermodynamic reversible process is a process that changes the state of a system in such a way that the net change in the combined entropy of the system and its surroundings is zero. The system and the surroundings ***can*** be restored to their initial states at the conclusion of a reversible process. No heat is wasted in a reversible process, therefore, the machine or engine's efficiency is maximized.

One of the attributes of a reversible process can be stated, mathematically, as follows:

$$\Delta \mathbf{S} = \mathbf{0}$$

Irreversible Process

A thermodynamic process that is not reversible is referred to as an irreversible process.

In addition to the fact that heat is or can be wasted in an irreversible process, there is a net change in entropy of the system. In other words:

$$\Delta \mathbf{S} \neq \mathbf{0} \,\big|_{\textbf{Irreversible}}$$

Ideal Heat Engine, Ideal Heat Engine Cycle and Energy Flow

Since the purpose of most engines is to convert one form of energy to another and perform work, an ideal heat engine's function can be simplified and understood through examination of heat/energy flow diagram in Figure 8-9.

As depicted in Figure 8-9, a heat engine performs the conversion of heat energy to mechanical work, driven by the temperature gradient between the higher heat content heat source and a lower heat content point referred to as the heat sink.

Ideal heat engine cycle and the flow of heat and energy in an ideal heat engine can be illustrated better through examination of the Heat Engine Energy Flow Diagram in Figure 8-9, and Heat Engine Process Flow Diagram in Figure 8-10, conjunctively.

As the heat energy is transferred from the heat source to a heat sink, through a mechanical device, such as a turbine, a substantial por-

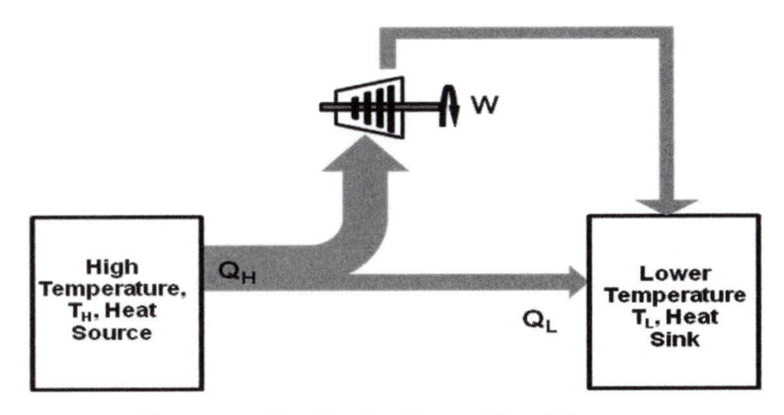

Figure 8-9. Heat Engine Energy Flow Diagram

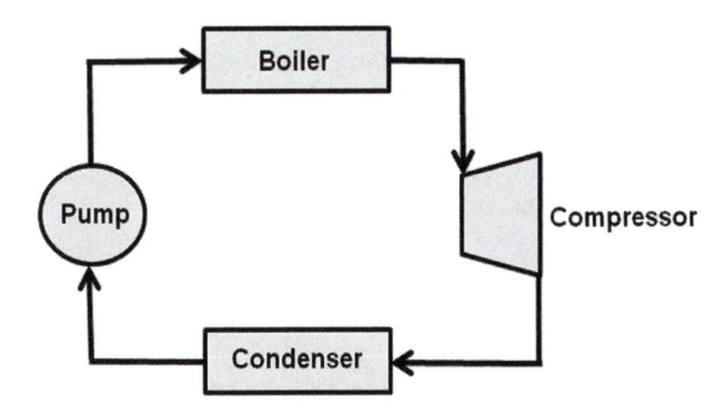

Figure 8-10. Heat Engine Process Flow Diagram

tion of the heat energy is transformed into mechanical energy or work. This is analogous to the flow of electrical current in an electrical circuit where current is driven by the electromotive force, or voltage difference between the positive terminal and the negative terminals of the voltage source. In the analogous electrical current scenario, when the current is routed through an electromechanical device like a motor, it ends up converting the electrical energy into mechanical energy, in form of work.

The transformation of heat energy to mechanical work through a heat engine is also analogous to the energy transformation that occurs in hydroelectric projects. In hydroelectric power generating systems, of course, the elevation head, or potential energy stored in the elevated water of the reservoir, is converted into the kinetic head or kinetic energy as the water is allowed to flow to the turbine. This kinetic energy is converted into mechanical energy of the hydroelectric turbine. The mechanical energy of the hydroelectric turbine, similar to a steam turbine, is then converted into electrical energy, or electric power, through electromagnetic transformation in the electric power generator.

A steam engine is a form of heat engine with the specific purpose of converting heat energy, derived through the combustion of some type of fuel, into mechanical energy or mechanical work. Of course, the complete process involves the transfer of heat energy, from combustion of fuel, to water. In the case of steam based power generating plants, the water, through sensible and latent process, is typically converted to, high enthalpy, superheated steam. Water in such a process is referred to as the ***working fluid.*** The heat energy contained in the superheated steam is then used to turn the turbine. The mechanical energy of the turbine is converted into electrical energy (kWh) through the coupled shafts of the turbine and the electric power generator. In the last stage, conversion of mechanical energy to electrical energy involves the rotation of generator's rotor within the magnetic field of the armature, thus inducing current flow in the armature or stator of the electric power generator.

Steam engines are commonly designed to operate in a heat cycle referred to as the ***Rankine Cycle.*** A ***Rankine Engine*** is simply an engine, or an energy conversion system, in which the working fluid experiences or follows a Rankine cycle. The vapor cycle pertaining to a Rankine cycle is a, relatively, simple and practical vapor cycle. Therefore, in this chapter, we will explore a basic Rankine cycle in detail, followed by a brief description of other variations of the Rankine cycle and an introduction to the Carnot cycle.

Rankine Cycle engines generate, approximately, 80% of all of the electric power consumed in the world. This includes the nuclear fission, fossil fuel, biomass and geothermal based electric power generation. Some notable renewable methods for generation of electric power, i.e. CSP, Concentrated Solar Power, geothermal power generation systems and biomass combustion systems, to name a few, employ the Rankine Cycle for conversion of heat energy into electrical energy. Figure 8-11 and Figure 8-12 show pictures and diagrams of examples of solar and geothermal projects.

Water is typically used as the working fluid in the most Rankine Cycle Engines, because of its favorable properties, such as, its relative chemical inertness—water is neither reactive nor toxic. Water is also abundant, accessible and available at reasonable costs. As shown in Figures 8-14 and 8-15, in Rankine Cycle Engines, water is used in a closed loop configuration.

Figure 8-14 and Figure 8-15 depict the Rankine Heat Engine and the associated vapor cycle. Figure 8-13 represents a pictorial or physical view of the Rankine cycle system. Figure 8-13 is based on symbolic mechanical drawings of the various components of this Rankine En-

Figure 8-11. Concentrated Solar Power Tower—Solucor/Estela Project.

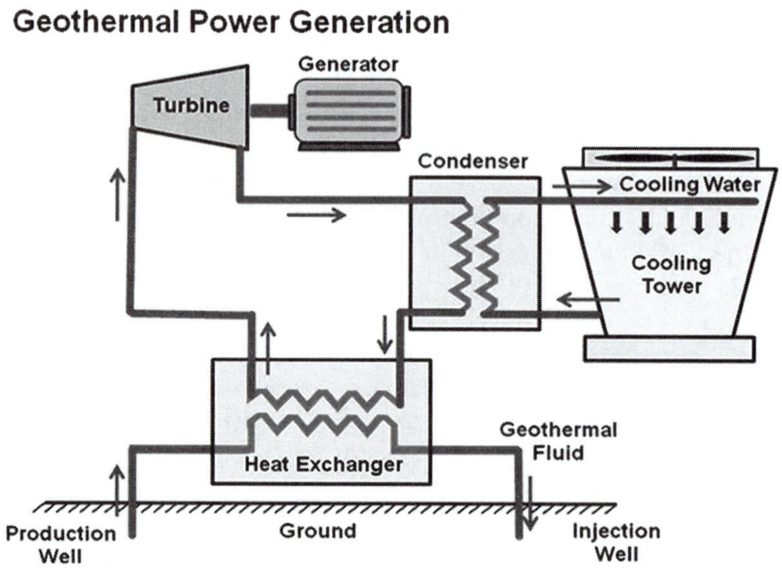

Geothermal Power Generation
Generator
Turbine
Condenser
Cooling Water
Cooling
Tower
Geothermal
Fluid
Heat Exchanger
Production
Well
Ground
Injection
Well

Figure 8-12. Geothermal Electrical Power Generation Process.

STEAM
$W_{turbine}$
BOILER
Heat, Q_{out}
Heat, Q_{in}
W_{pump}

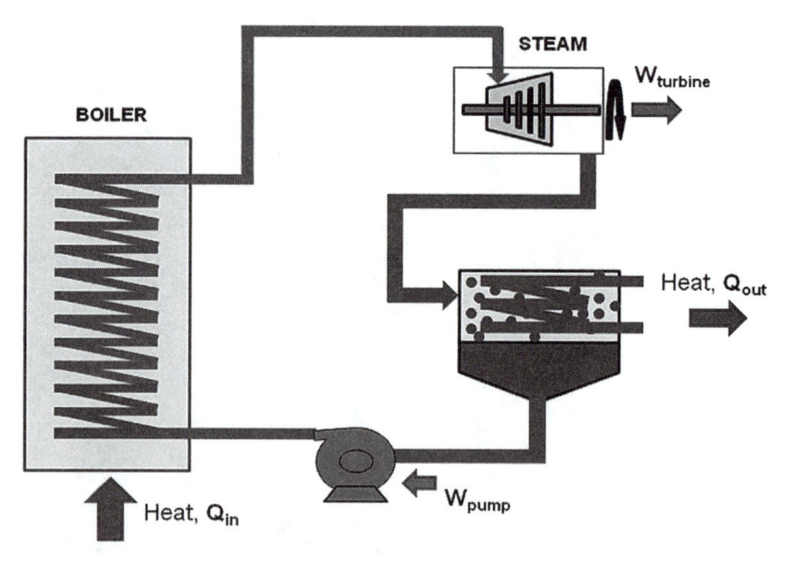

Figure 8-13. Heat cycle in a heat engine without superheat.

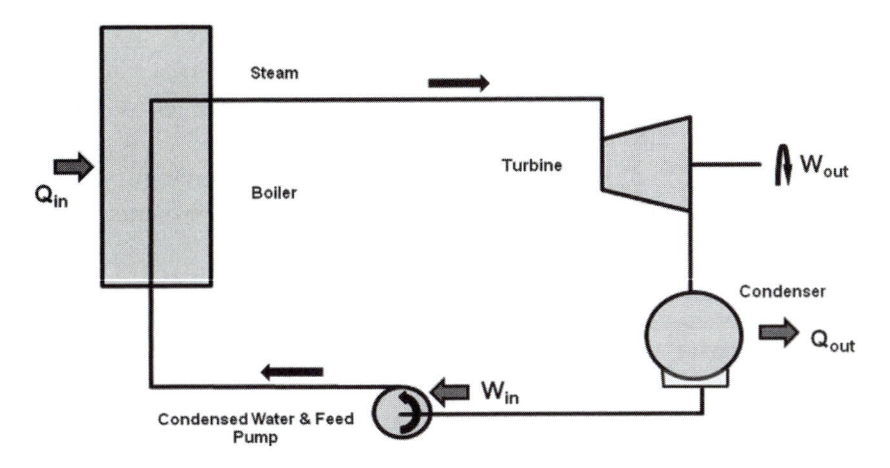

Figure 8-14. Heat cycle in a heat engine without superheat.

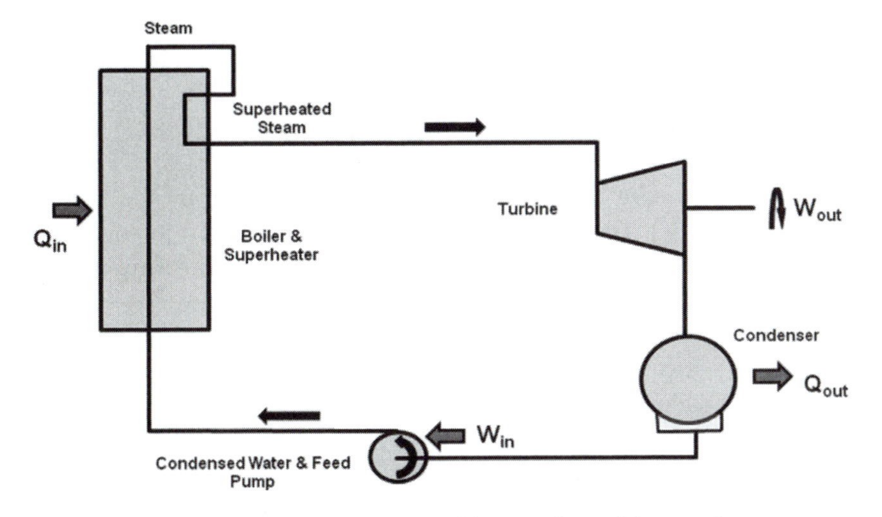

Figure 8-15. Heat cycle in a Rankine Engine with superheat

gine. Figure 8-14 shows the flow of water and vapor in a ***single line diagram*** form utilizing conventional symbols.

While Figure 8-14 represents a one line diagram of the Rankine cycle system that utilizes ***a simple*** water to vapor conversion stage, Figure 8-15 represents a Rankine cycle system that utilizes a water to vapor conversion stage coupled with a ***vapor superheating stage***.

Since the Rankine Cycle, with the superheat, represents a more common and practical Rankine cycle, we will explore it in depth in an effort to understand the process flow in Rankine Heat Engines, in general. Variations of the Rankine Engine, or Rankine Cycle will be described, briefly, later in this chapter.

Examination of Figure 8-15 shows that a basic Rankine Cycle, or a basic Rankine Engine consists of the following major components:

1) Boiler equipped with auxiliary superheating function.
2) Turbine
3) Condenser
4) Pump

Before we embark on a detail discussion of the Rankine Cycle, let us define the four major Rankine Cycle components stated above.

1) **Boiler with Superheating Function:** A boiler, sometimes referred to as a steam generator, is essentially a system designed to heat the feed water returned from the condenser when a heat engine recoups the condensate through a condenser in lieu of releasing it into the environment as effluent. This water is heated up to the saturated state through introduction of heat from the combustion burners. The saturated water is further heated to convert the saturated water into saturated steam. The superheating segment of the boiler converts the saturated steam and some of the lower temperature superheated steam into higher temperature superheated steam. The steam is heated to high temperature superheated steam in order ensure that vapor does not condense in the turbine. When steam condenses into liquid phase in the turbine segment of the heat engine, it tends to reduce the efficiency or energy output of the turbine, and it tends to cause pitting, corrosion and accelerated deterioration of the turbine blades.

2) **Turbine**: A turbine, in essence, converts the superheated steam's enthalpy, or heat content, into positive work, or work performed by the system on the surroundings. From basic function and operating principle perspective, a steam turbine is similar to a hydroelectric turbine or a wind turbine. In all three instances the energy contained in the fluid, gaseous or liquid, is converted into mechanical energy or ***brake horsepower***.

There are two major categories or types of steam turbines that can be differentiated on the basis of their operation. These two categories are as follows:

A Reaction Turbine
B. Impulse Turbine

Reaction Turbine

A reaction turbine is constructed in form of a drum equipped with nozzles or reaction jets. These jets or nozzles, as shown in Figure 8-16, are located around the circumference of the drum. As the steam exits from the nozzles at a certain velocity **V** and mass flow rate $\dot{m}$, a force equal and opposite to the force F, below, is produced rotating the drum in the opposite direction. In Figure 8-16, this force rotates the drum in the counter clockwise direction. The formula for this force is stated in form of Eq. 8-22.

$$F = \dot{m} \cdot \Delta V \qquad \textbf{Eq. 8-22}$$

Impulse Turbine

An impulse turbine system consists of stationary nozzles and vanes appointed on a wheel as shown in Figure 8-17. The high pressure high enthalpy steam released in a throttling process from the opposing nozzles impacts the vanes in a coupled moment fashion. This ***coupled*** moment, as depicted in Figure 8-18, rotates the wheel, thus transferring power to the wheel shaft as brake horsepower.

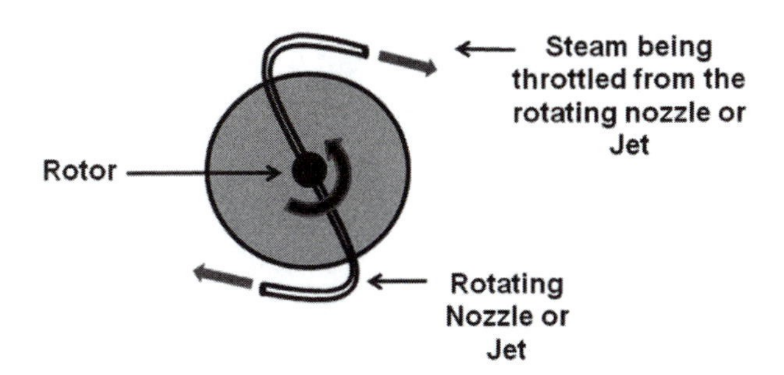

Figure 8-16. Reaction Turbine Design and Operating Principle

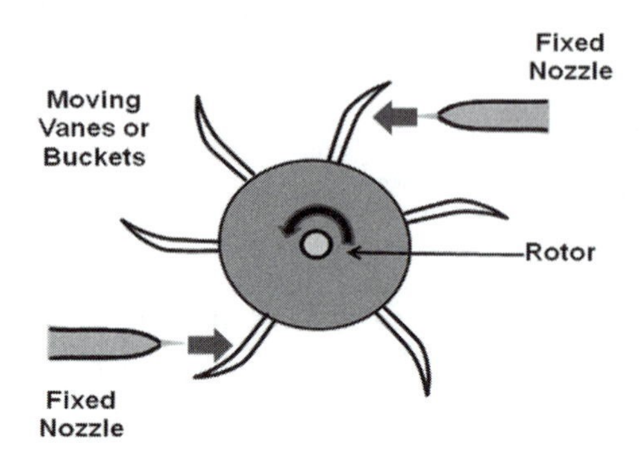

Figure 8-17. Impulse Turbine Design Principle

A coupled moment, or simply a couple, is comprised of two equal and opposite forces. The sum of these forces, as stated in Eq. 8-23, is zero.

$$\Sigma F = 0 \qquad \textbf{Eq. 8-23}$$

The coupled moment is shown in Figure 8-18 and represented mathematically in Eq. 8-24.

$$\text{Couple} = 2.\tau = 2(F.d) \qquad \textbf{Eq. 8-24}$$

Where,

- τ is the torque
- **F** is the magnitude of the equal and opposite forces
- **d** is the moment arm for each force, spanning from the center point of the vane to the center of the turbine wheel.

Figure 8-18.
Couple in an Impulse Turbine

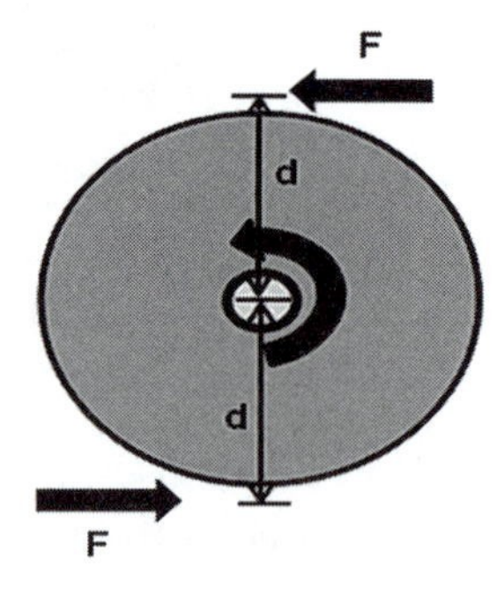

The throttling process that steam undergoes in the impulse turbine nozzles, as well as in the reaction turbine nozzles, is an adiabatic process; no heat is exchanged between the system and the environment. This can be physically explained on the basis of the fact that due to the high velocity of steam and short contact period between the steam and the nozzles, a negligible amount of heat is transferred from the steam to the nozzles. In other words, the enthalpy remains unchanged; $\mathbf{h_1 = h_2}$ and $\Delta h = 0$. This tenet is supported by the Bernoulli's law with appropriate assumptions and simplification. This concept is explained and discussed in more detail in Chapter 9.

3) **Condenser:** A condenser condenses the lower enthalpy vapor into water by facilitating extraction of heat "Q" from the saturated vapor. As explained in Chapter 7, when saturated vapor loses heat, it condenses into saturated liquid phase, or water. In that respect, a condenser can be referred to as a heat exchanger. A condenser is considered to be a special purpose non-adiabatic heat exchanger. A condenser is non-adiabatic because heat flows out of it as the working fluid, or vapor, is cooled. The heat flow rate out of a condenser is defined, mathematically, in Eq. 8-25. Since heat is lost by the vapor in the condenser, the final enthalpy will be lower than the initial enthalpy. In other words, $(h_2 - h_1)$ would be negative.

$$\dot{\mathbf{Q}} = \dot{\mathbf{m}}(h_2 - h_1) \qquad \textbf{Eq. 8-25}$$

In Eq. 8-25:

$\dot{\mathbf{Q}}$ represents the heat flow rate in Btu's per second, or kJ/second.
$\dot{\mathbf{m}}$ represents the mass flow rate in lbm per second or kg/second.
$\mathbf{h_2}$ is the final enthalpy.
$\mathbf{h_1}$ is the initial enthalpy.

4) **Feed Pump:** The feed pump simply receives the condensed water from the condenser, pressurizes it and transfers it to the boiler.

Process Flow in a Rankine Cycle with Superheat

In order to analyze and understand the processes or stages in the superheat equipped Rankine Cycle, the one line schematic from Figure 8-15 has been redrawn with appropriate stage or process annotations as shown in Figure 8-19. These stages or process are further addressed through pressure "**p**" versus specific volume "**υ**," temperature "**T**" versus entropy "**s**," and enthalpy "**h**" versus entropy "**s**" graphs; these graphs are depicted in Figure 8-20, Figure 8-21 and Figure 8-22, respectively.

The thermodynamic process flow stages in a superheated Rankine cycle, as shown in Figure 8-19, are numbered 1 through 6. In Figure 8-20, Figure 8-21 and Figure 8-22, these six stages are described in terms of the interplay between pressure (**p**), specific volume (υ), temperature (**T**), enthalpy, (**h**) and entropy (**s**).

Process Flow from Point 1 to 2

Point 1 lies in the sub-cooled water realm on all three graphs. In other words, it is located distinctly to the left of the bell shaped saturation curves on all three graphs. On the other hand, Point 2 lies directly on the saturation curve on all three graphs. This means that at point 2, the water is in saturated liquid state. Therefore, as heat is added to the subcooled water in the boiler, it transitions from point 1 to point 2 and changes states from subcooled form to saturated water form. As shown in the (**p**) versus specific volume (υ), diagram (Figure 8-20), this process is **isobaric**. Of course, since ***heat is being added*** to the system or the working fluid, the process flow from point 1 to point 2 is ***non-adiabatic.***

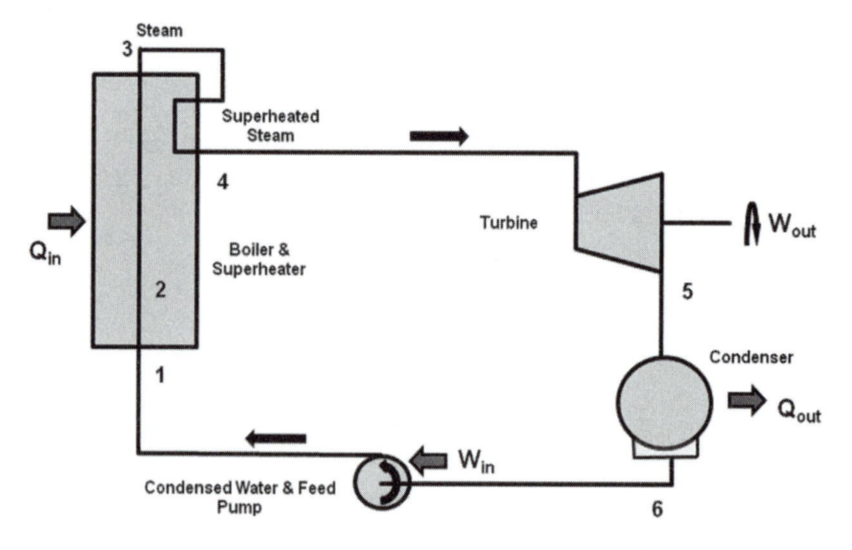

Figure 8-19. Heat Cycle in a Rankine Engine with Superheat

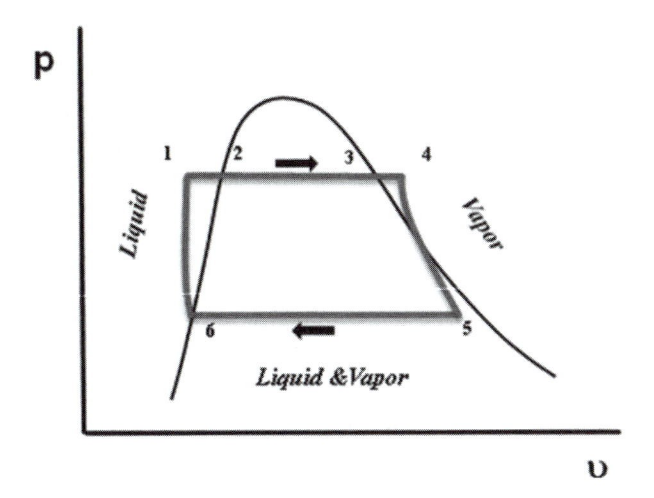

Figure 8-20. Heat Cycle in a Rankine Engine with Superheat, p versus υ.

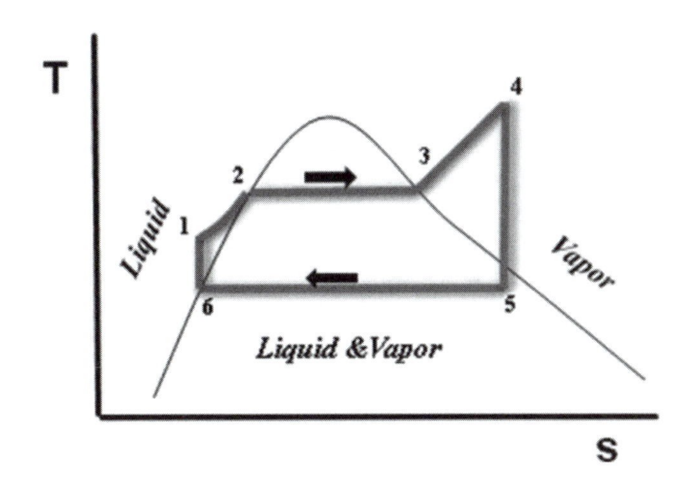

Figure 8-21. Heat Cycle in a Rankine Engine with Superheat, T versus s.

As we examine the temperature versus entropy graph, in Figure 8-21, we note that the process flow from point 1 to point 2 involves a **"sensible"** rise in temperature (**T**) accompanied by an increase in the entropy (**s**).

The transition from point 1 to point 2, from the enthalpy (**h**) and entropy (**s**) perspective, can be observed in Figure 8-22. This transition entails a small increase in the enthalpy and a more pronounced rise in the entropy.

Process Flow from Point 2 to 3

Point 2 lies directly on the saturation curve on all three graphs. As stated earlier, the water is in saturated liquid state at point 2. By definition, addition of ***any*** heat to the water in process stage 2 would result in evaporation.

Therefore, as heat is added to the saturated water in the boiler, it transitions from point 2 to point 3 and changes states from saturated water to saturated water vapor. As shown in the pressure **p** versus specific volume υ diagram (Figure 8-20), this process is ***isobaric***.

As we examine the temperature **T** versus entropy **s** graph, in Figure 8-21, we note that the process flow from point 2 to point 3 is an ***isothermal*** process coupled with an ***increase*** in the entropy.

The transition from point 2 to point 3, from the enthalpy **h** and entropy **s** perspective, as graphed in Figure 8-22, shows that the addition of heat in this stage not only raises the entropy but also raises the enthalpy as the heat is added to change states from saturated water to saturated vapor.

Process Flow from Point 3 to 4

Point 3 lies directly on the saturated vapor curve on all three graphs. The water is in saturated vapor state at point 3. By definition, ***removal*** of ***any*** heat from saturated vapor would result in condensation of the saturated vapor into liquid water.

However, as heat is ***added*** to the saturated vapor in the boiler, it transitions from point 3 to point 4 and changes states from saturated vapor to superheated vapor. All three graphs show that as additional heat is added to the saturated vapor at point 3, the vapor diverges off the saturation curve and into the superheated steam region. The pressure **p** versus specific volume υ graph in Figure 8-20 shows that this addition of heat that drives the water into the superheated state is also ***isobaric.*** Also, note that the high pressure of the superheated steam that is introduced into the turbine, is established in the preceding stages of the heat engine cycle.

The temperature **T** versus entropy **s** graph, in Figure 8-21, shows that the heat added in this stage escalates the temperature significantly; accompanied by a notable rise in the entropy. The increase in entropy can be observed in the enthalpy versus entropy graph in Figure 8-22, as well.

Process Flow from Point 4 to 5

The transition from point 4 to point 5 takes place in the turbine. This is the process stage in which the mechanical (fluid) energy is delivered from the superheated steam to the turbine blades. This energy is transferred to the turbine shaft in form of work. The work performed on the turbine blades $W_{Turbine}$, is represented mathematically by Eq. 8-28, in per unit mass form.

Point 4 lies clearly in the superheated steam region as depicted in all three graphs in Figures 8-20 through 8-22. However, as the superheated steam transfers its energy to the turbine, its temperature, enthalpy, and pressure drop to levels represented by point 5 on the pressure (**p**), specific volume (υ), temperature (**T**), enthalpy, (**h**) and entropy (**s**) diagrams.

As exhibited in the enthalpy versus entropy and the temperature versus entropy graphs, the entropy remains constant in this stage, making this stage isentropic. Note that the heat is neither added nor removed from the superheated vapor—the working fluid—in this stage. Therefore, this stage is not just isentropic but also adiabatic. Of course, this is based on the assumption that the turbine is well insulated.

The pressure, specific volume, temperature, enthalpy and entropy diagrams show that after delivering its energy to the turbine, the vapor and liquid mixture has significantly lower temperature, enthalpy and pressure, and is therefore no longer capable of performing useful work. However, as is the case with some of the multiple stage Rankine Cycles, if additional heat were added to the system and the steam raised to the superheated state, additional work production or power delivery could be accomplished through the turbine.

At point 5, the vapor and liquid mixture is ready to be condensed fully to saturated liquid phase, represented by point 6.

Process Flow from Point 5 to 6

The transition from point 5 to point 6 constitutes the condensation stage of this Rankine Cycle. This transition takes place in the condenser. In this process stage, heat is removed from the vapor and liquid mixture in order to transform the mixture into a saturated liquid.

Transition from point 5 to point 6 is a non-adiabatic process; in that, heat is removed from the system, or the working fluid, by the condenser.

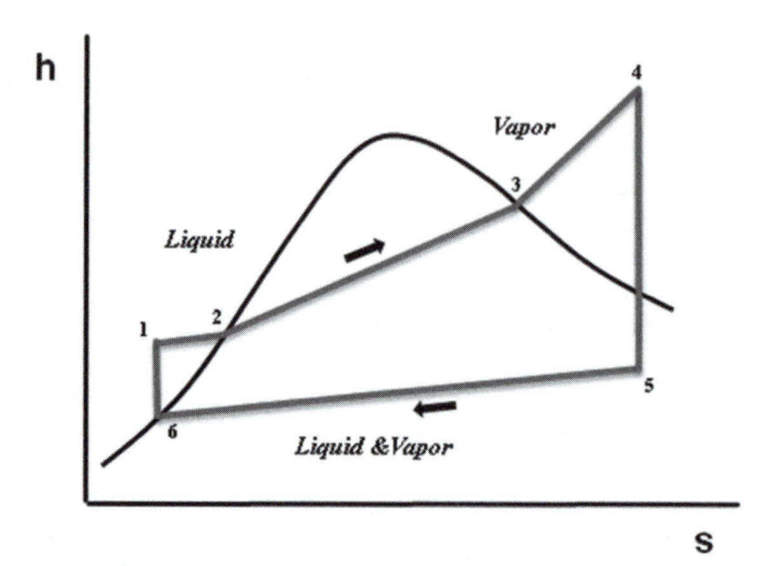

Figure 8-22. Heat Cycle in a Rankine Engine with Superheat, h versus υ.

As evident through the pressure versus specific volume graph in Figure 8-20, the transition from point 5 to point 6 occurs at a constant pressure; in other words, this portion of the Rankine Cycle is ***isobaric.*** In addition, the transition from point 5 to point 6, results in substantial reduction of the specific volume. And, since specific volume is the inverse of a density, as the specific volume decreases, the density increases. This observation is supported by the physical transformation of the working fluid from a lower density vapor liquid mixture to a, significantly, denser saturated water.

The temperature versus entropy graph in Figure 8-21 shows that as this Rankine cycle transitions from point 5 to point 6, the temperature stays constant, or this process is ***isothermal***. This comports with the physics associated with general condensation of vapors. Condensation of vapor involves "latent" heat removal and the temperature remains constant as the heat is removed.

Since heat is removed in this stage, there is a change in **Q** and it occurs at a constant temperature. According to Eq. 8-26, restated below, change in heat at a constant temperature results in net change in entropy:

$$\Delta s = q/T_{abs} \qquad \textbf{Eq. 8-26}$$

Therefore, in the process transition from point **5** to point **6,** there is a net change in the entropy. This is supported by the **T** versus **s** graph in Figure 8-21.

Figure 8-22 shows that the transition from point 5 to point 6 also involves some reduction in enthalpy, **h.** This is typically the case when there is a significant change in heat, **Q**.

Process Flow from Point 6 to 1

The transition from stage or point 6 to point 1 is the last stage in this Rankine Cycle. This stage of the Rankine Cycle boosts the pressure of the working fluid to the level of pressure in the boiler. This increase in the overall pressure head or pressure energy of the fluid is accomplished through the feed pump. The fact that the pressure of the working fluid is enhanced substantially is illustrated through the pressure versus specific volume graph in Figure 8-20. Note that the as the pressure rises significantly in the transition stage between point 6 and 1, the specific volume stays practically constant. Since the Rankine Cycle considered in this analysis recycles the condensate, it could—for most practical purposes—be construed as a ***closed system***. As we learnt earlier, in a closed thermodynamic system, the mass remains constant. Then, by definition, as represented mathematically in Eq. 8-27, the specific volume would remain constant in this last stage of the Rankine Cycle.

$$\upsilon = 1/\rho \qquad \textbf{Eq. 8-27}$$

Where,

υ = Specific Volume

ρ = Density = Mass / Volume

Note that the working fluid had already arrived at the saturation line—the line on the left half of Figure 8-20 graph that represents saturated liquid—at point 6. So, in the transition from point 6 to 1, as the pressure is raised by pump, the working fluid, or water, is driven into the ***subcooled*** liquid state. As we noted in the steam table discussion in chapter 6, as the pressure of saturated water is increased, the saturation temperature elevates, as well. Therefore, when the Rankine Cycle is repeated, the first stage—from point 1 to point 2—would require commensurate amount of heat and higher temperature (saturation temperature) to boil the water.

Since no heat is added or removed from the water in this stage (6 to 1), $\Delta Q = 0$, and according to Eq. 8-26, $\Delta s = 0$. In other words, as evident from the **T** versus **s** and **h** versus **s** graphs, this stage is ***isentropic***. Note that despite the fact that this stage is isentropic and that no heat is added to the system, transition from point 6 to 1 is ***not*** isoenthalpic (constant enthalpy). The small rise in temperature and enthalpy in this stage, as exhibited in the temperature versus entropy and enthalpy versus entropy graphs, is due to the addition of pump head or pressure energy.

Rankine Cycle Equations

Some of the mathematical relationships or equations that can be used to analyze and define various parameters in a Rankine cycle are listed below:

$$W_{Turbine} = h_3 - h_4 \qquad \textbf{Eq. 8-28}$$

$$W_{Pump} = v_6(P_1 - P_6) \qquad \textbf{Eq. 8-29}$$

$$Q_{in} = h_4 - h_1 \qquad \textbf{Eq. 8-30}$$

$$Q_{out} = h_5 - h_6 \qquad \textbf{Eq. 8-31}$$

$$\eta_{Thermal} = \frac{Q_{in} - Q_{out}}{Q_{in}} = \frac{W_{Turbine} - W_{Pump}}{Q_{in}} \qquad \textbf{Eq. 8-32}$$

Where,

$W_{Turbine}$ = Work performed on the turbine by the steam, in Btu's or kJ

W_{Pump} = Work performed by the pump on the vapor, in Btu's or kJ

$\eta_{Thermal}$ = Thermal efficiency of the entire Rankine cycle

Q_{in} = Heat added by the boiler to the working fluid, in Btu's or kJ

Q_{out} = Heat removed from the working fluid by the condenser, in Btu's or kJ

h_1 = Enthalpy at point 1 in Btu/lbm or kJ/kg, from steam tables

h_3 = Enthalpy at point 3 in Btu/lbm or kJ/kg, from steam tables

h_4 = Enthalpy at point 4 in Btu/lbm or kJ/kg, from steam tables

h_5 = Enthalpy at point 5 in Btu/lbm or kJ/kg, from steam tables
h_6 = Enthalpy at point 6 in Btu/lbm or kJ/kg, from steam tables
p_1 = Pressure at point 1, in lbf/ft^2, or Pa, i.e. N/m^2
p_6 = Pressure at point 6, in lbf/ft^2, or Pa, i.e. N/m^2

CASE STUDY 8-1. RANKINE ENGINE

A thermodynamic system consists of a Rankine engine with superheat function. The enthalpy versus entropy graph for this system is shown in Figure 8-23. The mass flow rate of the system or working fluid is 200 lbm/sec. As an energy engineer, you are to explore the answers to the following questions based on the data provided:

a) If the enthalpy of the fluid is 1500 Btu/lbm and the entropy of the system is approximately 1.8900 Btu/(lbm-°R), what phase or stage is the system in:

(i) Steam generation stage
(ii) Steam superheating stage
(iii) Condensation stage
(iv) Work producing turbine stage

b) If the enthalpy of the superheated steam entering the turbine is 1860 Btu/lbm and the discharged steam has an enthalpy of 1320 Btu/lbm, what is approximate amount of power delivered to the turbine shaft? Assume that the heat loss is negligible in this system.

(i) 1 MW
(ii) 10 kW
(iii) 50 MW
(iv) 80 MW

c) If the enthalpy of the working fluid is 1200 Btu/lbm and the entropy is 1.8100 Btu/(lbm-°R), what is the phase of the working fluid?

(i) Saturated vapor
(ii) Superheated vapor
(iii) A mixture of vapor and liquid
(iv) Sub-cooled liquid

d) If the enthalpy of the working fluid is 1500 Btu/lbm and the entropy is 1.8650 Btu/(lbm-°R), what is the phase of the working fluid?

(i) Saturated vapor
(ii) Superheated vapor
(iii) Saturated liquid
(iv) Sub-cooled liquid

Analyses/Solutions:

a) If the enthalpy of the fluid is 1500 Btu/lbm and the entropy of the system is approximately 1.8900 Btu/(lbm-°R), what phase or stage is the system in?

Answer:

To address this question, we must first locate the point on the graph where **h** = 1500 Btu/lbm and **s** = 1.8900 Btu/(lbm-°R). This point

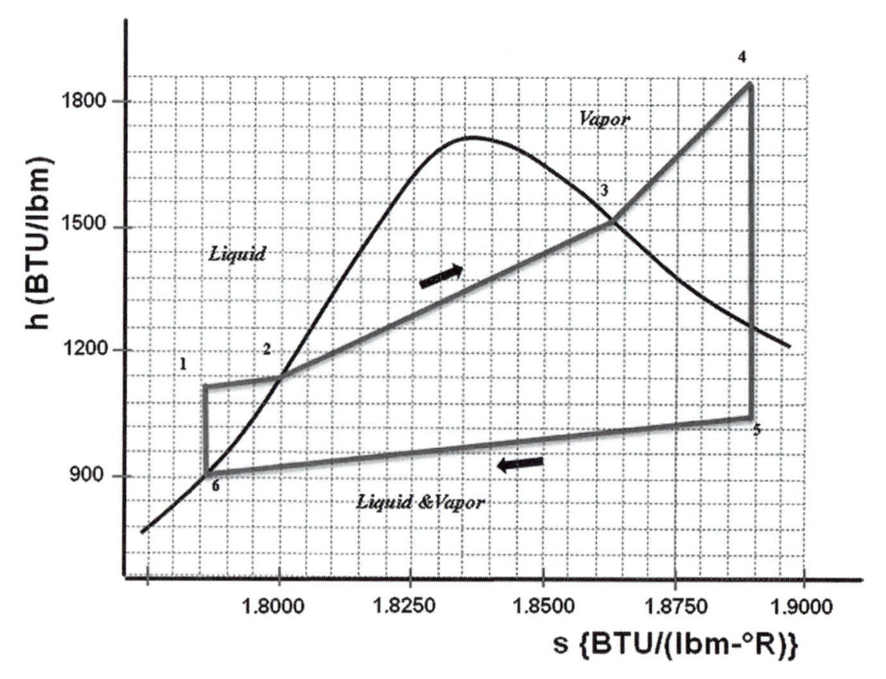

Figure 8-23. Heat Cycle in a Rankine Engine with Superheat, h versus υ. Case Study 8-1

is labeled as point "**a**" on the enthalpy versus entropy graph in Figure 8-24. Point **a** lies in the region that is clearly outside the saturated vapor curve. In other words, at point **a**, the working fluid is clearly ***superheated steam***. Furthermore, point **a** lies on the transition path between points **4** and **5**. As discussed earlier, the transition path between points **4** and **5** represents the work producing stage in the turbine

Therefore the correct answer would be:

(iv) Work producing turbine stage

b) If the enthalpy of the superheated steam entering the turbine is 1860 Btu/lbm and the discharged steam has an enthalpy of 1320 Btu/lbm, what is approximate amount of power delivered to the turbine shaft if the efficiency of the turbine is 70%? Assume that the heat loss is negligible in this system.

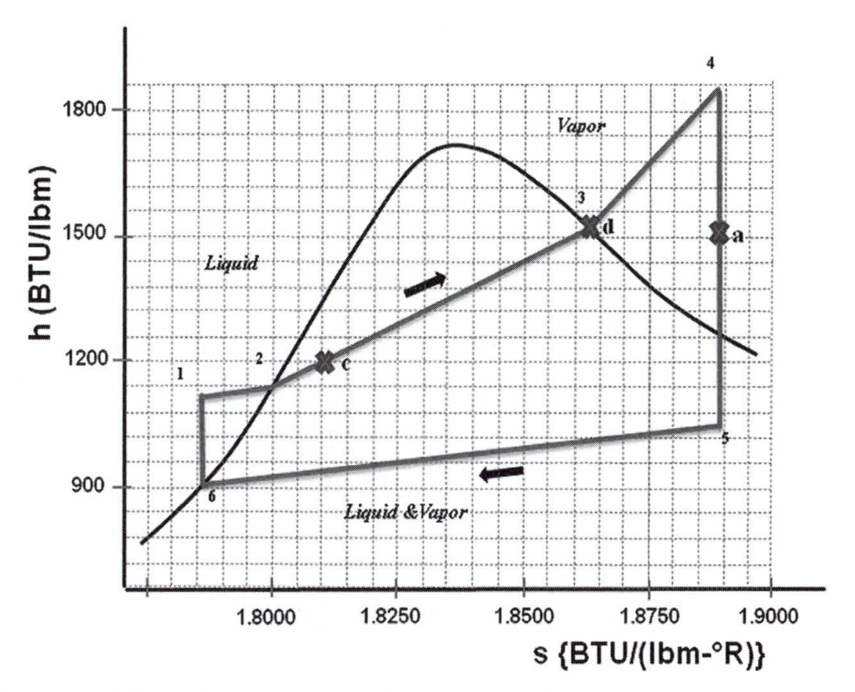

Figure 8-24. Heat Cycle in a Rankine Engine with Superheat, h versus υ; Case Study 8-1

Power delivered by the turbine to the shaft, or generator
= ($\mathbf{Power_{Steam}}$) x (η) **Eq. 8-33**

Where,

η = Efficiency of the turbine = **70%** {Given}

$\mathbf{Power_{Steam}}$ = Power Delivered by the Steam, to the Turbine Blades

Since we are allowed to assume that there is no heat loss in the turbine, the difference between the enthalpy of the superheated steam entering the turbine and the discharged steam would be equal to the energy delivered by the steam to the turbine.

$\therefore$ **Power delivered to turbine in Btu's/sec = ($h_i - h_f$) . ($\dot{m}$)** **Eq. 8-34**

Where,

$\dot{m}$ = Mass flow rate of the working fluid or water = **200 lbm/sec** {Given}

h_i = Enthalpy of steam entering the turbine = **1860 Btu/lbm.**{Given}

h_f = Enthalpy of steam discharged = **1320 Btu/lbm** {Given}

Then, application of Eq. 8-34 would yield:

Power delivered to turbine, in Btu's/sec
= **(1860** Btu / lbm **– 1320** Btu / lbm**) . (200 lbm/sec)**

Or

Power delivered to turbine, in Btu's/hr
= **(1860** Btu / lbm **– 1320** Btu / lbm**) . (200 lbm/sec) . (3600 sec/hr.)**
= 388,800,000 Btu's/hr

Since there are 3,413 Btu's/hr per kW,

Power delivered to turbine, in kW
= **(388,800,000 Btu's/hr)/3413 (Btu's/hr/kW**
= 113,917 kW or 13.9 MW

Then, by applying **Eq. 8-33**:

Power delivered by the turbine to the shaft, or the generator
= (**Power$_{\mathbf{Steam}}$**) x (η)
= **(113.9 MW) x (0.7)**
= **79.74 MW**

∴ **The correct answer is (iv) 79.74 MW or approximately 80 MW**

Ancillary to Example 8-1 (b):

As apparent upon examination of Figure 8-23 and Figure 8-24, the change in enthalpy stated in part (b)—from 1860 Btu/lbm to1320 Btu/lbm—occurs while the working fluid ***remains in superheated realm.*** If the final enthalpy, in this turbine phase of the engine cycle, had dropped below the saturated vapor curve, i.e. point 5, where h = 1020 Btu/lbm, the solution for this part of the problem would have required more steps. This is because at point 5, the working fluid would be a mixture of liquid and vapor phases and each phase would have a different or distinct enthalpy value. The total final enthalpy in such a case would be a sum of the saturated vapor and liquid enthalpies, added in proportion determined by the humidity ratio, ω.

c) If the enthalpy of the working fluid is 1200 Btu/lbm and the entropy is 1.8100 Btu/(lbm-°R), what is the phase of the working fluid?

Answer:

To address this question, we must locate the point on the graph where **h** = 1200 Btu/lbm and **s** = 1.8100 Btu/(lbm-°R). This point is labeled as point "**c**" on the enthalpy versus entropy graph in Figure 8-24. Point **c** lies in the region that falls between **saturated liquid line** and the **saturated vapor line**. In other words, at point **c**, the working fluid is in a phase that consists of a mixture of liquid and vapor.

Therefore the correct answer would be:

(iii) A mixture of vapor and liquid

d) If the enthalpy of the working fluid is 1500 Btu/lbm and the entropy is 1.8650 Btu/(lbm-°R), what is the phase of the working fluid?

Answer:

To address this question, we must locate the point on the graph where **h** = 1500 Btu/lbm and **s** = 1.8650 Btu/(lbm-°R). This point is labeled as point "**d**" on the enthalpy versus entropy graph in Figure 8-24. Point **d** lies directly on the **saturated vapor line**. In other words, at point **d**, the working fluid is in a phase that consists of entirely of saturated vapor. This occurs as the working fluid begins its transitions from points **3** to **4**; just before it is superheated to point **4**.

Therefore the correct answer would be:

(i) Saturated vapor

Carnot Cycle:

The Carnot cycle, similar to the Rankine cycle, is a heat engine cycle that converts thermal energy into mechanical and electrical energy. However, unlike the Rankine cycle, the Carnot cycle is an ideal power cycle. In strict context, a Carnot cycle is impractical to implement.

In order to understand the sequential transition of processes or stages within the Carnot cycle, we will examine the **pressure** versus **specific volume**, **temperature** versus **entropy** and **enthalpy** versus **entropy** graphs shown in Figure 8-25, Figure 8-26, and Figure 8-27, respectively. Note that these graphs provide a simplified view of the process mechanics in a Carnot cycle. A less simple but more realistic insight into the Carnot Cycle is presented in Figure 8-28.

Process Flow from Point 1 to 2:

Point 1 lies on the saturated water line on all three graphs; in Figure 8-25, Figure 8-26, and Figure 8-27. In other words, at point 1, the working fluid is in saturated liquid state on all three graphs. On the other hand, Point 2 lies directly on the saturated vapor curve and the working fluid is in the saturated vapor state on all three graphs Therefore, as heat is added to the saturated water at point 1, it transitions to point 2. As evident from Figure 8-25 and Figure 8-26, this transition from point 1 to point 2 is ***isobaric*** and ***isothermal***. Note that this transition or thermodynamic process from point 1 to point 2, in the Carnot cycle, is similar to the transition from point 2 to point 3 in the Rankine cycle. Of course, since ***heat is being added*** to the system or the working fluid, the process flow from point 1 to point 2 is ***non-adiabatic.*** This segment of the Carnot cycle can also be considered as isothermal expansion of the saturated liquid to saturated vapor.

As we examine the **T** vs. **s** graph, in Figure 8-26, we note that, while the process flow from point 1 to point 2 is isothermal or constant temperature, there is a distinct increase in the entropy (**s**).

The transition from point 1 to point 2, from the enthalpy (**h**) and entropy (**s**) perspective, can be observed in Figure 8-27, to entail the pronounced rise in the entropy accompanied by a notable increase in the enthalpy.

Process Flow from Point 2 to 3:

The transition from stage or point 2 to point 3 would, theoretically, take place in the work producing segment of the Carnot cycle. Note that, as apparent from Figures 8-25, 8-26 and 8-27, unlike the corresponding point in the Rankine cycle, point 2 in the Carnot cycle is ***not in the superheated steam*** realm. As the saturated steam transfers its energy to the turbine, its temperature, enthalpy, and pressure drop to levels represented by point 3 on the pressure (**p**), specific volume (υ), temperature (**T**), enthalpy, (**h**) and entropy (**s**) diagrams.

As exhibited in the enthalpy versus entropy (Figure 8–27) and the temperature versus entropy graphs (Figure 8–26), the entropy remains constant in this stage, making this stage isentropic. Note that the heat is neither added nor removed from the vapor—the working fluid—in this stage. Therefore, this stage is not just ***isentropic*** but also ***adiabat-***

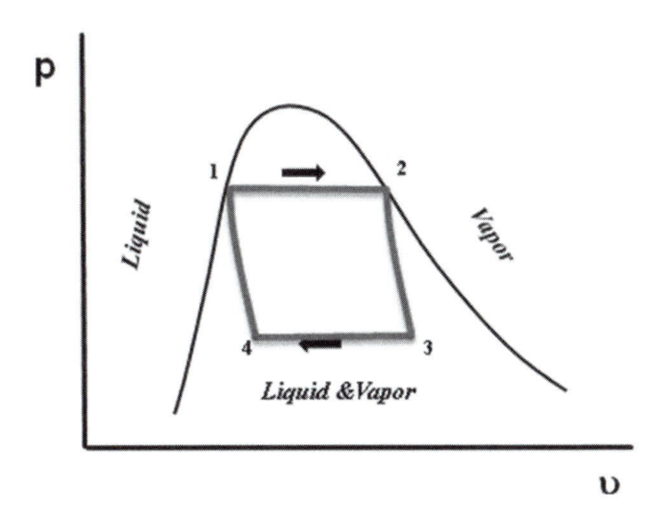

Figure 8-25. Heat Cycle in a Carnot Engine, p versus υ.

ic. This segment of the Carnot cycle can also be considered as ***isentropic expansion*** of vapor. This is because the entropy stays constant and the pressure drops substantially as the working fluid transitions from point 2 to point 3. The thermodynamic process diagram in Figure 8-26 shows that, as the isentropic expansion of the vapor in this stage of the Carnot cycle performs work on the surroundings and transfers energy in form of work, it results in lower vapor temperature, $\mathbf{T_{Low}}$.

Process Flow from Point 3 to 4:

The transition from point 3 to point 4 consists of isothermal compression of the lower temperature vapor. The fact that this stage of the Carnot cycle is isothermal is obvious from Figure 8-26. However, the fact that vapor is compressed in this stage is not clearly supported by the simplified pressure versus specific volume graph shown in Figure 8-25. The graph in Figure 8-25 implies that the pressure is constant in this stage. Examination of the more realistic pressure versus specific volume graph, in Figure 8-28, shows that the pressure of the working fluid does rise in this stage of the Carnot cycle.

In the transition from point 3 to point 4, some heat is removed from the working fluid or the system, thus rendering this segment of the Carnot cycle non-adiabatic. In other words, Q is negative in this stage. Heat is removed from system in order to maintain the vapor

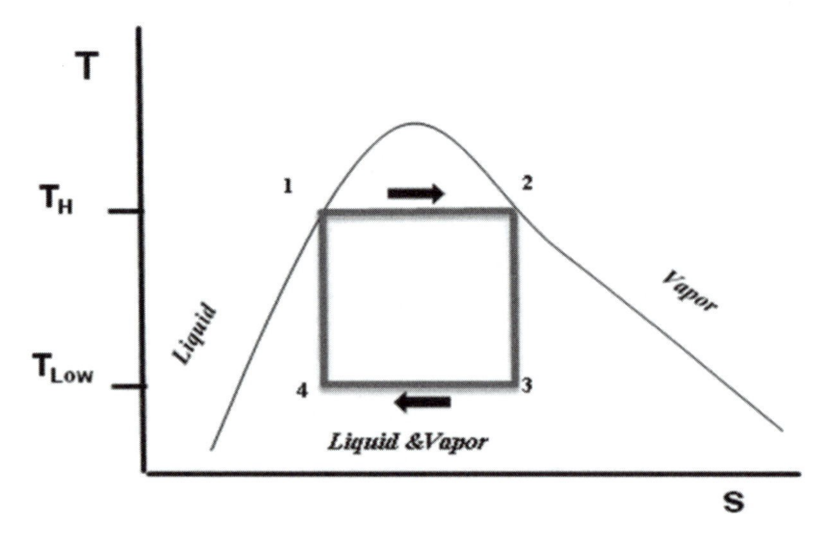

Figure 8-26. Carnot Heat Cycle, T versus s.

at a constant temperature as work is performed by the surroundings, namely the compressor pump, onto the system. The removal of heat in this stage comports with the small drop in the enthalpy, as shown in Figure 8-27.

Since heat is removed in this stage, there is a change in **Q** and it occurs at a constant temperature. According to Eq. 8-26, restated below, change in heat at a constant temperature results in net change in entropy:

$$\Delta s = q/T_{abs} \qquad \textbf{Eq. 8-26}$$

Therefore, in the process transition from point **3** to point **4,** there is a net change in the entropy. This is supported by the **T** versus **s** graph in Figure 8-26.

Process Flow from Point 4 to 1:

The transition from stage or point 4 to point 1 is the last stage in this Carnot cycle. This stage represents an isentropic compression process. As apparent from Figures 8–26 and 8–27, the transition from point 4 to 1 occurs at a constant entropy **s.** This stage of the Carnot cy-

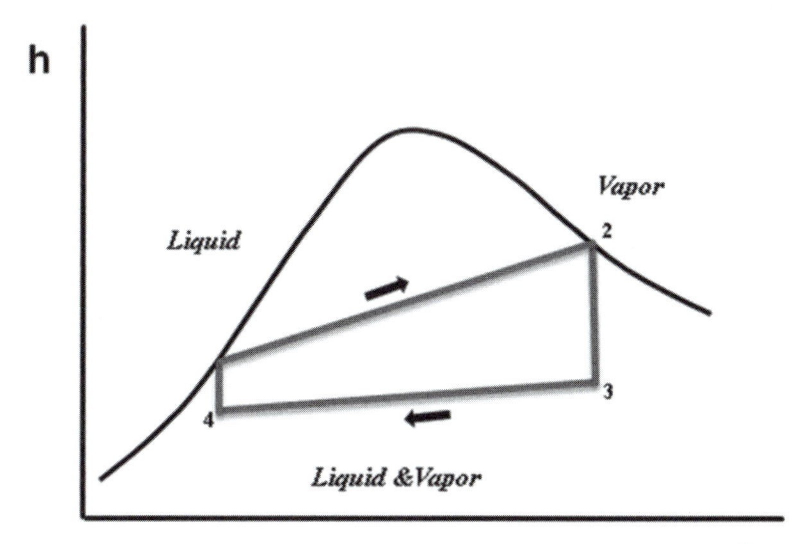

Figure 8-27. Carnot Heat Cycle, h versus s.

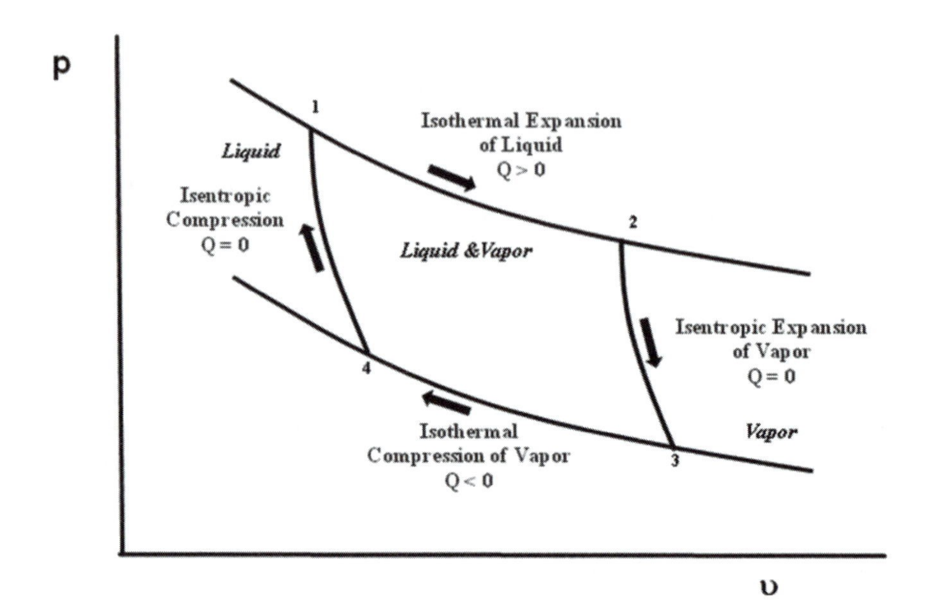

Figure 8-28. Alternative Representation of Carnot Heat Cycle, p versus υ.

cle boosts the pressure of the working fluid to the level of pressure in the boiler. The fact that the pressure of the working fluid is enhanced substantially is illustrated through the pressure versus specific volume graphs in Figures 8-25 and 8-28. In an ideal system, it is assumed that no heat is either added or removed from the system in the transition path from 4 to 1. Therefore, this process from point 4 to 1 is an adiabatic process. In other words, Q = 0 in the last stage of the Carnot cycle.

Carnot Cycle Equations:

Some of the mathematical relationships or equations that can be used to analyze and define various parameters in a Carnot cycle are listed below:

$$W_{Turbine} = h_2 - h_3 \qquad \textbf{Eq. 8-35}$$

$$W_{Pump} = h_1 - h_4 \qquad \textbf{Eq. 8-36}$$

$$Q_{in} = h_2 - h_1 \qquad \textbf{Eq. 8-37}$$

$$Q_{out} = h_3 - h_4 \qquad \textbf{Eq. 8-38}$$

$$\eta_{Thermal} = \frac{Q_{in} - Q_{out}}{Q_{in}} = \frac{W_{s-Turbine} - W_{s-Pump}}{Q_{in}}$$ **Eq. 8-39**

Where,

$W_{Turbine}$ = Work performed on the turbine by the steam, in Btu's or kJ

W_{Pump} = Work performed by the pump on the vapor, in Btu's or kJ

$W_{s\text{-}Turbine}$ = Work performed on turbine by the steam, in Btu's or kJ, during the isentropic process from point 2 to 3.

$W_{s\text{-}Pump}$ = Work performed by the pump on the vapor, in Btu's or kJ, during the isentropic process from point 4 to 1.

$\eta_{Thermal}$ = Thermal efficiency of the entire Carnot cycle

Q_{in} = Heat added by the boiler to the working fluid, in Btu's or kJ

Q_{out} = Heat removed from the working fluid by the condenser, in Btu's or kJ

h_1 = Enthalpy at point 1 in Btu/lbm or kJ/kg, from steam tables

h_2 = Enthalpy at point 2 in Btu/lbm or kJ/kg, from steam tables

h_3 = Enthalpy at point 3 in Btu/lbm or kJ/kg, from steam tables

h_4 = Enthalpy at point 4 in Btu/lbm or kJ/kg, from steam tables

Comparison between Rankine and Carnot Cycles:

A basic Rankine Cycle is similar to the Carnot Cycle with the following exceptions:

1) In the Rankine cycle, the compression process occurs in the liquid region. The compression process in a Carnot cycle takes place in the region where working fluid exists in vapor and liquid mixture form. This is apparent from examination of the two, comparative, temperature versus entropy graphs shown in Figure 8-29. Compression of the working fluid in liquid form is more efficient than the compression and pumping of the vapor and liquid mixture.

2) Rankine cycle is closely approximated and applied in steam turbine plants.

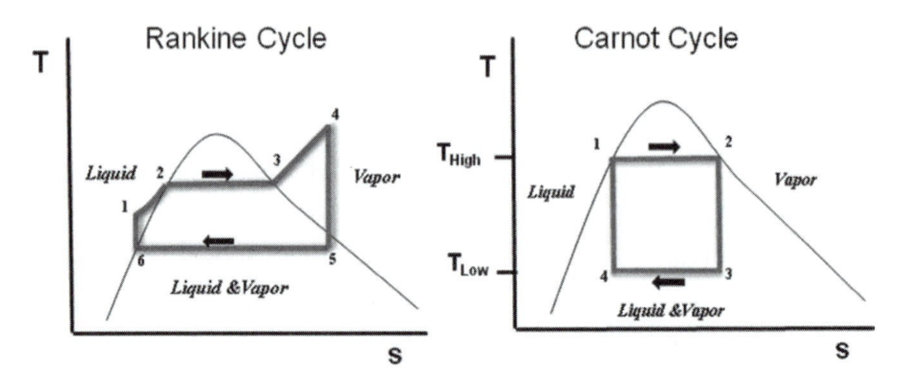

Figure 8-29. Rankine and Carnot Cycle Comparison, Temperature versus Entropy

3) The working fluid crosses through the saturated liquid line and is condensed completely into the subcooled phase in the Rankine cycle. In the Carnot cycle the working fluid is compressed only ***up to*** the saturated liquid line.

4) On the high enthalpy spectrum of the process, the working fluid is heated well into the superheated phase area in the Rankine cycle. While, in the Carnot cycle, the working fluid only heats ***up to*** the saturated vapor line.

Other Major Types of Cycles:

Aside from the simple Rankine cycle, Rankine cycle with superheating and the Carnot cycle discussed above, some other cycles, that find common applications are as follows:

1) Rankine Cycle with Superheat and Reheat

This type of heat engine or cycle is similar to the Rankine cycle with superheating function, except that it includes a reheat feature, as well. See Figure 8-30. The reheat feature allows for the steam to be reheated after its first pass through the turbine. This creates a second enthalpy differential and an additional opportunity to produce work or deliver energy to the turbine blades.

2) Rankine Cycle with Regeneration

This type of Rankine cycle allows the condensate to be preheated

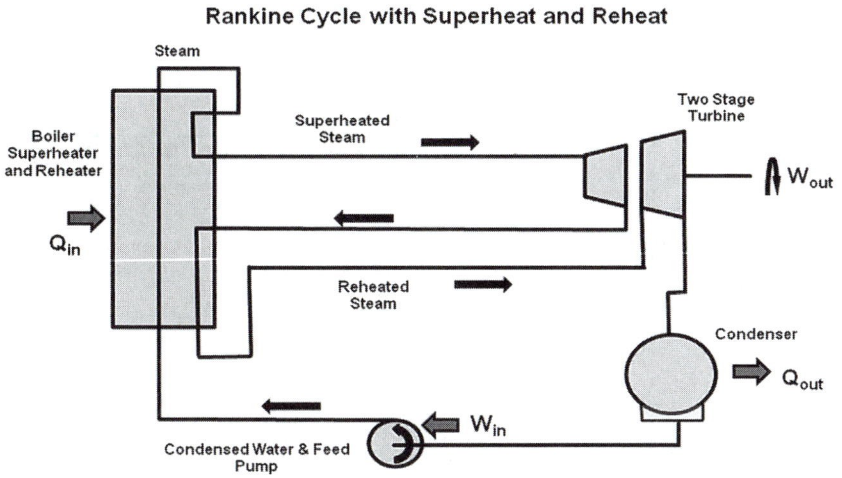

Rankine Cycle with Superheat and Reheat

Figure 8-30. Working Principle of a Rankine Cycle Engine with Superheat and Reheat

using bleed stream from multiple points in the turbine. This preheated condensate water is then pumped to the boiler. Rankine cycles with regeneration may or may not be equipped with a superheat stage. Regenerative Rankine cycles are typically equipped with multiple feedwater heaters.

3) Binary Cycle

A binary cycle engine is, in essence, a Rankine cycle engine that utilizes two different fluids. For example, in geothermal type binary cycle engines, as the one shown in Figure 8-31, these two fluids could be geothermal fluids, such as water and some type of organic working fluid, like pentane hydrocarbon or butane.

4) Cogeneration and Combined Cycles

Most heat engines operate at efficiencies that are less than 50%. So, if the heat engine efficiency is 35%, it would mean that 65% of the heat is ***not*** converted to work; instead it is discarded or dissipated into the ambient atmosphere as waste heat. The heat lost in the heat engines is referred to as ***rejected energy*** or ***rejected heat***. Cogeneration and combined cycle systems harness the rejected heat for various purposes, thus enhancing the efficiency of the overall system.

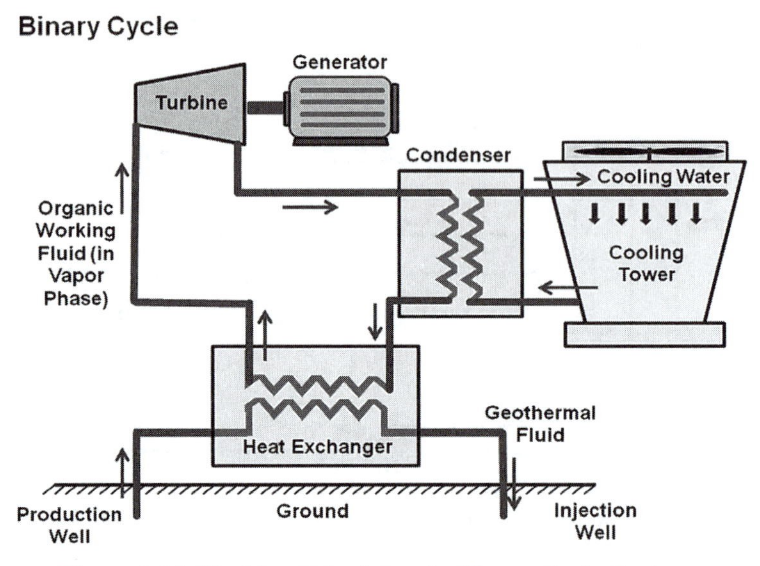

Figure 8-31. Working Principle of a Binary Cycle System

Cogeneration

When waste heat is recovered and used to produce hot water or for space heating purposes, also referred to as ***district heating***, the process is termed as cogeneration. In cogeneration systems the recovered heat is not directly used to generate electricity. If the heat recovered through a cogeneration system is in form of steam, steam that is a byproduct of an electric power generation system, then it is called a ***CHP***, or ***combined heat and power cycle***. The temperature of steam applied in cogeneration systems ranges from 80°C to 180°C. The cogeneration steam is not only used for space heating but can also be utilized for cooling purposes through the absorption process or absorption chillers.

Combined Cycle

If recovered waste heat from a power generating system is used to vaporize water and produce steam, and if that steam is subsequently used to generate electric power (once again), the process is referred to as a combined cycle.

Working principle behind a typical combined cycle is illustrated in Figure 8-32. As apparent from this diagram, generator No. 2 is powered by the steam produced through heat transfer in the primary generator's effluent stack.

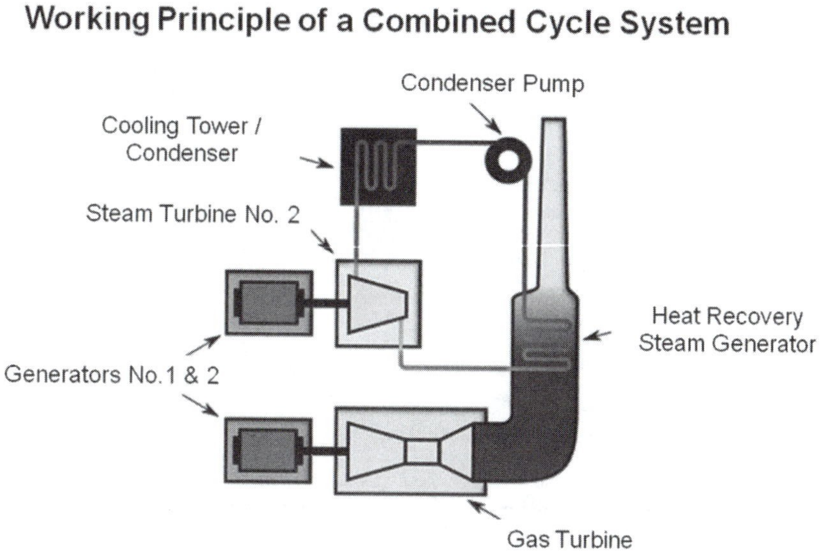

Figure 8-32. Working Principle of a Combined Cycle System

Chapter 8—Self Assessment Problems and Questions

1. Ideal heat engines always include a boiler with superheating function.

   A. True
   B. False

2. In the heat engine represented by the enthalpy versus entropy graph in Figure 8-22, the heat is added to working fluid in:

   (i) Steam generation stage
   (ii) Steam superheating stage
   (iii) Condensation stage
   (iv) Steam generation stage and the Steam superheating stage

3. In the heat engine represented by the enthalpy versus entropy graph in Figure 8-22, the energy contained in the superheated working fluid is converted into the rotational kinetic energy in:

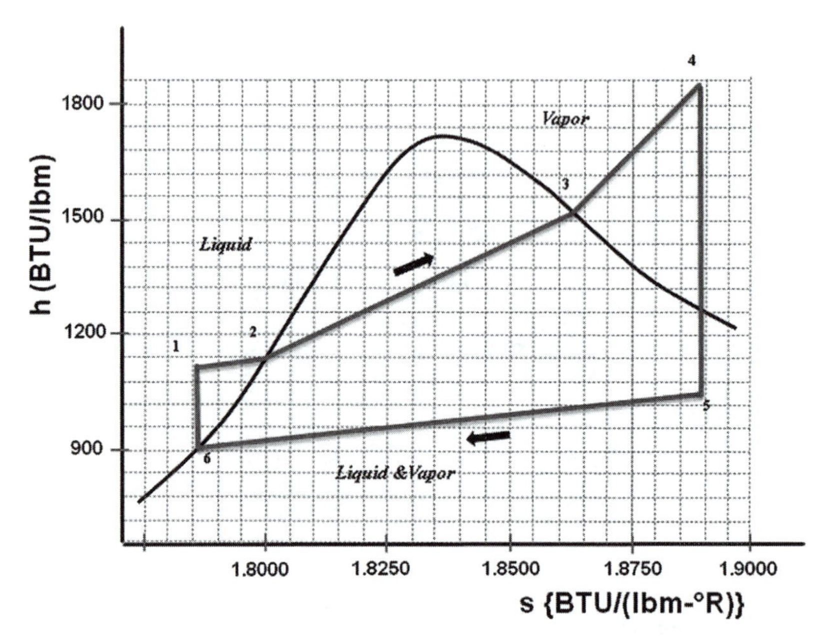

Figure 8-33. Heat Cycle in a Rankine Engine with Superheat, h versus s.

(i) Process transition from point 2 to 3
(ii) Process transition from point 4 to 5
(iii) Process transition from point 6 to 1
(iv) Process transition from point 1 to 2

4. A thermodynamic system consists of a Rankine engine with superheat function. The enthalpy versus entropy graph for this system is shown in Figure 8-33.

   The mass flow rate of the system or working fluid is 100 lbm/sec. Answer the following questions based on the data provided:

   a) If the enthalpy of the fluid is approx. 1850 Btu/lbm and the entropy of the system is approximately 1.8900 Btu/(lbm-°R), what phase or stage is the system in:

(i) Work producing turbine stage
(ii) Steam superheating stage

(iii) Condensation stage
(iv) Steam generation stage

b) If the enthalpy of the subcooled water entering the boiler is 900 Btu/lbm and the enthalpy of the water at a downstream point in the boiler is 1080 Btu/lbm, what is approximate amount of heat added in **MMBtu per hour**? Assume that there is no heat loss.

(i) 1 MMBtu/hr
(ii) 1.08 MMBtu/hr
(iii) 64.8 MMBtu/hr
(iv) 1.3 MMBtu/hr

c) If the enthalpy of the working fluid is 1440 Btu/lbm and the entropy is 1.8500 Btu/(lbm-°R), what is the phase of the working fluid?

(i) Saturated vapor
(ii) Superheated vapor
(iii) A mixture of vapor and liquid
(iv) Sub-cooled liquid

Chapter 9

Gas Dynamics

INTRODUCTION

This chapter is devoted to introduction of ***Gas Dynamics*** and topics within the realm of gas dynamics that are more common from practical application point of view. Gas dynamics constitutes the study of gases moving at high velocity. By most standards, a gas is defined as a high velocity gas when it is moving at a velocity in excess of 100 m/s or 300 ft/s. Traditional fluid dynamics tools such as the Bernoulli's equation, and the momentum and energy conservation laws—traditionally applied in mechanical dynamics study—do not account for the role internal energy plays in gas dynamics; therefore, they cannot be applied in a comprehensive study of high velocity gases. In this chapter, we will examine the behavior of high speed gases on the basis of key thermodynamic entities, such as enthalpy, **h,** and internal energy, **u**. The gas dynamics discussion is premised largely on the fact that high velocity of a gas is achieved at the expense of internal energy; where the drop in internal energy, **u**—as supported by equation **Eq. 9-1**—results in the drop in the enthalpy, **h**.

$$\mathbf{h = u + p.v} \qquad \textbf{Eq. 9-1}$$

STEADY FLOW ENERGY EQUATION

Consider the high velocity flow scenario depicted in Figure 9-1 below. We will use this illustration to explain important characteristics and components of high velocity gas flow system.

As shown in Figure 9-1, a high pressure reservoir is located on the extreme right. The properties of gas in this reservoir are referred to as the ***stagnation properties, chamber properties, or total properties.*** The gas possesses kinetic, potential, and thermal energy in all segments of the high velocity gas system. The thermal energy possessed by the gas is in form of internal energy and enthalpy. The pressure and temperature of the gas in the reservoir are denoted by P_o and T_o, respectively. The gas in the reservoir is high pressure gas. This gas travels through the mid segment, referred to as the ***duct***. In the duct, the gas continues to be considered as high pressure and low velocity. The duct leads to the segment called the ***throat*** where the pressure drops and velocity escalates; thus transforming the gas into high velocity gas.

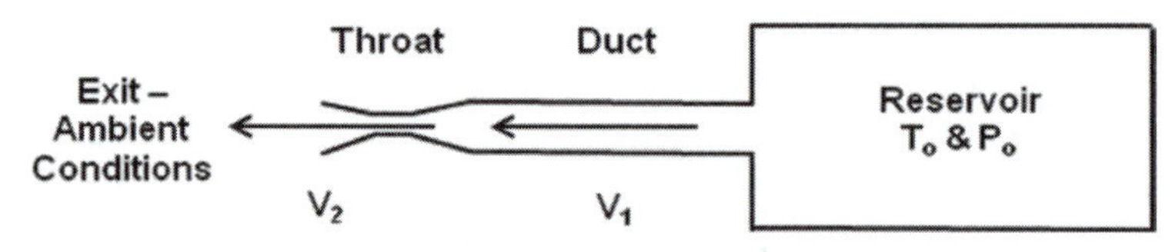

Figure 9-1. High Velocity Flow

The flow of gas in the high velocity gas system is considered to be adiabatic because the high speed of gas—due to its short residence time in the throat—does not allow significant amount of heat exchange. In addition, in a simplified scenario, the length of the duct is considered to be short enough, such that no significant frictional head loss occurs. Also, as seen in Figure 9-1, the high velocity gas exits out to ambient atmosphere; signifying its "**open-flow**" characteristic.

Since the high velocity gas system described above is an adiabatic open flow system, the **SFEE, steady flow energy equations**, as stated below, would apply:

In the SI (Metric) Unit System:

$$g.z_1 + \tfrac{1}{2}.v_1^2 + h_1 = g.z_2 + \tfrac{1}{2}.\ v_2^2 + h_2 \qquad \textbf{Eq. 9-2}$$

In the US (Imperial) Unit System:

$$(g/g_c).z_1 + \tfrac{1}{2}.(v_1^2/g_c) + J.h_1 = z_2.(\ g/g_c) + \tfrac{1}{2}.(v_2^2/g_c) + J.h_2 \qquad \textbf{Eq. 9-3}$$

Where,

h_1 = Enthalpy of gas entering the throat, in kJ/kg (SI) or Btu/lbm (US).

h_2 = Enthalpy of gas exiting the throat, in kJ/kg (SI) or Btu/lbm (US).

v_1 = Velocity of the gas entering the throat, in m/s (SI) or ft/s (US).

$\mathbf{v_2}$ = Velocity of the gas exiting the throat, in m/s (SI) or ft/s (US).

$\mathbf{z_1}$ = Elevation of the gas entering the throat, in m (SI) or ft (US).

$\mathbf{z_2}$ = Elevation of the gas exiting the throat, in m (SI) or ft (US).

$\mathbf{g}$ = 9.81 m/s^2 in the SI realm and 32.2 ft/s^2 in the US unit realm.

$\mathbf{g_c}$ = Gravitational constant, 32.2 lbm-ft/lbf-sec^2

$\mathbf{J}$ = 778 ft-lbf/Btu

Often, in practical gas dynamics scenarios, the exit velocity of the gas is the desired objective of analysis. Therefore, in practical situations, Equations 9-2 and 9-3 can be simplified to compute the exit velocity $\mathbf{v_2}$. Since reservoir area of cross section is inordinately larger than the orifice or throat area of cross section, the velocity, $\mathbf{v_1}$, of gas in the reservoir is considered to be negligible; or, $\mathbf{v_1 \cong 0}$. Since the density of gas is small, the potential energy component in Equations 9-2 and 9-3 can be disregarded. With these practical assumptions, Equations 9-2 and 9-3 can be distilled down to the following, simpler, practical forms:

$$v_2 = \sqrt{2(h_0 - h_2)} \quad \text{\{SI Unit System\}} \quad \textbf{Eq. 9-4}$$

$$v_2 = \sqrt{2g_c J(h_0 - h_2)} \quad \text{\{US Unit System\}} \quad \textbf{Eq. 9-5}$$

CASE STUDY 9-1

A nozzle is fed from a superheated steam reservoir, shown in Figure 9-2. The duct or hose connecting the nozzle to the reservoir is short and the frictional head loss in the hose is negligible. Based on these practical assumptions, the velocity of the superheated steam in the hose can be neglected. The steam exits the nozzle at 150°C (300°F) and 0.15 MPa (21.76 psia). Determine the exit velocity of the steam at the nozzle.

Solution:

Given:

$\mathbf{T_o}$ = 300°C or 572°F

$\mathbf{P_o}$ = 2.0 MPa or 290 psia

$\mathbf{v_1 = v_0 = 0}$

T_2 = 150°C or 300°F
P_2 = 0.15 MPa or 21.76 psia

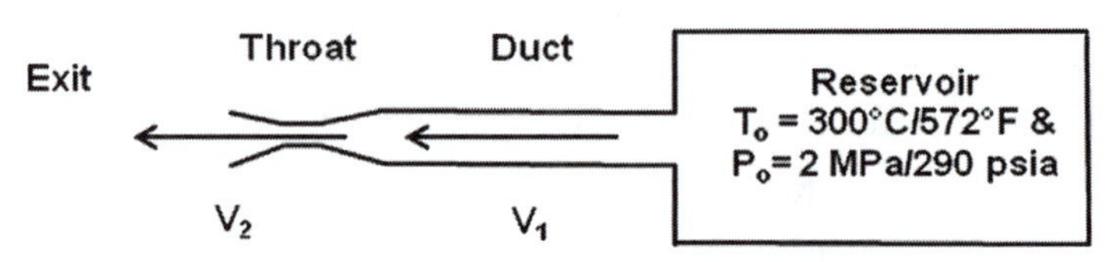

Figure 9-2. High Velocity Flow, Case Study 9-1

SI Unit System:

Apply Eq. 9-4 do calculate the exit velocity of the superheated steam in the SI units:

$$v_2 = \sqrt{2(h_0 - h_2)} \quad \text{\{SI Unit System\}} \qquad \textbf{Eq. 9-4}$$

From the steam tables in Appendix B, in the SI units:

$\mathbf{h_o = 3024\ kJ/kg}$
$\mathbf{h_2 = 2773\ kJ/kg}$

Then, by applying **Eq. 9-4**:

$$v_2 = \sqrt{(2).(3024 - 2773kJ/kg).(1000J/kJ)}$$

Note: The multiplier 1000J/kJ, in the equation above, is used to convert kJ to Joules as Eq. 9-4 is premised on Joules and not kilo Joules

$$v_2 = 709m/s$$

US Unit System:

Apply **Eq. 9-5** do calculate the exit velocity of the superheated steam in the US units:

$$v_2 = \sqrt{2g_cJ(h_0 - h_2)} \quad \text{\{US Unit System\}} \qquad \textbf{Eq. 9-5}$$

From the steam tables in Appendix B, in the US units, and through double interpolation:

$\mathbf{h_o = 1299\ Btu/lbm}$
$\mathbf{h_2 = 1191Btu/lbm}$

Note: These enthalpies can also be read from the Mollier diagram, without interpolation; albeit, the results might differ, slightly.

Then, by applying **Eq. 9-5**:

$$v_2 = \sqrt{2.\left(32.2\,\frac{lbm-ft}{lbf-s^2}\right).\left(778\,\frac{ft-lbf}{Btu}\right).\left(1299-1191\,\frac{Btu}{lbm}\right)}$$

v2 = **2324** ***ft/s***

ISENTROPIC FLOW

In gas dynamics, flow of gas is said to be ***isentropic*** when the process is adiabatic, frictionless, reversible, and when the ***change in entropy is negligible***. In many, practical, high velocity gas flow scenarios, there is a small entropy change due to nozzle and discharge loss coefficients.

Critical Point

A gas in flow is said to be at the critical point when its speed equals the speed of sound, i.e. Mach 1, or M=1, or 1130 ft/s (344 m/s) at 70°F/20°C, 1 Atm, or 1090 ft/s (331 m/s) at STP. At the critical point, parameters such as velocity, density, temperature, pressure, etc., are called sonic properties and are annotated by an asterisk, *; for instance; v*, ρ*, T*, and P*, respectively. The ratios of sonic properties to reservoir properties are referred to as ***critical constants*** or ***critical ratios***. For instance, the critical pressure ratio $\mathbf{R_{cp}}$ is represented, mathematically, as:

$$R_{cp} = \left[\frac{P^*}{P_o}\right] = \left(\frac{2}{k+1}\right)^{\frac{k}{k-1}} \qquad \textbf{Eq. 9-6}$$

Where,

$\mathbf{P^*}$ = Sonic pressure

$\mathbf{P_o}$ = Reservoir pressure

k = Ratio of specific heats; e.g., **k = 1.4** for air

$\mathbf{R_{cp}}$ = Critical pressure ratio

Shock Waves

Shock waves are thin layers of gas, several molecules in thickness that have substantially different thermodynamic properties. Shock waves develop when a gas moving at supersonic speed slows to subsonic speed. Shock waves propagate or travel ***normal to the direction of flow of gas***. Shock waves represent an ***adiabatic*** process and the total temperature of the system stays constant. However, the total pressure does decrease and the process is ***not isentropic.***

Chapter 9—Self Assessment Problems and Questions

1. A nozzle is fed from a superheated steam reservoir. The superheated steam in the reservoir is at 500°C (932°F) and 2.0 MPa (290 psia). The duct or hose connecting the nozzle to the reservoir is short and the frictional head loss in the hose is negligible. Based on these practical assumptions, the velocity of the superheated steam in the hose can be neglected. The steam exits the nozzle at 1.0 bar (14.5 psia) and 95% quality. Determine the exit velocity of the steam at the nozzle in SI units.

2. Solve Problem 1 in US Units, use Mollier diagram for all enthalpy identification and compare the resulting steam speed with results from computation conducted in SI units, in Problem 1.

3. The SFEE Equation 9-2 can be applied to compute the exit speed of gas, in high speed gas applications under which of the following conditions?

   A. When data are available in US units
   B. When data are available in SI units
   C. When the reservoir is large enough such that $\mathbf{v_0 = 0}$, applies.
   D. Both B and C.
   E. Both A and B.

4. Which of the following statements is true about shock waves?

   A. Shock waves require superheated steam
   B. Shock waves travel parallel to the direction of the flow of gas.
   C. Shock waves travel perpendicular to the direction of the flow of gas.
   D. Both A and B.

Chapter 10

Psychrometry and Psychrometric Analysis

INTRODUCTION

The scope of this chapter is to introduce energy engineers to psychrometry and to provide an understanding of psychrometric concepts, principles, tools and techniques available for analyzing existing and projected psychrometric conditions in an air conditioned environment. Psychrometry, like many other aspects of thermodynamics, deals with the basic elements of thermodynamics such as air, moisture, and heat. Our discussion in this chapter will focus heavily on the use of psychrometric chart as an important tool for evaluating current psychrometric conditions, defining transitional thermodynamic processes and projecting the post transition psychrometric conditions.

THE PSYCHROMETRIC CHART

A psychrometric chart is a graph of the physical properties of moist air at a constant pressure (often equated to an elevation relative to sea level). This chart graphically expresses how various physical and thermodynamic properties of moist air relate to each other, and is thus a graphical equation of state. See Figures 10-1, 10-2 and 10-3. Psychrometric charts are available in multiple versions. Some versions are basic and allow analysis involving only the basic parameters, such as the dry bulb, wet bulb, enthalpy, relative humidity, humidity ratio and the dew point. The detailed version psychrometric charts include additional parameters, like the specific volume, sensible heat ratio and higher resolution relative humidity scale for RH level below 10%. Psychrometric charts are available in the US or imperial units as well as the SI or metric units. Psychrometric charts are published by various sources including the major refrigerant and refrigeration systems manufacturers like Dupont, York, Carrier and Trane. Moreover, several tools are available, on line, for psychrometric analysis.

The versatility of the psychrometric chart lies in the fact that by knowing three independent properties of moist air (one of which is the pressure), other unknown properties can be determined.

The thermophysical properties and parameters found on most psychrometric charts are as follows:

Dry-bulb Temperature (DB)

Dry bulb temperature of an air sample is the temperature measured by an ordinary thermometer when the thermometer's bulb is dry; hence the term "Dry-bulb." Dry bulb temperature can also be measured using electronic or electrical instruments such RTD, Resistance Temperature Devices and thermocouples. When RTD's or thermocouples are employed for dry bulb measurement, the temperature sensing tips or junctions of these devices are simply exposed to ambient air. The units for dry bulb temperature are °F (US/Imperial domain) or °C (SI/Metric domain).

Wet-bulb Temperature (WB)

Wet bulb temperature is the temperature read by a thermometer whose sensing bulb is covered with a wet sock evaporating into a rapid stream of the sample air. When the air is saturated with water, the wet bulb temperature is the same as the dry bulb temperature and the psychrometric point lies directly on the saturation line. Similar to the dry bulb temperature, the units for wet bulb temperature are °F (US/Imperial domain) or °C (SI/Metric domain).

Dew-point Temperature (DP)

Dew point is the temperature at which water vapor begins to condense into liquid. The dew point temperature serves as an adjunct to and supports other psychrometric properties of moist air, such as the wet bulb and the relative humidity. Similar to the dry bulb and wet bulb temperatures, the units for dew point are °F (US/Imperial domain) or °C (SI/Metric domain).

Relative Humidity (RH)

Relative humidity of a sample of moist air—air that holds some measurable quantity of water vapor— is the ratio of the mole fraction of water vapor to the mole fraction of saturated moist air at the same temperature and pressure. Relative humidity is dimensionless, and is usually expressed as a percentage.

Humidity Ratio

Humidity ratio is the proportion of the mass of water vapor per unit mass of dry air under given set of dry bulb, wet bulb, dew point and relative humidity conditions. Humidity ratio is denoted by the symbol "ω." Humidity ratio is dimensionless. However, it is typically expressed in as grams of water per gram of dry air (in SI units) or grains of moisture per pound of dry air (in US units).

Specific Enthalpy

Specific enthalpy of a substance is defined as heat content of the substance per unit mass. In psychrometry, enthalpy represents the heat content of moist air. Enthalpy is measured in kilo Joules per kilogram of dry air (in SI units) or Btus per pound (in US units) of dry air. In the SI or metric unit realm, specific enthalpy is, sometimes, also stated in Joules/gram. Of course, enthalpy amounting to 1 kJ/kg of dry air is equivalent to an enthalpy of 1 J/gm of dry air. Specific enthalpy, as alluded to earlier in this text, is denoted by the symbol "**h.**"

METHOD FOR READING THE PSYCHROMETRIC CHART

Psychrometric chart reading guide shown in Figure 10-2 illustrates the general method for reading various psychrometric parameters on a typical psychrometric chart. Navigation to some of the basic psychrometric parameters, utilizing the guide in Figure 10-2 and a simple psychrometric chart shown in Figure 10-1, is outlined below:

- **Dry Bulb:** On the psychrometric chart, the dry bulb temperature scale appears horizontally, along the x-axis, See Figures 10-1 and 10-2. As apparent in these two diagrams, the dry bulb temperature increments from left to right. The scale for dry bulb temperature is graduated in °F (US/Imperial domain) or °C (SI/Metric domain).

- **Wet Bulb**: The wet bulb lines are inclined with respect to the horizontal. In other words, the wet bulb lines emanate diagonally from

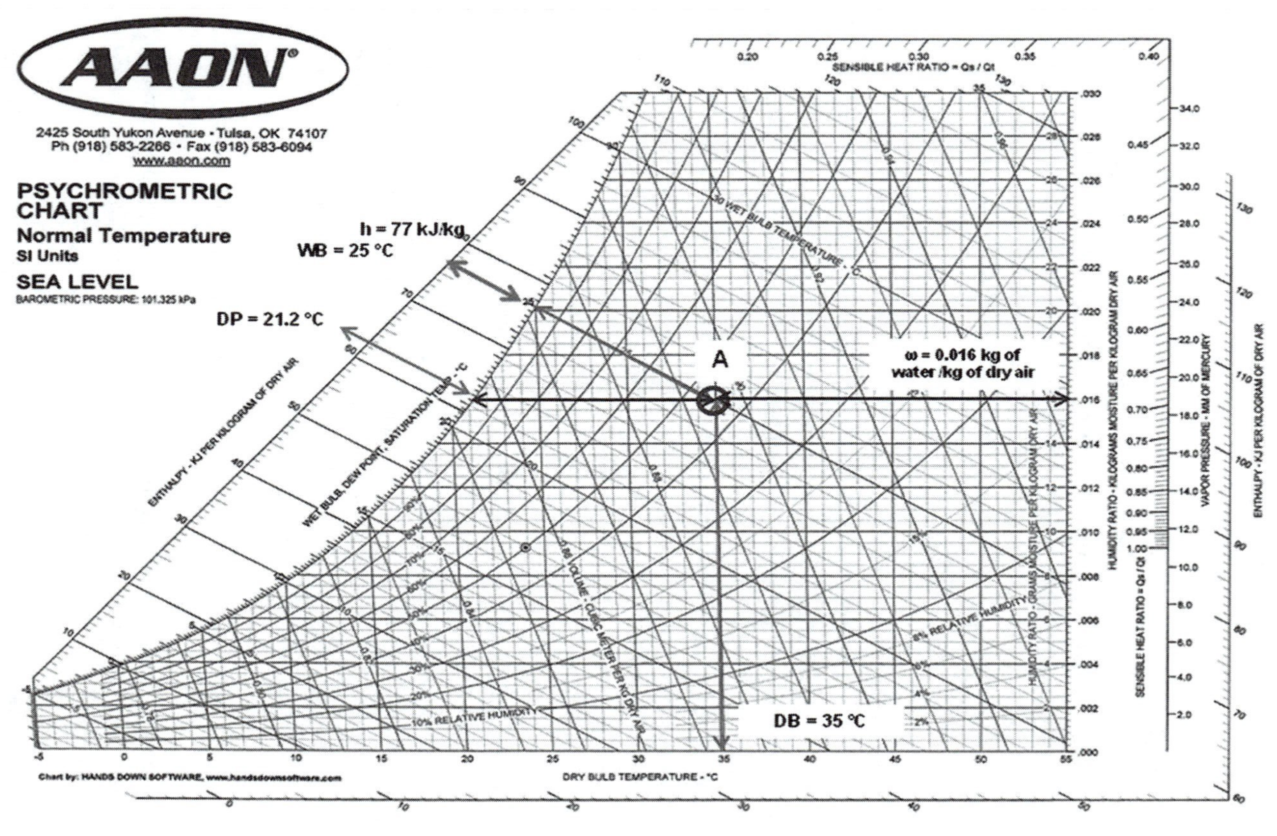

Figure 10-1. Psychrometric chart—Copyright and Courtesy AAON

the psychrometric point and intersect with the saturation curve on the left as they run parallel to the enthalpy lines. The wet bulb temperature shares its scale, on the saturation line, with the dew point. See Figures 10-1 and 10-2. Like the dry bulb temperature, the scale for the wet bulb temperature is graduated in °F (US/Imperial domain) or °C (SI/Metric domain).

- **Dew Point:** To read the dew point, follow the horizontal line from the psychrometric point to the 100% RH line. The100% RH line is also known as the ***saturation curve***. Note that the psychrometric point is a point on the psychrometric chart where wet bulb (slanted) and dry bulb (vertical) lines meet, or where dry bulb line (vertical) and the line/curve representing a given %RH intersect. The dew point is located where the horizontal dew point line intersects with the100% relative humidity line on the left. The dew point temperature shares its scale, on the saturation line, with the wet bulb temperature. Like the dry bulb and wet bulb temperatures, the scale for the dew point temperature is graduated in °F (US/Imperial domain) or °C (SI/Metric domain).

- **Relative Humidity Line**: Relative humidity is depicted in form of positively sloped curves, or lines, spanning from the bottom left corner of the psychrometric chart to the top right portion of the chart. These relative humidity lines are half parabolic asymptotic lines that are drawn to the right of the saturation curve. The relative humidity lines are typically graduated in 10% increments on most conventional psychrometric charts; ranging from 10% to 100%. The relative humidity scale is graduated in finer 2% increments below the 10% RH level. See Figures 10-1.

- **Humidity Ratio—**ω: Humidity ratio is read off the graduated vertical line, on the right side of the psychrometric chart, representing the humidity ratio scale. See Figures 10-1 and 10-2. The horizontal humidity ratio lines span from the saturation line side of the psychrometric chart to the extreme right side, intersecting on the right with the vertical humidity ratio scale. The humidity ratio scale ranges from 0.000 to 0.030—defined in kg of moisture per kg of dry air on the metric (SI) psychrometric charts or in pounds (lbm) of moisture per pound (lbm) or dry air on the US unit psychrometric charts.

- **Specific Enthalpy—h:** As shown in Figures 10-1 and 10-2, the specific enthalpy lines run parallel to the wet bulb lines on the psychrometric charts. In other words, the enthalpy lines emerge diagonally from the psychrometric point and intersect with the saturation curve on the left. In the commonly used segment of the psychrometric chart, the specific enthalpy scale ranges from 10 to 55 Btu/lbm of dry air, in the US unit realm, and 10 to 110 kJ/kg in the metric (SI) unit realm.

- **Specific volume—**υ: Specific volume lines appear on the psychrometric chart as equally spaced parallel lines representing specific volume ranging from 0.5 to 0.96 m^3/kg of dry air, in increments of 0.01m^3/kg of dry air, in the SI (metric) unit system. These lines span diagonally from the bottom left corner of the psychrometric chart to the top right corner. See Figure 10-1. On the US or imperial system psychrometric charts, the specific volume lines range from 13.0 to 15.0 ft^3/lbm of dry, in 1.0 ft^3/lbm of dry air increments. Specific volumes for psychrometric points that do not lie on the designated specific volume lines must be derived through interpolation, as illustrated in Case Study 10-2.

Example 10-1

A basic illustration of the method employed for reading and analyzing psychrometric charts can be seen in Figure 10-1, where, psychrometric point A is identified on the basis of the following two parameters:

a) The given dry bulb temperature of 35°C.
b) The given wet bulb temperature of 25°C.

Once point A is located on the psychrometric chart, the following additional psychrometric properties and attributes are read off the chart:

I. The dew point is read horizontally off to the left on the wet bulb and dew point scale, as 21.2°C.

II. The enthalpy is read diagonally to the left, parallel to the wet bulb lines, off the enthalpy scale as 77 kJ/kg.

III. The humidity ratio, ω, is read off the humidity ratio scale on the right as 0.016 kg of moisture per kg of dry air.

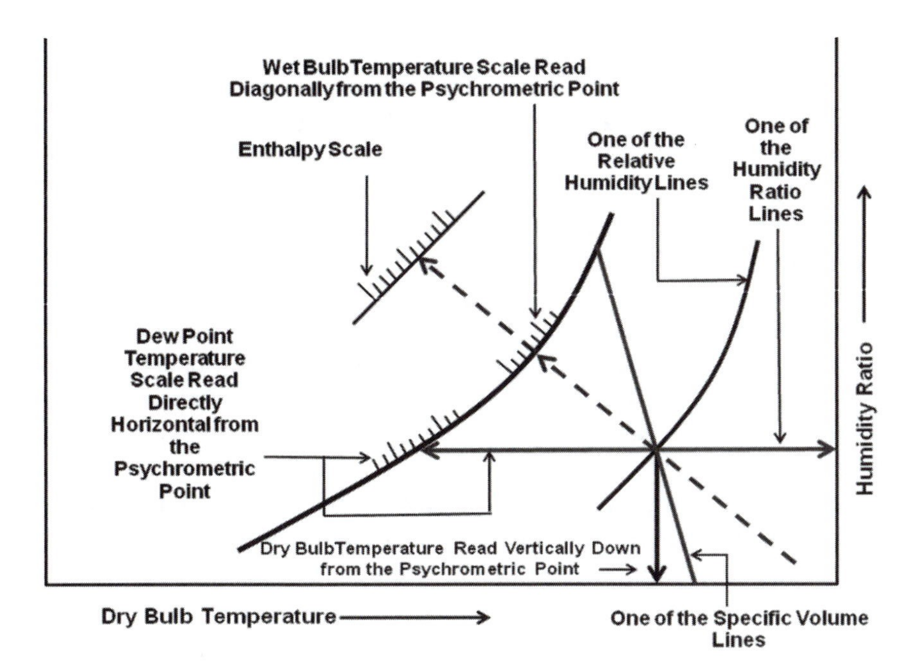

Figure 10-2. Psychrometric Chart Reading Guide

PSYCHROMETRIC TRANSITION PROCESS

Psychrometric transition process is the process involving changes in dry bulb, wet bulb, dew point, relative humidity, humidity ratio, and enthalpy, whereby, a psychrometric point representing a set of psychrometric conditions moves from one point on the psychrometric chart to another.

These psychrometric processes are illustrated in Figure 10-3. The paths representing eight of these processes and the psychrometric significance of each of these processes are as follows:

- **Path O-A:** Path O-A, in its absolute vertical configuration as depicted in Figure 10-3, represents a psychrometric transition or process involving an increase in relative humidity or humidity ratio. Since this path is vertical, the dry bulb stays constant.

- **Path O-B:** Since path O-B is at an upward diagonal attitude, relocation of psychrometric points along this path would involve an increase in dry bulb, enthalpy and humidity.

- **Path O-C:** Path O-B, in its absolute horizontal direction to the right, as depicted in Figure 10-3, represents a psychrometric transition or process involving a increase in sensible heat only. Since this path is horizontal, the dry bulb changes while the humidity ratio stays constant.

- **Path O-D:** Path O-D, with its diagonally downward attitude, represents dehumidification with some decline in the dry bulb temperature. This psychrometric process path can be implemented through chemical dehumidification systems; which are ideally suited for ice skating rinks where dehumidification is desired without an increase in the dry bulb temperature.

- **Path O-E:** Path O-E, in its direct downward vertical configuration, as depicted in Figure 10-3, represents a reduction in relative humidity or humidity ratio with no change in the dry bulb temperature.

- **Path O-F:** As obvious from the diagram in Figure 10-3, path O-F with its diagonally downward attitude to the left represents a simultaneous cooling and dehumidification process. This path is ideal for situations where lower dry bulb temperatures and lower dew points are desired.

- **Path O-G:** While relocation of a psychrometric point from "O" directly to the left, as depicted in Figure 10-3, would result in some increase in the relative humidity, the predominant impact is in form of reduction of sensible heat and the dry bulb temperature.

- **Path O-H:** This path is a classic representation of the evaporative cooling process where, typically, the dry bulb temperature is reduced through forced air evaporation of water. However, while the latent evaporative process extracts heat from the system—thus lowering the dry bulb temperature—the evaporated moisture increases the RH level and the humidity ratio.

When performing psychrometric analysis pertaining to a scenario that entails the transition from an initial set of psychrometric conditions to a final set of psychrometric conditions, a suitable approach is to begin with the identification or location of the initial and final psychrometric points on the psychrometric chart.

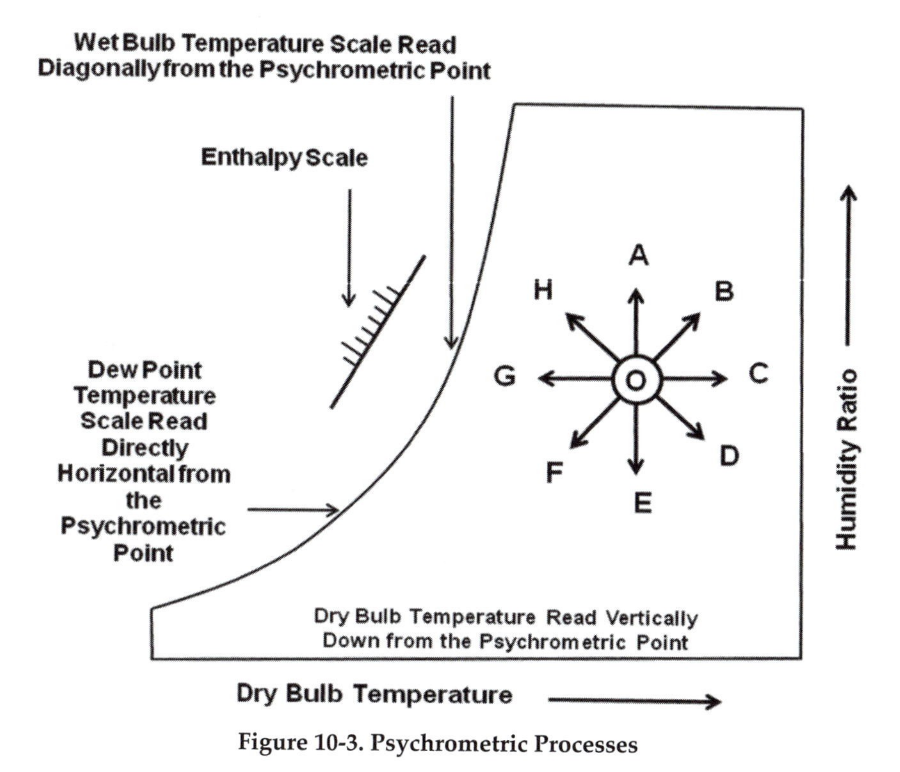

Figure 10-3. Psychrometric Processes

The location or identification of the initial and final psychrometric points on the psychrometric chart requires the knowledge—or field measurement—of at least two of the following important parameters associated with each point:

I. The dry bulb temperature
II. The wet bulb temperature
III. The % relative humidity
IV. The dew point
V. The enthalpy
VI. The humidity ratio

Among the six parameters listed above, the more conventional, measurable and more "likely to be known" parameters are the dry bulb temperature, the wet bulb temperature, the % relative humidity, and the dew point. The enthalpy and the humidity ratio are listed merely as theoretical possibilities.

Once the initial and final psychrometric points have been located on the psychrometric chart, other unknown parameters associated with these two points can be read off the psychrometric chart. Interpolation between known or graphed lines and points may be necessary in certain cases to locate the psychrometric points in question.

Also, once the initial and final psychrometric points have been located on the psychrometric chart, advanced analyses, such as the determination of SHR, sensible heat ratio, mass of water removed, amount heat involved, specific volume of the air, etc. can be determined through graphical or geometric analyses performed on the psychrometric chart.

CASE STUDY 10-1: PSYCHROMETRICS—SI UNIT SYSTEM

In an environment that is estimated to contain, approximately, 450 kg of air, the dry-bulb is measured to be 35 °C and the wet-bulb is at 25 °C. Later, the air is cooled to 13 °C and, in the process of lowering the dry-bulb temperature, the relative humidity drops to 75%. As an energy engineer, you are to perform the following psychrometric analysis on this HVAC system:

a) Find the initial humidity ratio, ωi.
b) Find the final humidity ratio, ωf.
c) Find the total amount of heat removed.
d) Find the amount of sensible heat removed.
e) Find the amount of latent heat removed.
f) Find the final wet-bulb temperature.
g) Find the initial dew point.
h) Find the final dew point.
i) Find the amount of moisture condensed/removed.
j) Can the amount of electrical power consumed by the A/C System be determined on the basis of the data provided in this case study?

Solution:

General approach to the solving this psychrometric case study problem and others similar ones is premised on the psychrometric transition process paths illustrated in Figure 10-3 and the psychrometric chart interpretation guide shown in Figure 10-2.

As explained in the discussion leading to this case study, we need to begin the analyses-associated with this case study—with the identification or location of the initial and final psychrometric points on the psychrometric chart.

Location of the initial psychrometric point can be established using the following two parameters associated with this point:

- Dry-bulb temperature of 35 °C
- Wet-bulb temperature of 25 °C

This point is shown on the psychrometric chart in Figure 10-4 as point A.

Location of the final psychrometric point can be established using the following two pieces of data associated with this point:

- Dry bulb temperature of 13 °C
- Relative humidity of 75%.

Relative humidity line, representing an RH of 75% is placed through interpolation between the given 70% and 80% RH lines on the psychrometric chart. The final point, thus identified, is shown as point B on the psychrometric chart in Figure 10-4.

a) Find the initial humidity ratio, ω_i.

To determine the initial humidity ratio, draw a horizontal line from the initial point to the vertical humidity ratio scale on the psychrometric chart as shown in Figure 10-4.

The point of intersection of this horizontal line and the humidity ratio scale represents the humidity ratio for the initial psychrometric point, ω_i.

As read from the psychrometric chart in Figure 10-4:

ω_i = 0.016 kg of moisture per kg of dry air

b) Find the final humidity ratio, ω_f.

Similar to part (a), the humidity ratio for the final psychrometric point can be determined by drawing a horizontal line from the final point to the vertical humidity ratio scale on the psychrometric chart in Figure 10-4.

The point of intersection of this horizontal line and the humidity ratio scale represents the humidity ratio, ω_f, for the final psychrometric point.

As read from the psychrometric chart in Figure 10-4:

ω_f = **0.007** kg of moisture per kg of dry air

c) Find the total amount of heat removed.

The first step is to identify the enthalpies, on the psychrometric chart, at the initial and final points. See Figure 10-4.

At the initial point, the dry-bulb temperature is 35 °C, the wet-bulb is 25 °C, and as shown on the psychrometric chart, h_i = 77 kJ/kg of dry air.

At the final point, dry-bulb is 13 °C, with RH at 75%. The enthalpy at this point, h_f = 32 kJ/kg of dry air.

$$\therefore \Delta \mathbf{h} = \mathbf{h_f - h_i}$$
$$= 32 \text{ kJ/kg} - 77 \text{ kJ/kg}$$
$$= -45 \text{ kJ/kg of dry air.}$$

And,

$$\Delta \mathbf{Q} = (\Delta \mathbf{h}) \cdot \mathbf{m_{air}} \qquad \textbf{Eq. 10-1}$$
$$= (-45 \text{ kJ/kg}) \cdot m_{air}$$

Where, the mass of dry air, m_{air}, needs to be derived through the given combined mass of moisture and air, 450 kg, and the humidity ratio, ω.

Humidity ratio is defined as:

ω = **mass of moisture (kg)/mass of dry air (kg)**

And, as determined from the psychrometric chart, earlier, in part (a):

ω = **0.016** kg of moisture per kg of dry air, at the initial point

$$\omega = \mathbf{m_{moisture}/m_{dry\ air}}$$

Or,

$$\omega = \mathbf{(m_{moist\ air} - m_{dry\ air})/m_{dry\ air}}$$

Through algebraic rearrangement of this equation, we get:

$$\mathbf{(1 + \omega) = m_{moist\ air}/m_{dry\ air}}$$

Or,

$$\mathbf{m_{dry\ air} = mass\ of\ dry\ air = m_{moist\ air}/(1 + \omega)}$$

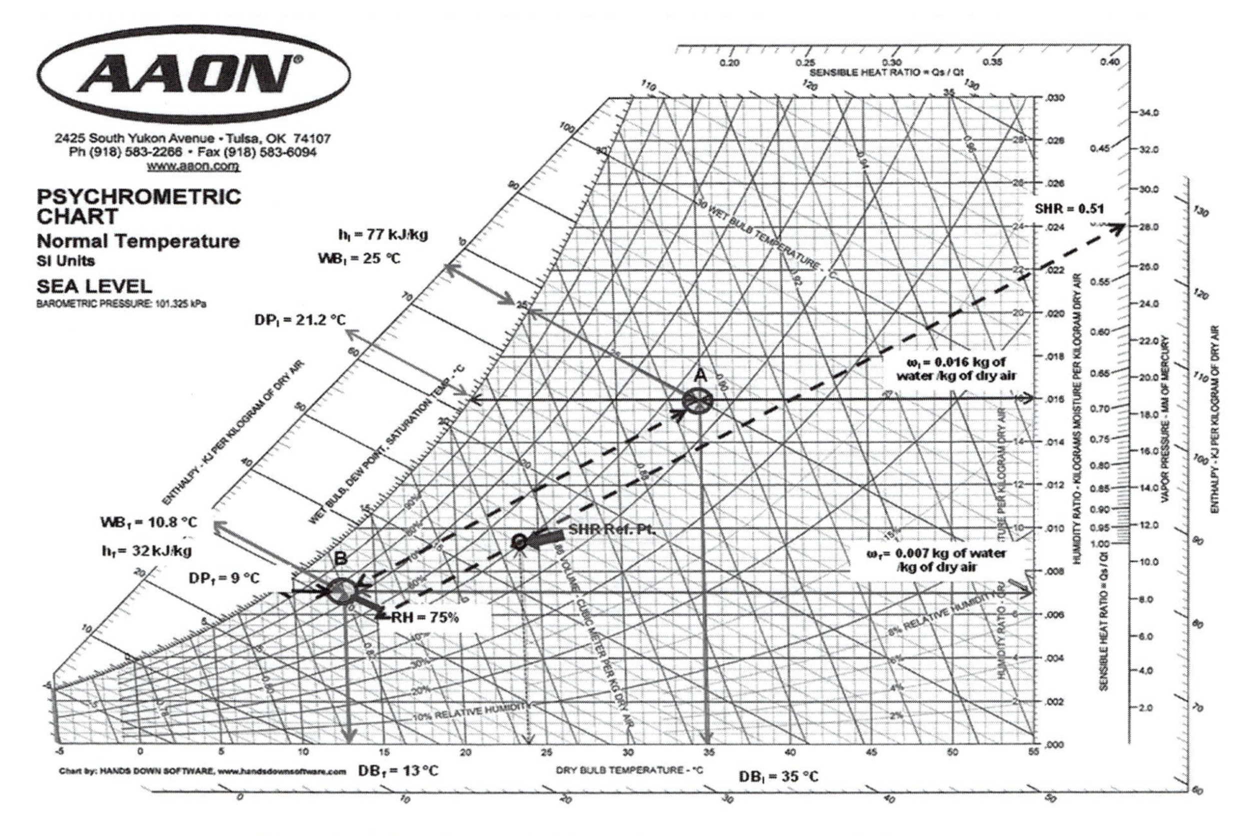

Figure 10-4. Psychrometric Chart—Case Study 10-1, SI Unit System

Where the total combined of mass of the moisture and the dry air is given as 450 kg.

$\therefore m_{\text{dry air}} = 450 \text{ kg}/(1 + 0.016)$
$= 443 \text{ kg}$

Then, by applying Eq. 10-1:

ΔQ = Total Heat Removed
$= (\Delta h) \cdot m_{\text{dry air}}$
$= (-45 \text{ kJ/kg}) \cdot (443 \text{ kg})$

Or,

ΔQ = Total Heat Removed = **- 19,935 kJ**

The negative sign, in the answer above, signifies that the heat is extracted or that it exits the system as the air conditioning process transitions from the initial psychrometric point to the final psychrometric point.

d) Find the amount of sensible heat removed.

The first step in determining the amount of sensible heat removed is to identify the SHR, Sensible Heat Ratio, from the psychrometric chart. This process involves drawing a straight line between the initial and final points. This line is shown as a dashed line between the initial and final points. Then draw a line parallel to this dashed line such that it intersects with the SHR Reference Point and the vertical scale representing the SHR. The point of intersection reads, approximately, 0.51. **See Figure 10-4.**

Note: For additional discussion on the significance of SHR, refer to Case Study 10-2, part (g) and the self assessment problem 10-2 (g).

SHR, Sensible Heat Ratio, is defined, mathematically, as:

SHR = Sensible Heat/Total Heat

In this case,

$SHR = Q_s/Q_t = 0.51$

Or,

Q_s = Sensible Heat = $(0.51) \cdot (Q_t)$

And, since $Q_t = \Delta Q$ = Total Heat Removed = - 19,935 kJ, as calculated earlier,

Q_s = **Sensible Heat** = (0.51). (-19,935 kJ)
= **-10,167 kJ**

e) Find the amount of latent heat removed.

The total heat removed consists of sensible and latent heat components.

Or,

$Q_t = Q_s + Q_l$
Q_l = **Latent Heat** = $Q_t - Q_s$
= - 19,935 kJ – (- 10,166 kJ)
∴ Q_l = **Latent Heat Removed = - 9,768 kJ**

f) Find the final wet-bulb temperature.

As explained in the discussion associated with the psychrometric chart interpretation guide in Figure 10-2, wet bulb is read diagonally from the psychrometric point along the wet bulb temperature scale. The wet bulb lines run parallel to the enthalpy lines on the psychrometric chart.

The diagonal line emerging from the final point intersects the wet bulb and dew point scale at approximately 10-8 °C as shown on the psychrometric chart in Figure 10-4. Therefore, the wet bulb temperature at the final point is **10.8 °C.**

g) Find the initial dew point.

To determine the initial dew point, read the dew point temperature for the initial point on the psychrometric chart in Figure 10-4 using the psychrometric chart interpretation guide in Figure 10-2.

Follow the horizontal "dew point" line drawn from the initial point to the left, toward the saturation curve. The point of intersection of the saturation line and the dew point line represents the dew point. This point lies at 21.2°C.

Therefore, the dew point at the initial point, as read off from Figure 10-4, is **21.2°C.**

h) Find the final dew point.

To determine the final dew point, follow the horizontal "dew point" line drawn from the final point to the left, toward the saturation curve. The point of intersection of the saturation line and the dew point line represents the dew point. This point lies at 9 °C.

Therefore, the dew point at the initial point, as read off from Figure 10-4, is **9°C.**

i) Find the amount of moisture condensed/removed.

In order to calculate amount of moisture condensed or removed, we need to find the difference between the humidity ratios for the initial and final points.

Humidity Ratio, ω = mass of moisture (kg)/mass of dry air (kg)

From the psychrometric chart, in Figure 10-4:

ω_i = **0.016** kg of moisture per kg of dry air, at the initial point
ω_f = **0.007** kg of moisture per kg of dry air, at the final point
$\therefore \Delta\omega$ = Change in the Humidity Ratio
= **0.016 - 0.007**
= **0.009** kg of moisture per kg of dry air

The amount of moisture condensed or removed
= **($\Delta\omega$) . (Total mass of Dry Air)**
= **($\Delta\omega$) . ($m_{dry\ air}$)**

Where, $m_{dry\ air}$ = 443 kg of dry air, as calculated in part (a)

$\therefore$ **The amount of moisture condensed or removed**
= (0.009 kg of moist./kg of dry air) . (443 kg of dry air)
= **3.987 kg**

j) Can the amount of electrical power consumed by the A/C System be determined on the basis of the data provided in this case study?

Calculation of the electrical power consumed by the A/C System would require data pertaining to the brake horsepower demanded by the A/C compressor.

The brake horse power can be calculated from the efficiency of the pump, differential pressure, head added and the volumetric flow rate of the refrigerant system. However, since none of these parameters are known, determination of the electrical power consumed is **not feasible** due to insufficient data.

CASE STUDY 10-2: PSYCHROMETRICS—US UNIT SYSTEM

As an energy engineer, you have been assigned to perform psychrometric analysis on an air conditioned environment. The results of measurements performed are as follows:

Estimated mass of *dry air*: **900 lbm**
Initial Dry Bulb Temperature: **81 °F**
Initial Wet Bulb Temperature: **70.4 °F**
The air is cooled to a final temperature of: **75 °F**
Final Dew Point: **48°F**

a) What is the RH, Relative Humidity, at the initial point?
b) What is the RH, Relative Humidity, at the final point?
c) What is the initial Dew Point?
d) What is the final point Wet Bulb?
e) Find the initial Humidity Ratio, ωi.
f) Determine the SHR for the change in conditions from the initial to the final point.
g) What is the significance of the low SHR in this scenario as compared to the scenario analyzed in Case Study 10-1?
h) What is the estimated specific volume at the initial point?
i) Estimate the total volume of the air in the system.

Solution:

As explained in Case Study 10-1, we need to begin our analyses of this case study with the identification or location of the initial and final psychrometric points on the psychrometric chart.

Location of the initial psychrometric point can be established using the following two parameters associated with this point:

- Dry-bulb temperature of **81 °F**
- Wet-bulb temperature of **70.4 °F**

This point is shown on the psychrometric chart in Figure 10-5 as point **A**.

Location of the final psychrometric point can be established using the following two pieces of data:

- Dry bulb temperature of **75 °F**
- Final Dew Point: **48°F**

The final point, thus identified, is shown as point **B** on the psychrometric chart in Figure 10-5.

a) Relative Humidity at the initial point:

Locate the initial point, as described above, on the psychrometric chart shown in Figure 10-5. As evident from the psychrometric chart in Figure 10-5, the 60% RH line passes directly through the initial point. Therefore, the RH at the initial point is **60%.**

b) Relative Humidity at the final point:

Similar to the approach used in part (a), locate the final point, as described above, on the psychrometric chart shown in Figure 10-5. Use the psychrometric chart interpretation guide from Figure 10-2 for clarification and review, as needed. As shown on the psychrometric chart in Figure 10-5, the final point lies directly on the 40% RH line. Therefore, the RH at the final point is **40%.**

c) Initial Dew Point:

To determine the initial dew point, read the dew point temperature for the initial point on the psychrometric chart in Figure 10-5 using the psychrometric chart interpretation guide in Figure 10-2.

Follow the horizontal "dew point" line drawn from the initial point to the left, toward the saturation curve. The point of intersection of the saturation line and the dew point line represents the dew point. This point lies at 66°F.

Therefore, the dew point at the initial point, as read off from Figure 10-5, is **66°F.**

d) Final Point Wet Bulb Temperature:

The wet bulb is read diagonally from the psychrometric point along the wet bulb temperature scale. Note that the wet bulb lines are parallel to the enthalpy lines on the psychrometric chart.

The diagonal line emerging from the final point intersects the wet bulb and dew point scale at approximately 59°F. Therefore, the wet bulb temperature at the final point is 59°F.

e) Initial Humidity Ratio, ω_i:

As shown in Figure 10-5, draw a straight, horizontal, line from the initial point to the right until it intersects with the vertical scale labeled **Humidity Ratio, ω.** This point of intersection with the vertical humidity ratio line lies at $\omega = 0.0138$. Therefore, the humidity ratio at the initial point is **0.0138 lbm of moisture per unit lbm of dry air.**

f) SHR for the change in conditions from the initial to the final point.

Draw a straight line between the initial and final points as shown in Figure 10-5. This line is shown as a dashed line spanning between the initial and final points. Then draw a line parallel to this dashed line such that it intersects with the SHR Reference Point and the SHR scale, at the top. The point of intersection reads, approximately, 0.18. Therefore, the SHR for the change in conditions from the initial to the final point is **0.18.**

g) What is the significance of the low SHR in this scenario as compared to the scenario analyzed in Case Study 10-1?

The **SHR of 0.18,** for the scenario portrayed in this problem, is significantly lower than the **SHR of 0.51** for the scenario in Case Study 10-1 because the dry bulb change in this case study is significantly smaller than the dry bulb change in Case Study 10-1. The dry bulb drop in Case Study 10-1 is **22 °C** (or, 95°F – 54°F = 40°F) while the **dry bulb reduction** in this case study is only 6°F, or **3.3°C** (or, 27.2°C -23.9°C). In "°F," the dry bulb reduction in Case Study 10-1 is **40°F,** versus a rather small reduction of only **6°F** in this case study. While dry bulb change in this case study is relatively minute, the dew point change is substantial; an **18°F drop**, from 66°F to 48°F. In other words, in this case, while the dew point changes significantly, the dry bulb changes negligibly.

When the dry bulb change is small or negligible, the amount of sensible heat involved in the transition process is much smaller than the latent heat. This explains the reason behind SHR being only 0.18 in this case. Note that the SHR of 0.18 implies that only 18% of the total heat involved in this process transition is sensible heat, the remaining 82% of the heat extracted in this process transition is latent heat. This larger proportion for extracted latent heat explains the significant 18°F drop in the dew point

h) What is the estimated specific volume at the initial point?

Specific volume at the initial point can be estimated through interpolation between the given specific volume lines, in proximity of the initial psychrometric point, on a standard psychrometric chart. These two lines, as shown on the psychrometric chart in Figure 10-5, represent specific volumes of 13 cu-ft/lbm of dry air and 14 cu-ft/lbm of dry air. On the psychrometric chart in Figure 10-5, a diagonal, specific volume line is drawn such that it passes through the initial point and is parallel to the given specific volume lines. The specific volume represented by this initial point specific volume line is interpolated to be approximately **13.9 cu-ft/lbm of dry air.**

i) Estimate the total volume of the air in the system.

The total volume of the air in the system can be estimated on the basis of the specific volume determined in part (h) and the mass of dry air given in the problem statement.

Given:

Estimated mass of *dry air*: **900 lbm**
Specific volume: **13.9 cu-ft/lbm of dry air.**

Since Specific Volume, = Volume/Mass,

∴ **Volume of the air in the system = (v) . (Mass)**
= (13.9 cu-ft/lbm of dry air).(900 lbm)
= **12,510 cu-ft**

Chapter 10—Self Assessment Problems and Questions

1. In an environment that is estimated to contain, approximately, 400 kg of air, the dry-bulb is measured to be 40°C and the wet-bulb is at 27.3°C. Later, the air is cooled to 20°C and, in the process of lowering the dry-bulb temperature, the relative humidity drops to 47%. As an energy engineer, you are to perform the following psychrometric analysis on this system using the psychrometric chart in Figure 10-6:

    a) Find the initial humidity ratio, ω_i.
    b) Find the final humidity ratio, ω_f.
    c) Find the total amount of heat removed.
    d) Find the amount of sensible heat removed.

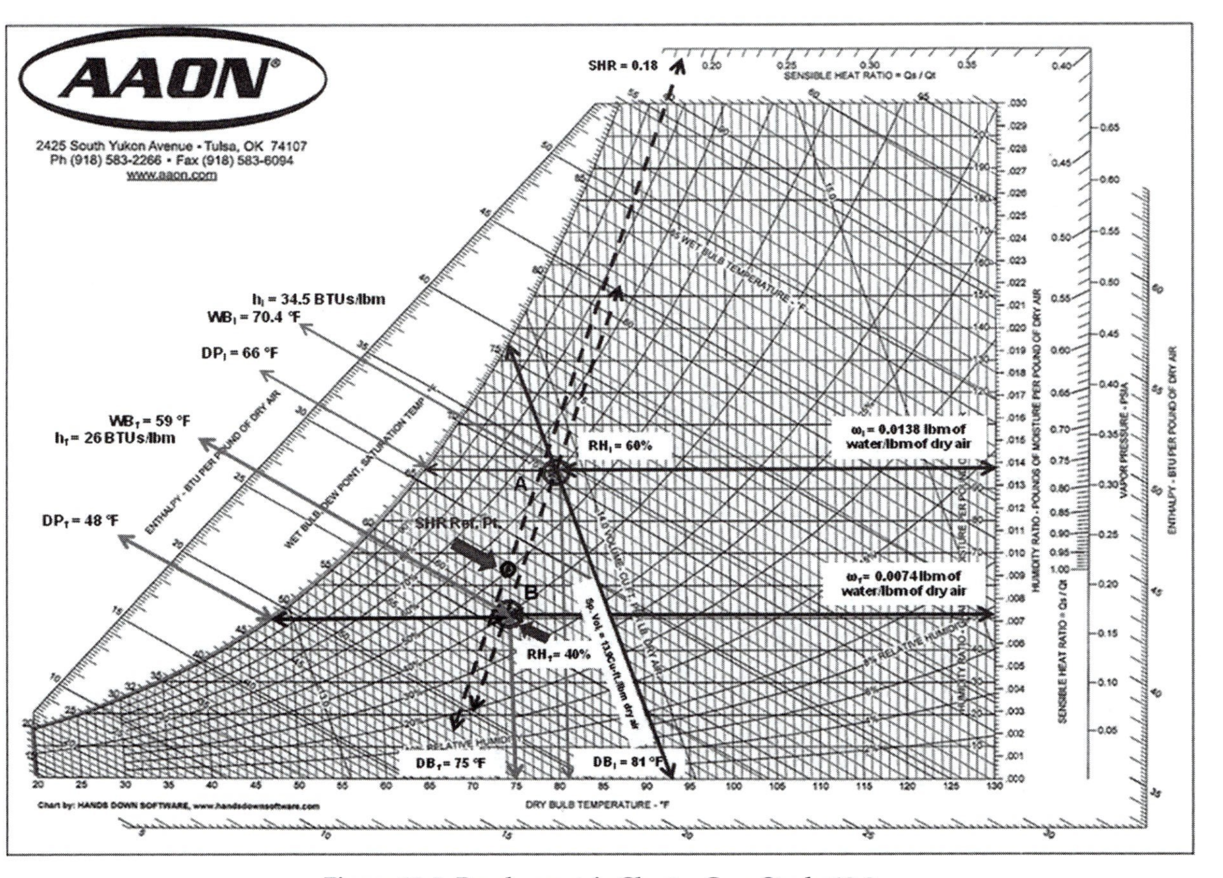

Figure 10-5. Psychrometric Chart – Case Study 10-2

e) Find the amount of latent heat removed.
f) Find the final wet-bulb temperature.
g) Find the initial dew point.
h) Find the final dew point.
i) Find the amount of moisture condensed / removed.

2. Psychrometric chart for the initial and final conditions in a commercial warehouse is shown in Figure 10-7. Dry Bulb and Wet Bulb data associated with the initial and final points is labeled on the chart. Assess the disposition and performance of the HVAC system in this building as follows:

a) What is the initial Dew Point?
b) What is the final Dew Point?
c) Based on the results of dew point determination in parts (a) and (b), define the type of thermodynamic process this system undergoes in the transition from initial point to the final point.
d) What is the RH, Relative Humidity, at the initial point?
e) What is the RH, Relative Humidity, at the final point?
f) Determine the SHR for the change in conditions from the initial to the final point.
g) Comment on why the SHR for this scenario is significantly higher than the scenario analyzed in Case Study 10-2?

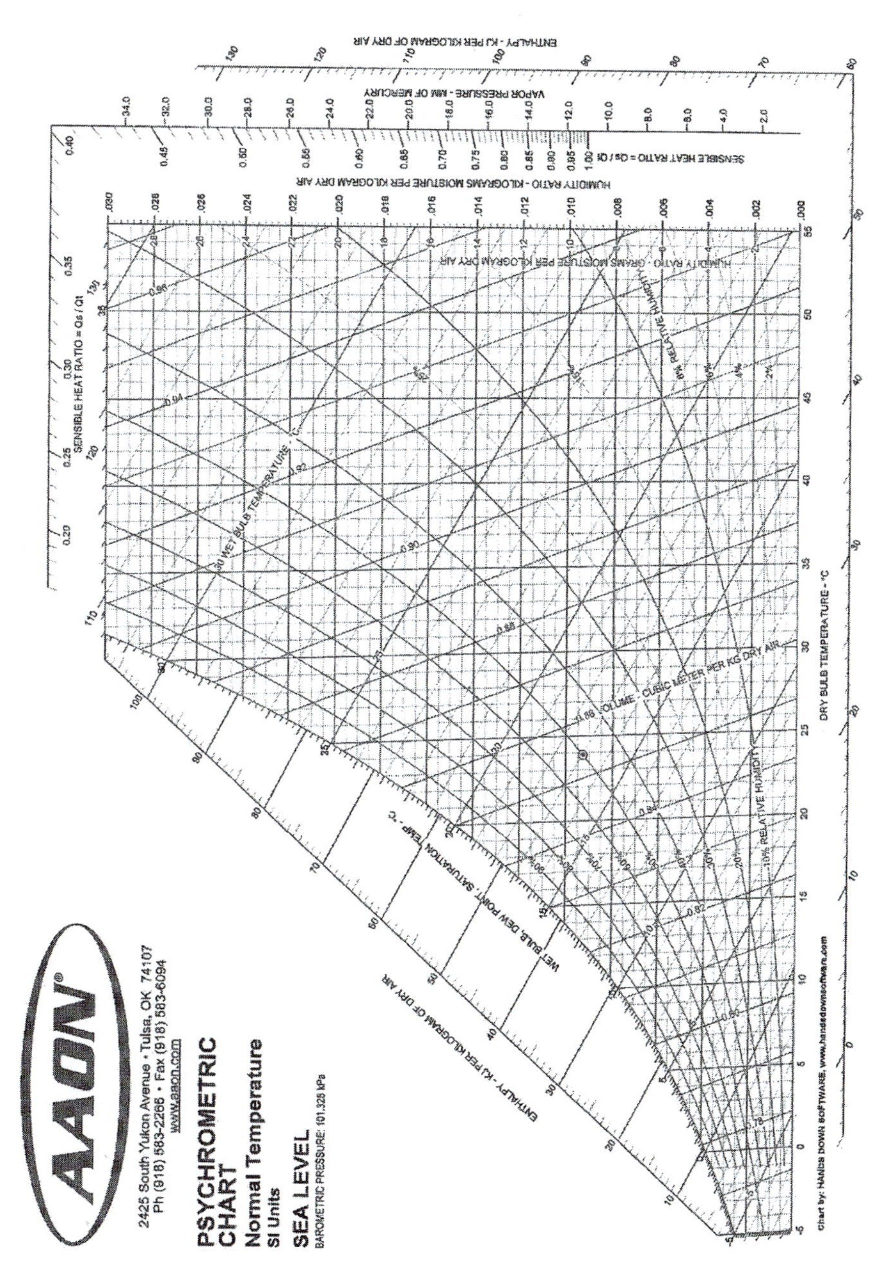

Figure 10-6. Psychrometric Chart – Problem 1

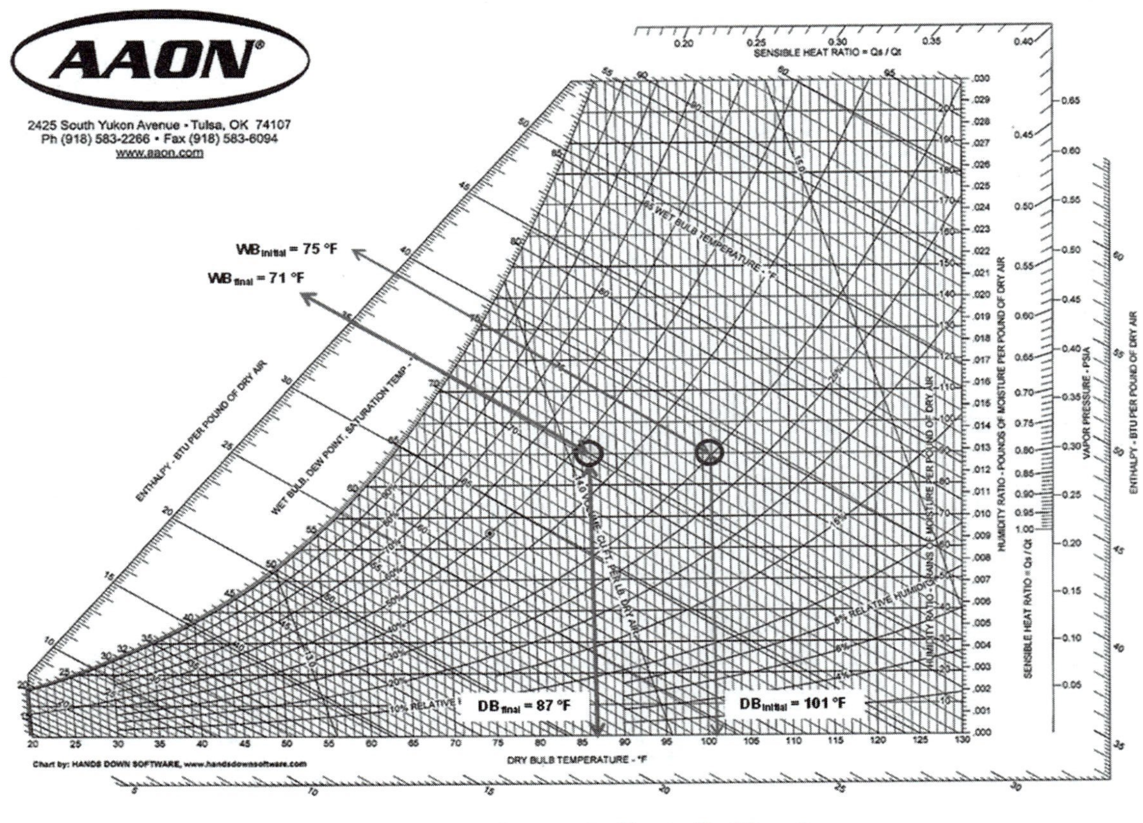

Figure 10-7. Psychrometric Chart – Problem 2

Chapter 11

Refrigeration Cycles and HVAC Systems

INTRODUCTION

Study and understanding of the Basic Refrigeration Cycle, HVAC Systems and Automated HVAC Systems is an essential an integral part of thermodynamics. This chapter provides the reader an opportunity to learn or review important fundamental concepts, principles, analysis and computational techniques associated with refrigeration and HVAC systems. The study and exploration of refrigeration cycles and HVAC system analysis is illustrated through practical examples, case study and end of the chapter self assessment problems—formulated with the energy engineer's role in mind.

This chapter includes definitions and explanation of HVAC terms, concepts and mechanical components not introduced before in this text. Definitions and explanation of several other important HVAC terms and concepts, such as, dry bulb, wet bulb, dew point, enthalpy, specific enthalpy, humidity ratio, SHR or Specific Heat Ratio, entropy, saturated liquid, saturated vapor and superheated vapor are covered under Chapter 10 and the preceding material.

TYPES OF AIR CONDITIONING SYSTEMS

There are several types of air conditioning systems. One could categorize air conditioning systems based on their application and size. The fundamental refrigeration system principles that govern functionality of a refrigerator or chiller versus a typical air conditioning system are the same. Therefore, most of our discussion and engineering analysis in this chapter would apply to all of these devices.

Within the air conditioning realm, differences between different types of air conditioning system are premised on their application and size. In large air conditioning systems, such as those pertaining to industrial and commercial applications, major components of the refrigeration systems are sizeable, somewhat independent, and are located separately. Some of the large industrial and air conditioning systems consist of large single unit chillers. A typical chiller for air conditioning applications is rated between 15 to 1500 tons. This would translate into180,000 to 18,000,000 Btu/h or 53 to 5,300 kW in cooling capacity. Chilled water temperatures in such systems can range from 35°F to 45°F or 1.5°C to 7°C, depending upon specific application requirements. Figure 11-1 draws a comparison between a large industrial or commercial chiller and a typical refrigerator compressor. The large chiller in the picture is rated over 700 hp, while the small compressor is rated, approximately 1-3 hp.

Large chiller-based air conditioning systems and process chilled water systems utilize water as a "secondary" working fluid. But, the primary working fluid is still a typical refrigerant, i.e., HFC-134a, ammonia, R-500, etc. Large chiller based air conditioning systems can be categorized as Open Air Conditioning Systems or Closed Air Conditioning Systems. An open air conditioning system utilizes a Freon based refrigerant, in a large chiller, to cool the water to 35°F to 45°F or 1.5°C to 7°C range. This chilled water is then conveyed to Open Air Washers, equipped with chilled water spray nozzles. See Figure 11-2. The return or outside air is passed through chilled water spray. The high moisture content and higher temperate return or outside air is thus cooled and dehumidified as it passes through the air washer. The supply air exiting the air washer is at lower dry bulb and lower

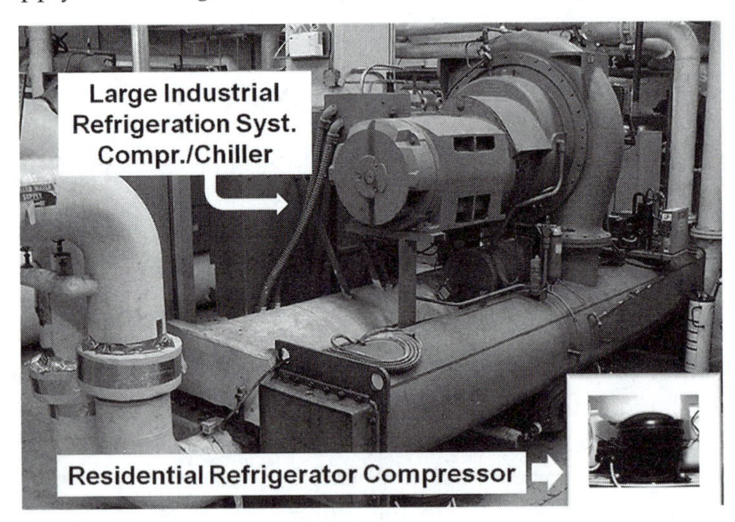

Figure 11-1. Large Refrigeration System Chiller vs. a Refrigerator Compressor

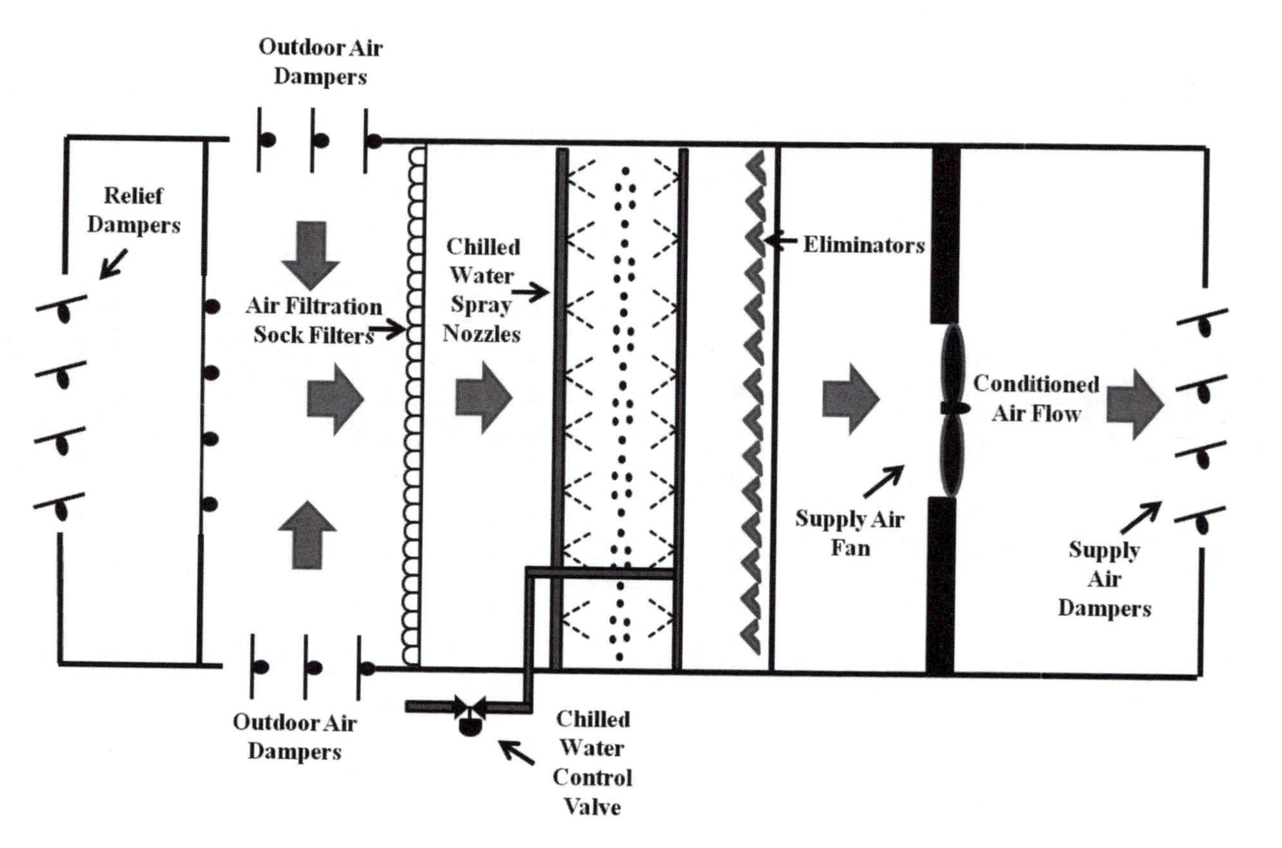

Figure 11-2. Open Air Washer System Architecture

dew point, with lower relative humidity. The supply air is then driven by supply air fan to work spaces, or occupied spaces in general, as conditioned air. A Closed Air Conditioning System, on the contrary, in most cases, does not use chilled water as a secondary working fluid to cool and condition the ambient air. Closed air conditioning systems are similar or equivalent to residential air conditioning systems where Freon or refrigerant is used as the working fluid.

Refrigeration System Compressors

There are five main types of air conditioner compressors:

1. Rotary compressor
2. Reciprocating compressor
3. Centrifugal compressor
4. Screw compressors
5. Scroll compressors

While the function and ultimate output of these different types of compressors is the same—high pressure refrigerant vapor—the mechanical components and principles employed to accomplish the compression differ. These differing mechanical approaches and principles are apparent from the names of compressors listed above.

The most common compressor is the reciprocating compressor. These compressors can be open type or hermetically sealed type. A typical refrigerator compressor is hermetically sealed (Figure 11-1).

Common refrigeration compressors range in sizes from less than 9 kW (approx. 9 hp) to 1 MW (approx. 1,000 hp), with condensing temperatures ranging from 15°C to 60°C, or higher. Mainstream refrigeration compressors power-source specifications, in terms of voltage, frequency and phases include: 12 VDC and 24 VDC, 115/60/1 (single phase AC), 230/50/1, 208.230/60/1, 208.230/60/3 (three phase AC), 380/50/3, 460/60/3 and 575/60/3.

Refrigeration System Condenser

Condensers, in essence, are heat transfer devices. They permit extraction of heat form the hot, high pressure, refrigerant vapor. Thus, allowing the vapor to condense into high pressure liquid phase. While the function of all condensers is the same, which is to condense high temperature, high pressure and high enthalpy refrigerant vapor, like the compressors and chillers, they differ based on their size and specific applications. For example, many large open air washer type air conditioning systems include large, water based, cooling towers for cooling the high pressure, high enthalpy, refrigerant vapor. On the other hand, some large closed air conditioning systems employ dry, forced air, type cooling towers such as the one shown in Figure 11-3.

Refrigerants

A refrigerant is a substance or medium used in a refrigeration system heat cycle. Refrigerants allow heat exchange and work to be performed in refrigeration systems as they undergo repetitive and cyclical phase changes from liquid to vapor and vapor to liquid, as illustrated in Figure 11-5.

Traditionally, fluorocarbons (FC) and chlorofluorocarbons (CFC) have been used as refrigerants. However, they are being phased out because of their ODP, ozone depletion Potential and, some cases, GWP, Global Warming Potential. They are being replaced by hydro-flourocarbons, i.e. HFC-134a. Other, non-CFC and non-HFC refrigerants used in various applications are non-halogenated hydrocarbons such as methane and non-hydrocarbon substances such as ammonia, sulfur dioxide.

Table 11-1 lists some of the commonly applied refrigerants and

Figure 11-3. Forced Air Type Condenser Cooling Tower for Refrigeration System

Table 11-1. Commonly used refrigerants and some of their important properties.

Refrigerant	Formula	Boiling temperature (°C)	Critical temperature (°C)	Properties	Applications
Amonia	NH_3	-33	133	Penetrating odor, soluble in water. Harmless in concentration up to 1/30%, non flammable, explosive	Large industrial plants
R11	CCl_3F	8.9	198	R11 is a single chlorofluorocarbon or CFC compound. ODP = 1 and GWP = 4000	Commercial plants with centrifugal compressors.
R12	CCl_2F_2	-29.8	112	Little odor, colorless gas or liquid, non flammable, non corrosive of ordinary metals, stable	Small plants with reciprocating compressors. Automotive, Medium Temperature Refrigeration
R22	$CHClF_2$	-40.8	96	R22 is a single hydrochlorofluorocarbon or HCFC compound.Low chlorine content and ozone depletion potential and only a modest global warming potential. R22 can still be used in small heat pump systems, but new systems cannot be manufactured for use in the EU after 2003. From 2010 only recycled or saved stocks of R22 can be used. It will no longer be manufactured. ODP = 0.05 and GWP = 1700 .	Packaged air-conditioning units where size of equipment and economy are important. Air Conditioning, Low and Medium

Table 11-1 (Cont'd). Commonly used refrigerants and some of their important properties.

Refrigerant	Formula	Boiling temperature (°C)	Critical temperature (°C)	Properties	Applications
R-134a	CH_2FCF_3			R134a is a single hydrofluorocarbon or HFC compound. No chlorine content, no ozone depletion potential. Modest global warming potential. ODP = 0 and GWP = 1300	Automotive replacement for R-12, Stationary A/C,
R417A				R417A is the zero ODP replacement for R22. Suitable for new equipment and as a drop-in replacement for existing systems.	Offers aprox. 20% more refrigeration capacity than R12 for same compressor.
R500	CCl_2F_2 (73,8%); CH3 CH F2 (26.2%)	-33		Similar to R12	Offers aprox. 20% more refrigeration capacity than R12 for same compressor.
R502	CCl F_2 (48,8%), CCl; F2-CF3 (51.2%)	-45.6	90.1	Non flammable, non toxic, non corrosive, stable	Capacity comparable to R22.

some of their important properties. These tables include chemical formulas, boiling points, critical temperatures, chemical properties, ozone depletion potential, global warming potential and likely application for the listed refrigerants.

***ODP** or ozone depletion potential of a refrigerant, or any other substance, is defined as the capacity of a single molecule of that refrigerant to destroy the ozone layer. All refrigerants use R11 as a datum reference, with R11 reference ODP of 1.0. The lower the value of the ODP, the less detrimental the refrigerant is to the ozone layer and the environment.

****GWP** stands for global warming potential. GWP is a measurement based over a 100-year period. It quantifies the effect a refrigerant will have on global warming relative to the GWP of carbon dioxide, CO_2. Carbon dioxide is assigned a GWP of 1. The GWP of all other substances or chemicals is assessed relative the carbon dioxide GWP of 1. The lower the value of GWP—the better the refrigerant is for the environment.

Note: Currently there are no restrictions on the use of R134A, R407C, R410A, and R417A in original equipment or for maintenance and repair

Expansion Valve

Expansion valve is an apparatus or component used in refrigeration systems to throttle the high-pressure refrigerant, in liquid phase, from high pressure liquid state to low pressure liquid state.

Common refrigeration system expansion valves are also referred to as Thermal Expansion Valves or TXV's. Operating principle of the thermal expansion valve is illustrated in Figure 11-4. A thermal expansion valve functions as a metering device for the high pressure liquid refrigerant; it allows small proportionate amounts of the high pressure liquid refrigerant into the discharge side. This permits the refrigerant to transform into low pressure liquid; ready to be converted to vapor phase as it absorbs heat from the warm ambient air passing through the heat exchanger coils. As the refrigerant evaporates to higher temperature, the temperature of the gas in temperature sensing bulb rises. The higher temperature gas develops higher pressure thus pushing the expansion valve open. The valve, in its metering function, stays open only until the temperature in the evaporator section drops. When the temperature in the evaporator

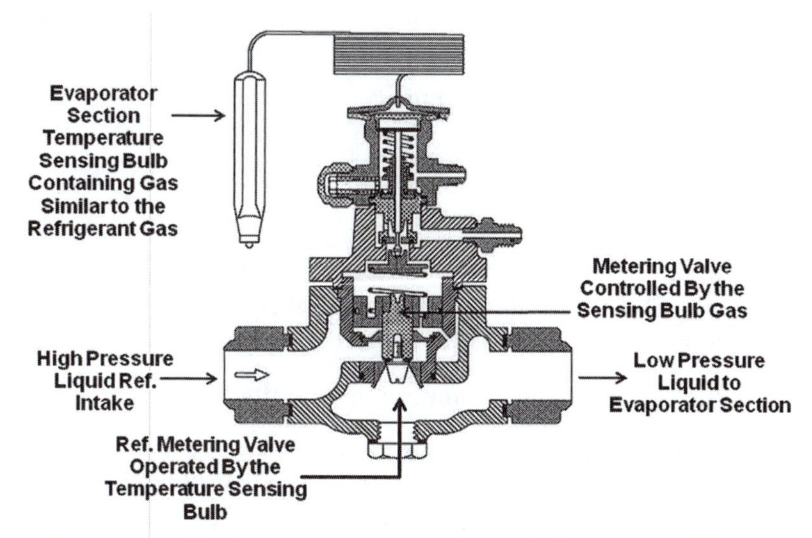

Figure 11-4. Refrigeration System Thermal Expansion Valve

section drops, the temperature and the pressure of the gas in the temperature sensing bulb drop, thus, resulting in the valve closure. This cycle repeats itself in a closed loop control fashion, continuously, in a typical refrigeration system.

Cooling Capacity of Refrigeration Systems

Cooling capacity of a refrigeration system is essentially the capacity of the refrigerant to exchange heat with the environment or ambient air. While in the absolute sense, cooling capacity represents the capacity of a refrigeration system to ***cool*** the environment or surroundings, in the case of heat pumps, cooling capacity of the system could broadly include the capacity of the system to heat the environment; explained on the basis of role reversal of the evaporator and condenser.

Refrigeration System Capacity Quantification in A/C Tons

The cooling capacity of refrigeration systems is often defined in units called "tons of refrigeration." One ton of refrigeration represents the rate of refrigeration required to freeze a ton of 32°F (0°C), water in 24-hr period. Stated alternatively, a ton of refrigeration is the rate of heat removal necessary to freeze a 2000 lbm of saturated water, at 32°F (0°C), within a period of 24 hours. For water, one ton of refrigeration amounts to 12,000 Btu/hr (12,660 kJ/hr). This is premised on heat of fusion of water being 143.4 Btu/lbm as illustrated below:

The amount of heat that must be extracted from 32°F (0°C) water to freeze it to 32°F (0°C) ice is equal to the amount of heat that must be added to 32°F (0°C) ice to melt it to 32°F (0°C) water.

Therefore,

Rate of refrigeration for one ton of 32°F (0°C) water
= (143.4 Btu/lbm).(2000 lbm)/24 hr
= 11,950, or approximately, 12,000 Btu/hr

Since 1 Btu = 1055 Joules in metric or SI units' realm:

Rate of refrigeration for one ton of 0°C water
= (12,000 Btu/hr).(1055Joules/Btu)/(3,600 sec/hr)
= 3,517 watts or 3.517 kW

In the metric unit realm, the counterpart unit to 1 ton (US) of refrigeration is a tonne (Metric/European). One tonne of refrigeration is based on freezing 1000 kg of 0°C water to 0°C ice in a 24 hour period. Calculation similar to the one illustrated in US units above equates one tonne of refrigeration to 3.86 kW. Note that most residential air conditioning units range in refrigeration capacity from about 1 to 5 tons.

Basic Refrigeration Cycle

As we describe the refrigeration cycle, let's begin tracking the process at the point in the cycle where the refrigerant enters the compressor in form of saturated or superheated low pressure vapor. Refer to Figure 11-5. The refrigerant, at this point, is high in enthalpy; albeit, the enthalpy is not as high as it is when it exits the compressor

The compressor compresses the saturated vapor and raises the pressure of the vapor to the maximum level. In doing so, the compressor packs the vapor or gaseous molecules of the working fluid closer together. The closer the molecules are together, the higher the energy and the temperature. The higher energy of the refrigerant vapor is manifested in its higher enthalpy.

The working fluid leaves the compressor as a hot, high pressure, gas with highest enthalpy and pressure. As the high pressure high enthalpy refrigerant, in vapor or superheated vapor form, enters the

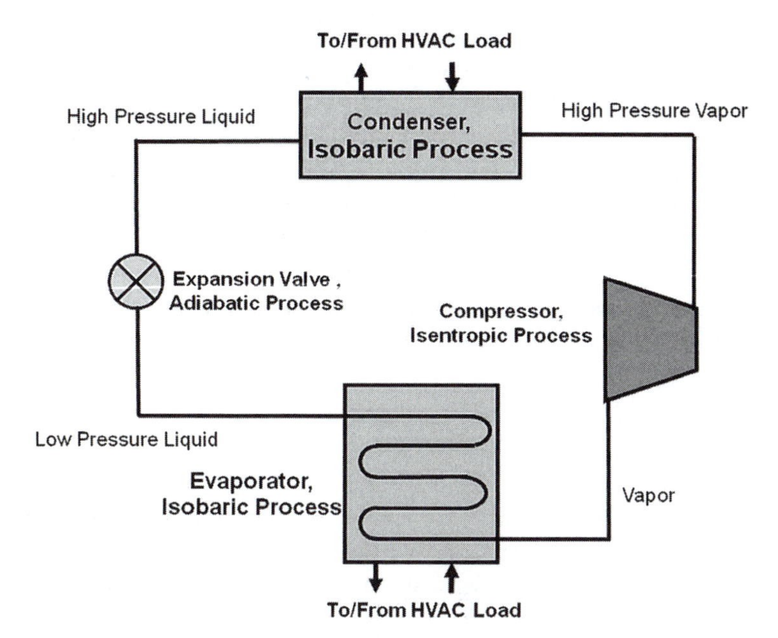

Figure 11-5. Refrigeration Cycle Process Flow Diagram

condenser, the cooling and heat exchanging process begins. In the condenser, the condenser's cooling system extracts the heat from the superheated refrigerant vapor. This loss of heat transforms the refrigerant into saturated liquid phase.

The next stage of the refrigeration cycle involves expansion or throttling of the high pressure liquid refrigerant. The expansion and the evaporation stages can be explained as two separate stages or they can be viewed as one contiguous stage. Expansion valve, as described in detail earlier and as illustrated in Figure 11-4, meters the high pressure liquid refrigerant out into the evaporator segment. This allows the refrigerant to expand into low pressure liquid form. The low pressure liquid refrigerant then readily evaporates into low pressure saturated vapor form. In doing so, the low pressure liquid refrigerant engages in heat exchange with the ambient air. The ambient air is forced past the refrigerant coils by the system return or supply fan. This heat exchange in the evaporator section results in a substantial increase in the enthalpy of the refrigerant. The end product is low pressure refrigerant vapor, ready to enter the compressor and repeat the refrigeration cycle.

The refrigeration cycle as described above is explained from mechanical or fluid dynamics perspective. The thermodynamic properties and thermodynamic changes associated with each segment or stage of the refrigeration cycle are shown on the Pressure-Enthalpy graph in Figure 11-6. As indicated on the pressure-enthalpy graph, the compression process is an isentropic, where the entropy stays constant, or, $\mathbf{\Delta s = 0}$. The condensation stage, as illustrated in the Pressure-Enthalpy graph, is a non-adiabatic isobaric process. The next stage involving expansion of the high pressure liquid refrigerant is an adiabatic and isenthalpic process, with $\mathbf{\Delta Q = 0}$ and $\mathbf{\Delta h = 0}$.

REFRIGERANT COMPRESSION

As the name implies, the compression segment of the refrigeration cycle involves transformation of the low pressure refrigerant vapor into high pressure refrigerant vapor. The low pressure refrigerant vapor can be transferred from the evaporator to the compressor in the following forms:

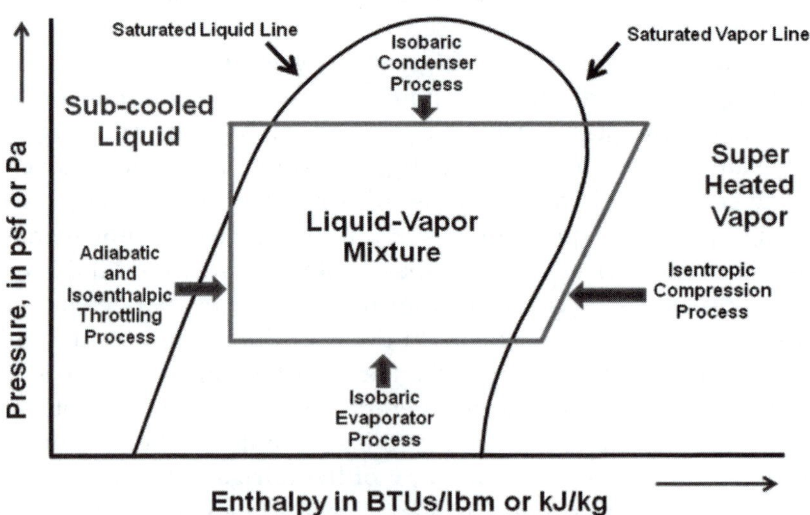

Figure 11-6 Refrigeration Cycle Pressure-enthalpy Graph

(a) Mixture of liquid and vapor; also referred to as wet vapor. See Figure 11-7.
(b) Saturated vapor
(c) Slightly superheated vapor

Wet Vapor Compression Process

Figure 11-7 depicts the temperature versus entropy diagram of a refrigeration cycle that is based on a ***wet compression cycle***. A wet compression cycle involves compression of a refrigerant before it has evaporated completely into saturated vapor or slightly superheated vapor form. This state of the refrigerant is represented by point 3 in Figure 11-7.The refrigerant, at point 3, exists in vapor and liquid mixture form and cannot be compressed or pumped as efficiently as it can be when it is in saturated vapor or slightly superheated form. In addition, wet compression results in compressor wear and performance problems.

Refrigerant Vapor Quality Ratio

The refrigerant vapor quality ratio is denoted by "ω," and is defined as the ratio of the mass of pure vapor to the total mass of vapor and liquid mixture. The vapor quality ratio can be defined, mathematically, as follows:

$$\omega = (m_{vapor})/(m_{vapor} + m_{liquid}) \qquad \textbf{Eq. 11-1}$$

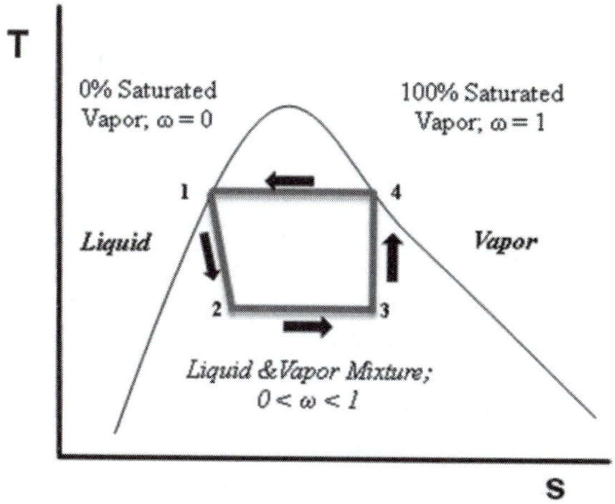

Figure 11-7. Wet Vapor Compression Cycle in Refrigeration Systems

Where,

$\mathbf{m_{vapor}}$ = Mass of refrigerant in vapor form
$\mathbf{m_{liquid}}$ = Mass of refrigerant in liquid form

The values of quality ratio ω at points 1, 2, 3 and 4, as noted on the graph in Figure 11-7, project the state and composition of the refrigerant as follows:

Point 1: This point lies directly on the saturated liquid line. There is no vaporized refrigerant along this line; $\mathbf{m_{vapor}} = 0$ and, therefore, $\omega = 0$. In other words, the refrigerant is in pure liquid phase all along this line.

Point 2: This point lies in the area of the graph where the refrigerant exists in vapor-liquid mixture form. The fact that point 2 is closer to the saturated liquid line than it is to the saturated vapor line implies that, at point 2, the percentage of liquid refrigerant is greater than the percentage of vapor. In other words, $\mathbf{m_{liquid}} > \mathbf{m_{vapor}}$. Note that all points on path 2-3 represent wet vapor state.

Point 3: This point also lies in the area of the graph where the refrigerant exists in vapor-liquid mixture form. However, point 3 is closer to the saturated vapor line than it is to the saturated liquid line. This means that, at point 3, the percentage of vaporized refrigerant is greater than the percentage of liquid. In other words, $\mathbf{m_{vapor}} > \mathbf{m_{liquid}}$. Refrigerant compression process begins at point 3 and extends up to point 4.

Point 4: This point lies directly on the saturated vapor line. There is no liquid refrigerant along this line; $\mathbf{m_{liquid}} = 0$ and, therefore, $\omega = 1$, or 100%. The refrigerant is in pure vapor phase all along this line. However, this vapor state is a saturated vapor state. Any loss of heat at this point would slide the refrigerant back into the condensed, or partially condensed, phase. On the other hand, if this point were to shift to the right of the saturated vapor line, it would be in superheated vapor phase. Superheated vapor is also referred to as ***"dry vapor."***

Dry Vapor Compression Process

The compression efficiency in a refrigeration cycle can be enhanced by extending the refrigerant evaporation process all the way to the saturated vapor line or beyond. This is precisely the strategy employed with dry vapor compression process as shown in Figure 11-8. In dry vapor compression refrigeration systems, the refrigerant leaves the evaporator section in either saturated vapor or superheated form as shown by point 3 in Figure 11-8. Furthermore, as obvious from Figure 11-8, the temperature of the refrigerant rises during the compression phase in dry vapor compression systems. As the temperature of the refrigerant rises, the refrigerant vapor becomes superheated. This is affirmed by the location of point 4 in Figure 11-8.

Following equations find common application in refrigeration cycles:

Computation of refrigerant compressor power utilizing the compression path 3-4:

$$P = \dot{W} = \dot{m}(h_4 - h_3) \qquad \textbf{Eq. 11-2}$$

Where, $\dot{W}$ represents "work flow rate" in Btu/sec or J/s.

Computation of refrigerant mass flow rate $\dot{m}$, utilizing the evaporation path 2-3:

$$\dot{m} = \frac{\dot{Q}_{in}}{(h_3 - h_2)} \qquad \textbf{Eq. 11-3}$$

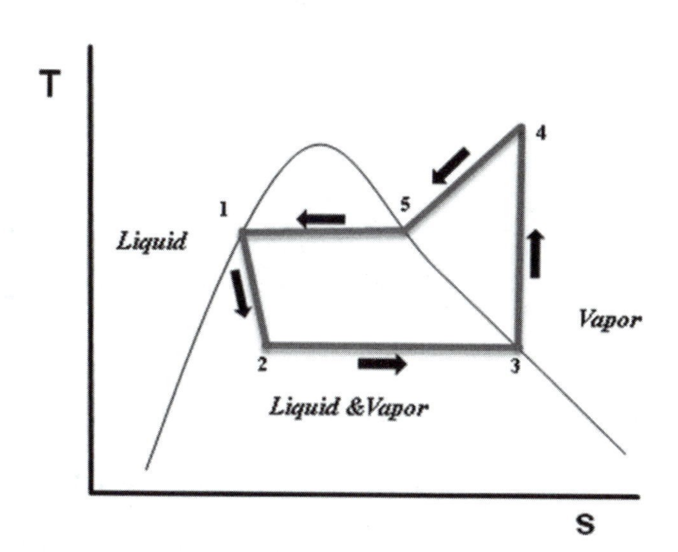

Figure 11-8. Dry Vapor Compression Cycle in Refrigeration Systems

Where, Q_{in} represents "heat flow rate" in Btu/sec or J/s.

Rearrangement of Eq. 11-3 yields the equation for heat flow rate calculation:

$$\dot{Q}_{in} = \dot{m}(h_3 - h_2) \qquad \text{Eq. 11-4}$$

These equations are premised on the refrigeration cycle depicted in Figure 11-8. The nomenclature used in these equations is specific to the refrigeration cycle shown in Figure 11-8. If letters are used to denote various points in the refrigeration cycle, the numerical subscripts are replaced with letters as illustrated in Case Study 11-1.

Coefficient of Performance, or COP, in Refrigeration Systems

Coefficient of performance of a refrigeration system is defined as the ratio of useful energy transfer to the work input. The system is considered to be the refrigerant in the computation of COP. The equations for the COP are as follows:

$$COP_{refrigerator} = \frac{Q_{in}}{(Q_{out} - Q_{in})} \qquad \textbf{Eq. 11-5}$$

Also,

$$COP_{refrigerator} = \frac{Q_{in}}{W_{in}} \qquad \textbf{Eq. 11-6}$$

And,

$$COP_{refrigerator} = COP_{heatpump} - 1 \qquad \textbf{Eq. 11-7}$$

$$\text{EER} = 3.41 \times \text{COP} \qquad \textbf{Eq. 11-8}$$

SEER, SEASONAL ENERGY EFFICIENCY RATIO

SEER or seasonal energy efficiency ratio is a rating based on the cooling output in Btu during a cooling season divided by the total electrical energy drawn from the utility, during the same period, in Watt-Hours. Therefore, the engineering units for SEER rating are

Btu/W-h. The higher the SEER rating of an air-conditioning system the more efficient it is.

Example 11-1

As an energy engineer, you have been asked by your client to determine the total annual cost of electrical energy consumed and the input power demanded by an air conditioning system with the following specifications:

SEER Rating: 10 Btu/W-h
Air Conditioning System Rating: 10,000 Btu/hr
Total, Annual, Seasonal Operating Period: 130 days, 8 hours per day.
Average, Combined, Electrical Energy Cost Rate: $0.18/kWh

Solution:

Annual Cost of Energy = ($0.18/kWh).(Total Energy Drawn From The Utility, Annually)

Total Power Demanded from the Utility
= (Air Conditioning System Rating, in Btu/hr)/(SEER Rating, in Btu/W-hr **Eq. 11-9**

Note: Both Btu values in **Eq. 11-9** are outputs, while the W-hr value represents the input energy drawn from the line (utility) side of the power distribution system.

∴ **Total Power Drawn From The Utility**
= (10,000 Btu/hr)/(10 Btu/W-hr)
= 1,000 Watts, or 1 kW

Then,

Total Energy Drawn From The Utility, Annually
= (1 kW).(Total Annual Operating Hours)
= (1 kW).(130 Days).(8 Hours/Day)
= 1,040 kWh

∴ **Total Annual Cost of Electrical Energy Consumed**
= (1,040 kWh)/($0.18/kWh)
= $187.20

CASE STUDY 11-1—REFRIGERATION CYCLE:

An air-conditioning system uses HFC-134a refrigerant. The refrigeration system is cycled between 2.0 MPa and 0.40 MPa. A pressure-enthalpy diagram for HFC-134a is presented on the next page.

a) Draw the refrigeration cycle on the given diagram. See Figure 11-9.
b) Determine the change in entropy during the throttling process
c) Determine the percentages of liquid and vapor at the end of the throttling segment of the refrigeration cycle.
d) How much enthalpy is absorbed by the system (refrigerant) in the evaporation (latent) phase?
e) How much enthalpy is extracted from the system (refrigerant) in the condensation (condenser) phase of the cycle?
f) In which leg of the refrigeration cycle would expansion be used?
g) If the refrigeration capacity of this system were sized based on the enthalpy extracted from the refrigerant, as calculated in part (e), what would the specification be in tonnes (Metric).

(a) Solution—The process involved in the drawing of the refrigeration cycle is as follows:

C - D:

See Figure 11-9. Locate the 2 MPa and 0.4 MPa points along the pressure (vertical) axis of the chart, name these points "C" and "D," respectively. This is the throttling portion of the refrigeration cycle. Note: HFC-134a, at point C is in, high pressure, saturated liquid phase.

Throttling process is adiabatic and Δ h = 0

∴ Draw a straight, vertical, line down from C to D. At point D, R-134a is in liquid-vapor mixture phase.

D - A:

See Figure 11-9. The next step involves complete transformation of the refrigerant from liquid to gaseous phase through absorption of heat, or Δ h. This is an, non-adiabatic, isobaric process; so, draw a straight, horizontal, line from D to A. This step is referred to as the ***evaporator*** segment of the refrigeration cycle. This is where the

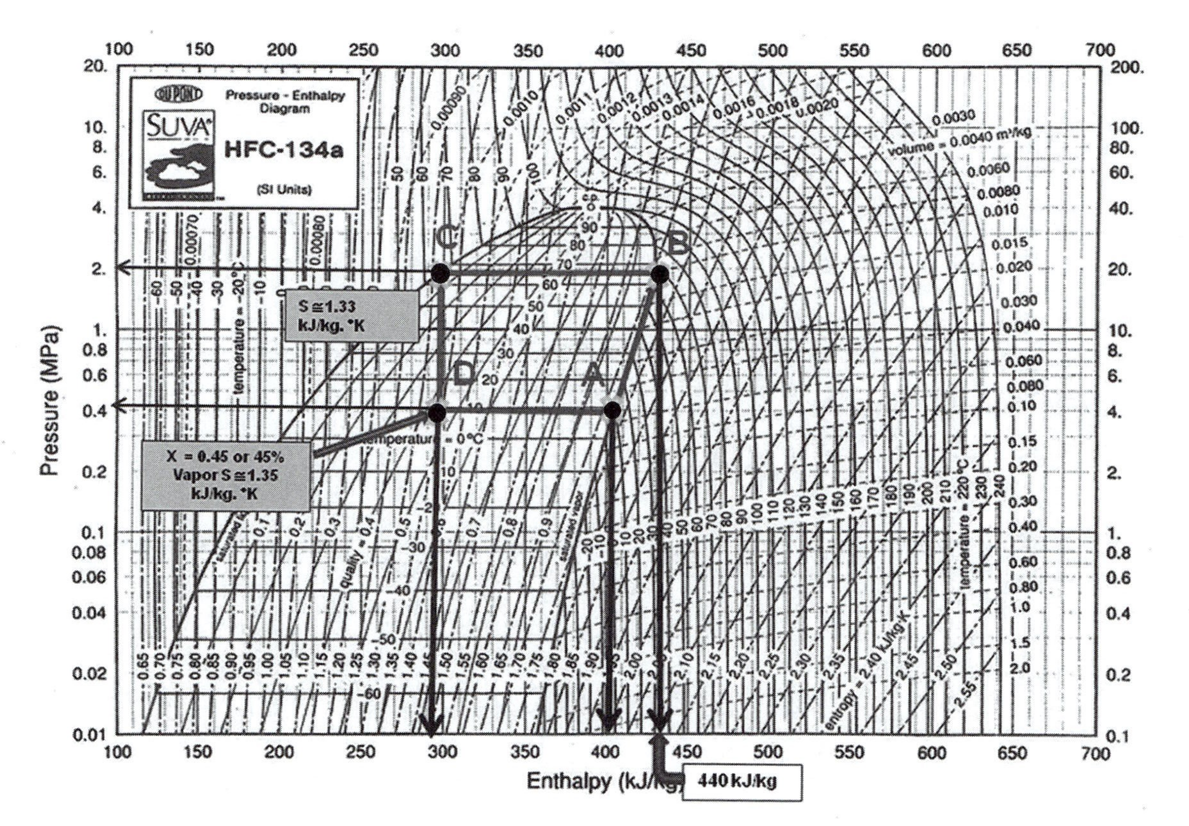

Figure 11-9. Pressure-Enthalpy Diagram, Case Study 11-1

system (refrigerant) performs cooling of the environment as its phase undergoes through latent transformation from liquid to gaseous phase.

A - B:

See Figure 11-9. The next step involves the transformation of HFC-134a from LOW pressure (0.4 MPa) gaseous phase to HIGH (2 MPa) pressure gaseous phase. This is the ***compressor*** segment of the refrigeration cycle. This phase is an isentropic process, Δ **s** = 0, therefore, draw a straight line from point A to B, asymptotic to S ≅ 1.73 kJ/kg. °K.

B - C:

See Figure 11-9. The next step involves the transformation of HFC-134a from high (**2.0 MPa**) pressure gaseous phase to high pressure, saturated, liquid phase. This segment constitutes the ***condenser*** segment of the refrigeration cycle. This is an **isobaric** process, Δ **P** = **0.** Therefore, draw a straight line from point B to C, along **P** = **2 MPa** line.

(b) Determine the change in entropy during the throttling process:
Solution: See Figure 11-9.

$$\Delta s = s_D - s_C \cong 1.35 - 1.33 \cong 0.02 \text{ kJ/kg. °K.}$$

(c) Determine the percentages of liquid and vapor at the end of the throttling segment of the refrigeration cycle.

Solution:

This involves reading the value of "x," the quality, at point "**D,**" from the ressure-enthalpy diagram. **See Figure 11-9.**

$$x = (m_{vapor})/(m_{vapor} + m_{liquid}) = 0.45 \text{ or } 45\%$$

In other words,

$$m_{vapor}(\%) = 45,$$

And since:

$$(\%m_{vapor} + \%m_{liquid}) = 100\%,$$

$$m_{liquid}(\%) = 100 - 45 = 55\%$$

(d) How much enthalpy is absorbed by the system (refrigerant) in the **evaporation** (latent) phase?

Solution:

This involves step **D - A,** see the Pressure – Enthalpy Diagram, **see Figure 11-9**:

$$\Delta h_{D-A\ Phase} = h_A - h_D = 400 - 300 = 100 \text{ kJ/kg}$$

(e) How much enthalpy is extracted from the system (refrigerant) in the **condensation** (condenser) phase of the cycle?

Solution:

This involves step **B** - C, **see Figure 11-9**

$$\Delta h_{B-C\ Phase} = h_B - h_C = 440 - 300 = 140 \text{ kJ/kg}$$

(f) In which leg of the refrigeration cycle would expansion occur?

Answer: The throttling leg, Step **C - D. See Figure 11-9.**

(g) Determination of the refrigeration capacity in tonnes based on the enthalpy extracted from the refrigerant, as calculated in part (e):

Solution:

Heat extracted from the refrigerant in part (e) = 140 kJ/kg.

Rate of refrigeration for this system, per tonne, where 1000 kg = 1 tonne

= (140 kJ/kg).(1000 kg)/24 hr
= 140,000 kJ/24 hr
= 140,000 kJ/(24 hr)/(3600sec/hr)
= 1.62 kJ/sec
= 1.62 kW

Since 1 tonne of refrigeration amounts to 3.86 kW, the refrigeration capacity of this system, in tonnes, would be:

= (1.62 kW)/(3.86 kW)
= 0.4 tonne.

DIRECT DIGITAL CONTROL OF HVAC SYSTEMS

Like many manufacturing operations and chemical process, nowadays, HVAC systems are taking advantage of automation and automated control systems. Automated closed loop control systems permit, operator-error-free, and reliable operation of HVAC systems. Prior to the 1990s and 1980s, proper and effective operation of many large HVAC systems in industrial and commercial domains required a sizeable crew of utilities engineers and technicians. Their sole responsibility, in many cases, was to monitor, track, audit and optimize the performance of chillers, compressors, air washers, cooling towers, fans and pumps. This approach required continuous manual monitoring, data recording frequent manual adjustments of HVAC controls.

The advent of industrially hardened computers and PLC's, programmable logic controllers transformed all that and ushered the era of automation into the HVAC realm. Central controllers, whether they consist of PC's or PLC's, are programmable, meaning, the control system code/program may be customized for the specific use. Major program features, within the overall application program, include synchronous controlled events, time schedules, set-points, control logic, timers, trend logs, regression analysis based forecasts, alarms, graphs, graphical depictions of the HVAC systems with ***live*** or ***real-time*** data points.

There are myriad alternative automated HVAC brands in the market to choose from. Some are relatively small concerns and some brands are well established and well supported subsidiaries of large firms. End users in the market for automated HVAC systems are advised to consider the following criteria in the formulation of their decision for a specific brand:

1) Is the automated HVAC system provider well established in the market with reasonable availability of technical support during start-up phase, commissioning phase and post installation operation?
2) Ensure that the technology—hardware, software and firmware—offered by the vendor is well beyond its "***infancy***" period, and therefore, vetted.
3) If possible, avoid technology that is proprietary and offers little compatibility with the mainstream PC's and PLC's.

4) Select brands and technology that is compatible with established, recognized, standard communication protocol—such as that which is established and sponsored by recognized entities like IEEE, Institute of Electrical and Electronic Engineers.
5) Choose field input an output hardware that is versatile and compatible with mainstream sensors and output transducers
6) Preference should be given to PC's and PLC that operate on—or are compatible with—mainstream operating system platforms, i.e. Microsoft Windows 7, Vista, and equivalent late generation systems. This is to ensure compatibility with useful application software packages, such as the Wonderware ® HMI, Human Machine Interface package and other equivalent software packages.

HVAC control systems are sometimes embedded into comprehensive EMS, energy management systems, or BMS, building management systems. This approach offers the following advantages:

1) Cost reduction through economies of scale.
2) More attractive return on investment and economic justification.
3) Obvious confluence of automated HVAC system projects with other energy productivity improvement projects. This could potentially provide additional advantage of financing such projects through ESCO/EPC programs. **(Reference: *Finance and Accounting for Energy Engineers*, By S. Bobby Rauf, The Fairmont Press)**
4) Central monitoring and control of all utilities in an industrial or commercial facility.
5) Computerized streamlining and scheduling of HVAC system PM programs along with the PM and Predictive Maintenance of all other plant equipment.

General approach to the architecture of most automated HVAC systems is illustrated in Figure 11-10. As shown in this system architecture, the core brain of the system is the CPU, central processing unit. This CPU can be a PLC, programmable logic controller, a DDC, direct digital controller, or a simply an industrially hardened PC. The language or code utilized to program the PLC or the PC would be proprietary and specific to the type of PLC or PC installed. Specifications such as the size of CPU memory, RAM, random access memory and the ROM, read only memory, would determine the length of code or program that can be written and the number of data points that can be monitored, tracked, trended, and controlled. The heart of the CPU is referred to as a ***"microprocessor."*** Controllers or CPU's that drive large HVAC systems, or combination of large HVAC and EMS Systems, are sometimes equipped with dual microprocessors in order to limit the program scan times. Program scan times are sometimes also referred to as cycle times. Reasonable scan time for an average automated HVAC system is approximately 30 milliseconds. Shorter scan times are desirable. Longer cycle times can, potentially, cause the control system to become ineffective and dysfunctional.

As shown in Figure 11-10, the CPU is sometimes connected to a monitoring terminal, which includes a PC, display/monitor, keyboard and a mouse. This PC System serves as an interface between the CPU and the programmers, maintenance technicians and certain qualified production personnel. Often, this is where the control program and application software reside.

The PC and the monitor shown at the bottom of Figure 11-10 serve as the process annunciation system, complete with HVAC graphics, real time data, alarms, trend charts, event logs, and production performance data diagnostics. This system, or sub-system, is referred to as the human machine interface or HMI system. This PC based system is equipped with a suitable operating system such as the Microsoft Windows®, or Vista®, and an HMI software package, such as, Wonderware®, or equivalent. The HMI system and/or the CPU Terminal are often connected, in network format, to other IT, Information Technology, and accounting computers for monitoring of productivity and production cost tracking purposes. As shown in the automated HVAC system architecture diagram, the HMI system and the CPU terminal are sometimes linked to remote or off-site locations, i.e. corporate offices, through Ethernet, wireless routers and modems.

As shown in Figure 11-10, automated HVAC systems also consist of other peripheral equipment, such as, input and output modules. Input modules receive different types of signals from peripheral devices/sensors and process them for presentation to CPU. Output modules take various outputs or commands from the CPU and package them such that they can be used to control equipment in the field.

Inputs and outputs, in essence are signals or commands. Main categories of inputs and outputs are as follows:

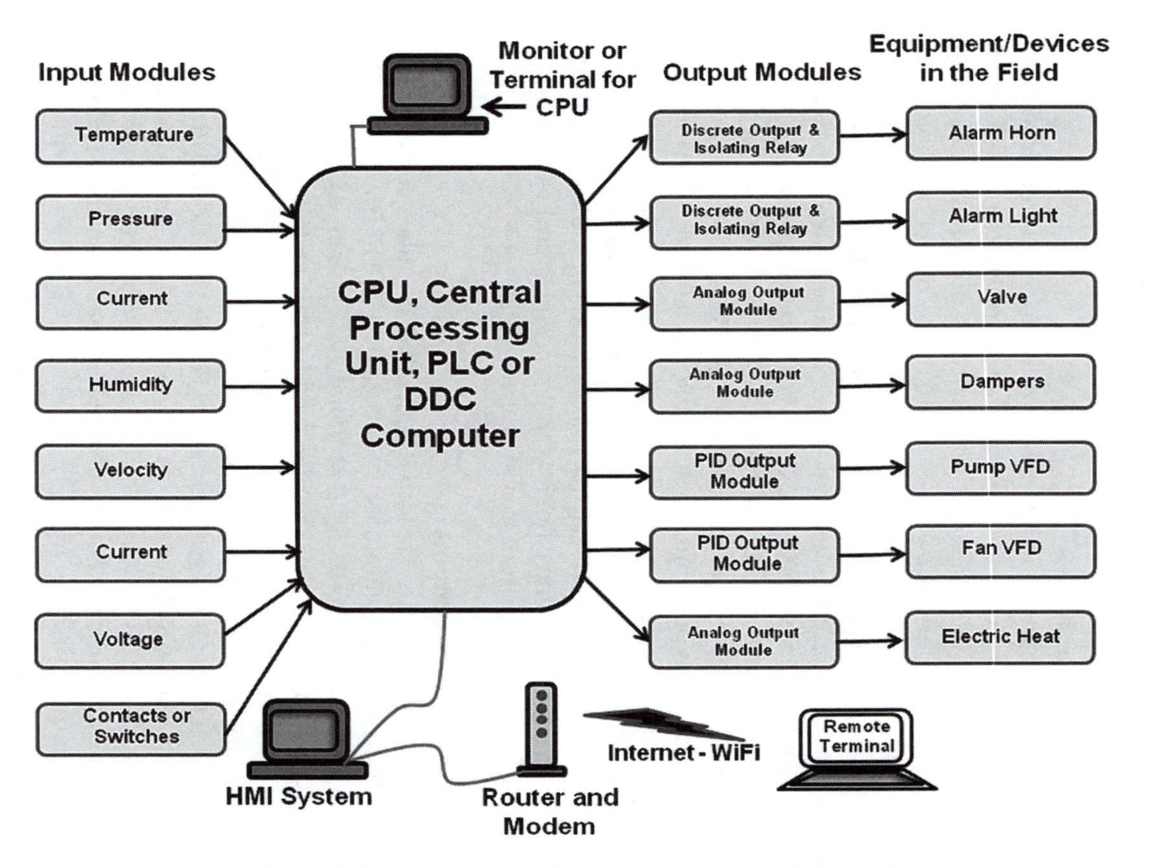

Figure 11-10. Automated HVAC Control System Architecture

1) Digital or discrete inputs
2) Digital or discrete outputs
3) Analog inputs
4) Analog outputs

Digital or Discrete Inputs

The digital inputs—sometimes referred to as discrete inputs—are simply closed or open contact signals that represent the closed or open status of switches in the field. These switches or contacts can be limit switches, safety interlock switches or auxiliary contacts of motors, sensors, etc. The normally closed safety interlock contact shown in Figure 11-11 presents 110 volts AC to the discrete input block. Closed contact type inputs can present other 'non-zero" voltages, such as, 5 volts DC, 10 volts DC and 24 volts AC/DC, or 110 volts AC to the discrete input modules. The end result is a logic level HIGH or "1" to the CPU for computation and decision making purposes. On the other hand, an open contact type input, such as the one shown to represent the normally open pressure switch in Figure 11-11, would be interpreted as a LOW logic level or "0" by the discrete input module and the CPU. Another perspective on the role of input modules would be to view them as intermediary devices that transform and isolate discrete signals being received from peripheral devices.

Digital or Discrete Outputs

Digital outputs are typically generated by relay contacts based in the discrete output modules. A "1" or a "HIGH" from the CPU is transformed into a 5 volts DC, 10 volts DC, 24 volts AC/DC, or 110 volts AC signal, which can then be used to start and stop equipment, as a part of the overall automated HVAC control scheme. For example, a "1" from the CPU is transformed into a 110 volt AC signal by the 110 Volt AC Discrete Output Module in Figure 11-11. This 110 Volt AC output from the output module is fed to the AC motor starter coil to turn on the motor, in response to the specific input conditions and the program or algorithm. Another discrete output from the 110 Volt AC Discrete Output Module, in Figure 11-11, is fed to an alarm horn to annunciate an alarm condition. Note that in both of these examples the second terminal of each controlled field device is connected to the power system neutral, designated as "N."

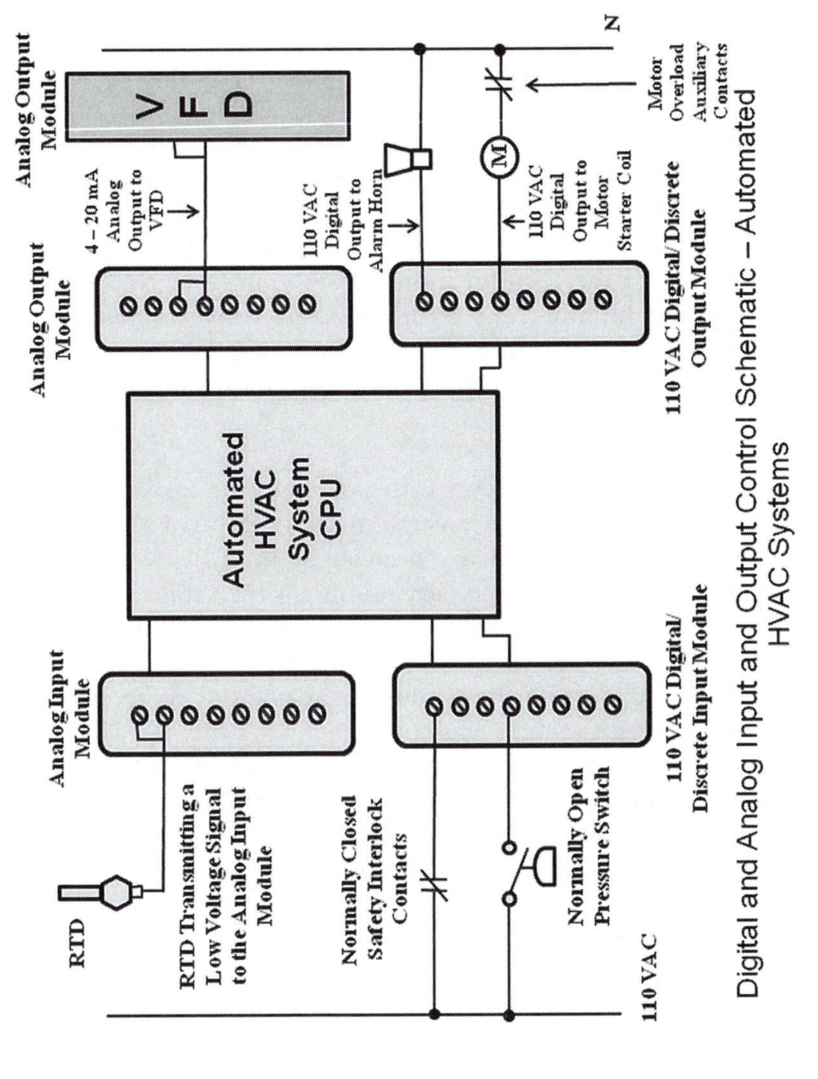

Figure 11-11. Automated HVAC Control System Architecture

Analog Inputs

Analog inputs represent gradually varying parameters or signals. Analog signals, in automated HVAC systems, include temperature, humidity, volume, pressure and even electrical current drawn by fan, blower or pump motors. Common analog current range is 4-20 mA. In certain cases, signals representing pressure, temperature, and volume are converted into analog voltage signals. These analog voltage signals range, commonly, from 0-10 Volts DC. Analog input modules can also convert signals from thermocouples, RTDs, resistance temperature detectors, pressures sources and other types of analog signal sources, into digital logic numbers for use by the CPU. See the RTD application example shown in Figure 11-11. The analog temperature measurement from the RTD, in form of low voltage signal, is fed to the analog input module. The analog module converts this voltage signal into an equivalent digital value. This digital value is presented to the CPU for computation or algorithm execution purposes.

Analog Outputs

Operation of analog outputs can, essentially, be explained as the operation of analog inputs in reverse. Analog output begins with a presentation of a digital or binary number by the CPU to the analog output module. This digital or binary number is then transformed into an equivalent analog signal such as 0 – 5 Volt DC, 0 – 10 Volt DC, 0 – 110 Volt AC or 4 – 20 mA DC signal. Any of such analog signals can then used to command the gradual or analog operation of HVAC control equipment like valves, variable frequency drives, heat sources, electric valve actuators, pneumatic actuators, etc. In Figure 11-11, the analog output module is shown feeding a 4-20 mA analog output to a VFD, Variable Frequency Drive. The drive can base its output power frequency on this 4 – 20 mA analog input to control the speed of an HVAC fan or pump motor.

Chapter 11—Self Assessment Problems and Question

1. An air-conditioning system uses HFC-134a refrigerant. The pressure-enthalpy diagram for this refrigerant is shown in Figure 11-12. The refrigeration system is cycled between 290 psia and 60 psia.

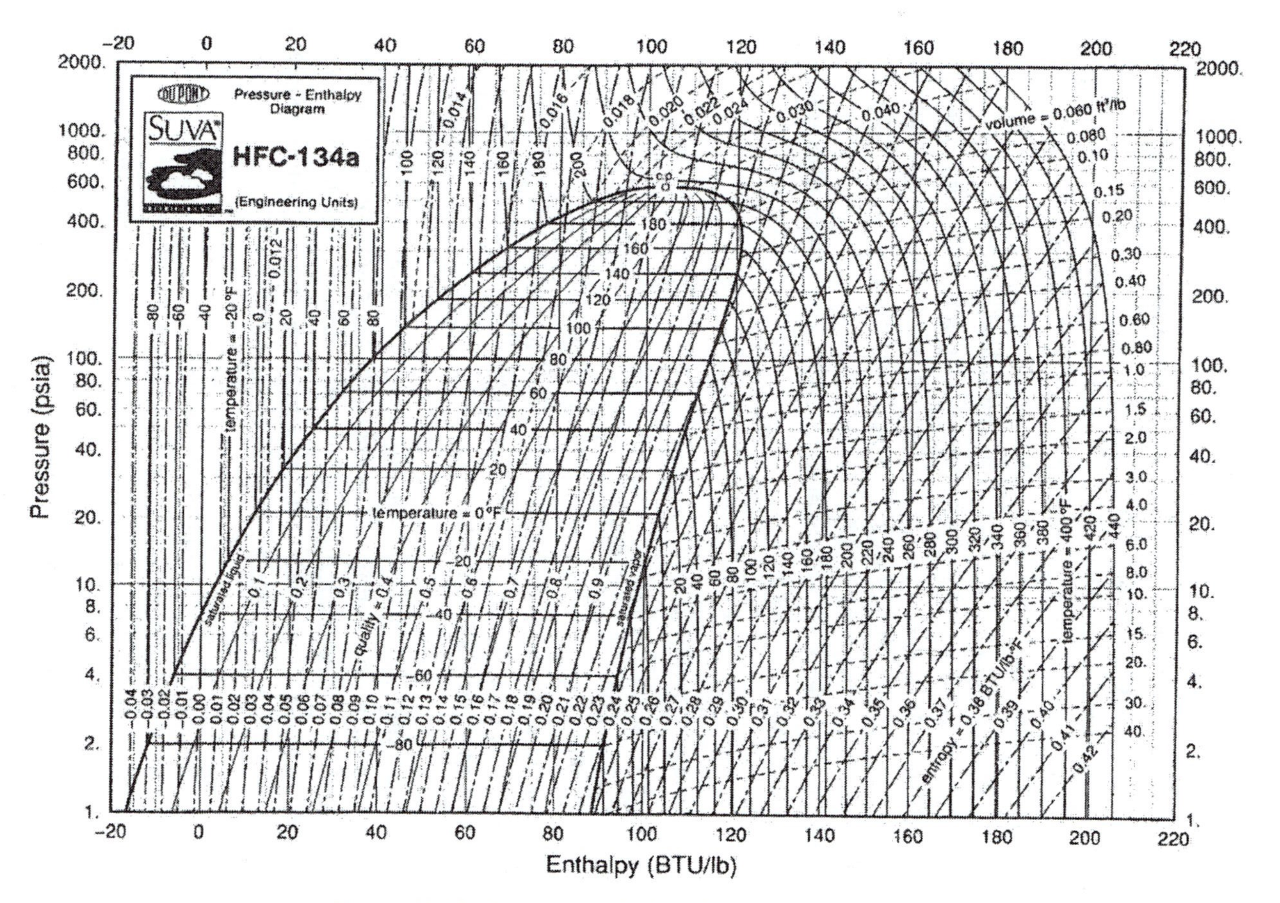

Figure 11-12. Pressure-enthalpy Diagram, HFC-134a

(a) Draw the refrigeration cycle on the given diagram
(b) What is the change in enthalpy during the expansion process
(c) Determine the percentages of liquid and vapor at the end of the throttling segment of the refrigeration cycle.
(d) How much enthalpy is absorbed by the system (refrigerant) in the evaporation (latent) phase?
(e) How much enthalpy is extracted from the system (refrigerant) in the condensation (condenser) phase of the cycle?
(f) Determine the percentages of liquid and vapor at B.
(g) Assume that the mass flow rate of refrigerant being cycled in this air-conditioning system is 10 lbm/min and the compressor efficiency is 70%. Determine the amount of electrical power demanded by the compressor motor if the compressor motor efficiency is 90%.
(h) Which leg of the refrigeration cycle would be considered isentropic.

I. A-B
II. B-C
III. C-D
IV. D-A

2. As stated in this chapter, 1 tonne (SI/Metric) of refrigeration capacity is equivalent to 3.86 kW of power. Provide the mathematical proof for this equivalence.

3. As an energy engineer, you are performing an energy cost assessment for operating a 20,000 Btu/hr air conditioner. Based on the data and specifications provided below, determine the total annual cost of electrical energy consumed and the input power demanded by the air conditioning system:

— SEER Rating: 12 Btu/W-h
— Air Conditioning System Rating: 20,000 Btu/hr
— Total, Annual, Seasonal Operating Period: 200 days, 10 hours per day.
— Average, Combined, Electrical Energy Cost Rate: $0.20/kWh

Appendices

Appendix A

This appendix includes the solutions and answers to end of chapter self assessment problems and questions

CHAPTER 1—SELF ASSESSMENT PROBLEMS AND QUESTIONS

1. Determine the amount of heat extracted by the quench water, per block, in **Case Study 1-2**, ***using the temperature rise of the water*** when the steel block is dropped into the quenching tank. The temperature of the block is **100°C** when it enters the quench water. The initial temperature of the water in the quench tank is **20°C** and volume of water is **6.038 m^3**. The final, equilibrium, temperature of the water and the block is **30°C**.

Note: This is an alternate approach to the **solution for part (g) in Case Study 1-2**. This solution is premised on the use of the **rise in water temperature** to calculate the amount of heat extracted by the quench water, per block.

Solution:
Given or known:

$\mathbf{c_{water} = 4.186\ kJ/kg.\ °K}$ {From **Table 1-5**}

$m_{water} = (1000\ kg/m^3)\ .\ (6.038\ m^3)$

$= \mathbf{6{,}038\ kg}$

$T_{water-i} = 20\ °C = 273 + 20°C = 293\ °K$

$T_{water-f} = 30\ °C = 273 + 30C = 303\ °K$

$\therefore \mathbf{\Delta T_{water} = 303\ °K - 293\ °K = +10°K}$

Since

$\mathbf{Q_{absorbed\ by\ water} = (m_{water})\ .\ (c).\ (\Delta T_{water})}$

$Q_{absorbed\ by\ water} = (6{,}038\ kg)\ .\ (4.186\ kJ/kg.\ °K)\ .\ (+10°K)$

$\mathbf{Q_{absorbed\ by\ water} = 252{,}750\ kJ}$

$\therefore \mathbf{Q_{absorbed\ by\ water} =}$

$\mathbf{- Q_{lost\ by\ the\ block} = 252{,}750\ kJ \cong 252{,}770\ kJ}$

CHAPTER 2—SELF ASSESSMENT PROBLEMS & QUESTIONS

1. As an energy engineer, you are charged with the task to estimate the amount of **electrical power produced, in MW**, by a steam based power generating plant. Assume that there is no heat loss in the turbine system and that difference between the enthalpies on the entrance and exit ends of the turbine is completely converted into work, minus the inefficiency of the turbine. All of the data available, pertinent to this project, is listed below:
   - Electrical Power Generator Efficiency, $\eta_{Generator}$: 87%
   - Steam Turbine Efficiency, $\eta_{Turbine}$: 67%
   - Mass flow rate for steam, $\dot{m}$: 20 kg/s (44 lbm/s)
   - Exit enthalpy, hf, of the steam: 2900 kJ/kg (1249 Btu/lbm)
   - Incoming superheated steam enthalpy, hi: 3586 kJ/kg (1545 Btu/lbm)

Solution:

Solution Strategy: As explained in Chapter 2, the power delivered by steam to the turbine blades, $\mathbf{P_{steam}}$, in a simplified, no heat loss, no kinetic head loss, no potential head loss and zero frictional head loss scenario can be represented by the mathematical relationship stated in form of **Eq. 2-1**. And, in the context of flow of energy from steam to electricity, functional relationship between electrical power, $\mathbf{P_{Electrical}}$, generator efficiency $\eta_{\mathbf{Generator}}$, steam turbine efficiency $\eta_{\mathbf{Turbine}}$, and $\mathbf{P_{steam}}$ can be expressed in form of **Eq. 2-2.**

$$\mathbf{P_{steam} = (h_i - h_f) \cdot \dot{m}} \qquad \textbf{Eq. 2-1}$$

$$\mathbf{P_{Electrical} = (P_{steam}) \cdot (\eta_{Turbine}) \cdot (\eta_{Generator})} \qquad \textbf{Eq. 2-2}$$

Solution in SI/Metric Units:

The power imparted by the steam onto the turbine blades, $\mathbf{P_{steam}}$, can be determined by applying **Eq. 2-1:**

$$\mathbf{P_{steam} = (3586\ kJ/kg - 2900\ kJ/kg) \cdot (20\ kg/s)}$$
$$\mathbf{= 13{,}720\ kJ/s}$$

Then, the electrical power produced by generator, $\mathbf{P_{Electrical}}$, can be determined by applying **Eq. 2-2:**

$$\mathbf{P_{Electrical} = (P_{steam}) \cdot (\eta_{Turbine}) \cdot (\eta_{Generator})}$$

$$P_{Electrical} = (13{,}720 \text{ kJ/s}) \cdot (0.67) \cdot (0.87)$$
$$\mathbf{= 7{,}997 \text{ kJ/s}}$$

Since 1.00 kJ/s = 1 kW,

$$P_{Electrical\ in\ kJ/s} = (7{,}997 \text{ kJ/s}) / (1.00 \text{ kJ/s/kW})$$
$$\mathbf{= 7{,}997 \text{ kW}}$$

Since 1,000 kW = 1 MW

$$P_{Electrical\ in\ MW} = (7{,}997 \text{ kW})/(1{,}000 \text{ kW/MW})$$
$$\mathbf{= 8 \text{ MW}}$$

Solution in US/Imperial Units:

The power imparted by the steam onto the turbine blades, $\mathbf{P_{steam}}$, can be determined by applying **Eq. 2-1:**

$$\mathbf{P_{steam} = (1545 \text{ Btu/lbm} - 1249 \text{ Btu/lbm}) . (44 \text{ lbm/s})}$$
$$\mathbf{= 13{,}024 \text{ Btu/s}}$$

Then, the electrical power produced by generator, $\mathbf{P_{Electrical}}$, can be determined by applying **Eq. 2-2:**

$$\mathbf{P_{Electrical} = (P_{steam}) \cdot (\eta_{Turbine}) \cdot (\eta_{Generator})}$$

$$P_{Electrical} = (13{,}024 \text{ Btu/s}) \cdot (0.67) \cdot (0.87)$$
$$\mathbf{= 7{,}592 \text{ Btu/s}}$$

Since 1.055 kJ/s = 1Btu/s,

$$P_{Electrical\ in\ kJ/s} = (7{,}592 \text{ Btu/s}) \cdot (1.055 \text{ kJ/s})$$
$$\mathbf{= 8010 \text{ kJ/s}}$$

Since 1 kJ/s = 1 kW

$$\mathbf{P_{Electrical\ in\ kW} = 8{,}010 \text{ kW}}$$

Since 1 MW = 1000 kW,

$$P_{Electrical\ in\ MW} = (8{,}010 \text{ kW})/(1000 \text{ kW/MW})$$
$$\mathbf{= 8 \text{ MW}}$$

2. Consider the scenario described in Problem (1). Your client has informed you that the power generating plant output requirement has now doubled. Based on the concepts and principles learned in Chapter 2, what is the most suitable alternative for doubling the power output if the exit enthalpy, $\mathbf{h_f}$, of the steam must be kept constant at the original 2900 kJ/kg (1249 Btu/lbm) level?

A. Double the mass flow rate, $\dot{m}$, only.
B. Double the incoming superheated steam enthalpy, hi only.
C. Double the efficiency of the turbine.
D. Double the efficiency of the generator.
E. Increase mass flow rate, $\dot{m}$, incoming superheated steam enthalpy, hi and increase the efficiency specification on the turbine.

Answer:

As apparent from inspection of Eq. 2-1 and Eq. 2-2, theoretically, the power output can be doubled by doubling the mass flow rate, $\dot{m}$. However, doubling the mass flow rate is ***not practical under normal circumstances.***

$$P_{steam} = (h_i - h_f) \cdot \dot{m} \qquad \text{Eq. 2-1}$$

$$P_{Electrical} = (P_{steam}) \cdot (\eta_{Turbine}) \cdot (\eta_{Generator}) \qquad \text{Eq. 2-2}$$

It is also apparent from examination of Eq. 2-1 that the relationship between $\mathbf{h_i}$ and $\mathbf{P_{Electrical}}$ or $\mathbf{P_{steam}}$ is not linear. Therefore, doubling $\mathbf{h_i}$ is not only impractical from superheating capacity point of view, but could escalate power to a level that is more than twice the original level.

Doubling the efficiency of the turbine or the generator is not possible since that would exceed 100% in each case; and efficiency cannot exceed 100%.

Hence, the most practical and reasonable answer would be "E." A suitable combination of an increase in mass flow rate, $\dot{m}$, a commensurate increase in incoming superheated steam enthalpy, $\mathbf{h_i}$, and increase in the efficiency of the turbine would be optimum and more practical.

CHAPTER 3—SELF ASSESSMENT PROBLEMS AND QUESTIONS

1. Calculate the volume **1 kg** of vapor would occupy under the following conditions:

h = **2734 kJ**
u = **2550 kJ**
p = **365.64 kPa = 365.64 kN/m^2**
V = ?

Solution:

By definition, the volume that **1 kg** of vapor occupies is the ***specific volume*** of the vapor. Specific volume, as defined in detail later in this text, is volume per unit mass. In this case, specific volume would be volume (in **m^3**) on per kg basis.

Apply **Eq. 3-2,** re-arrange it and substitute the given values:

$$\mathbf{h = u + p.V} \qquad \textbf{Eq. 3-1}$$

$$\mathbf{V = (h - u)/p} \qquad \textbf{Eq. 3-1a}$$

$$\mathbf{V} = (2734\ \text{kJ} - 2550\ \text{kJ})/365.64\ \text{kN/m}^2$$

$$\mathbf{V} = (2734\ \text{kN-m} - 2550\ \text{kN-m})/365.64\ \text{kN/m}^2 = \mathbf{0.503\ m^3}$$

Therefore, each kg of vapor would occupy 0.503 m^3 under the conditions stated above.

2. In a certain solar system there are four (4) planets oriented in space as shown in Figure A3-1. As apparent from the orientation of these planets, they are exposed to each other such that heat transfer can occur freely through radiation. All four (4) planets are assumed to be massive enough to allow for the interplanetary heat transfer to be an isothermal phenomenon for each of the planets. ***Perform all computation in the US Unit System.***

a. If the 1,300 Btu/lbm of radiated heat transfer occurs from planet X to planet Y, what would be the entropy changes at each of the two planets?

b. If a certain radiated heat transfer between Planets Y and Z causes an entropy change of -2.9 Btu/lbm.°R at Planet Y and an entropy change of 3.1 Btu/lbm.°R at Planet Z, what would be the overall, resultant, entropy of this planetary system?

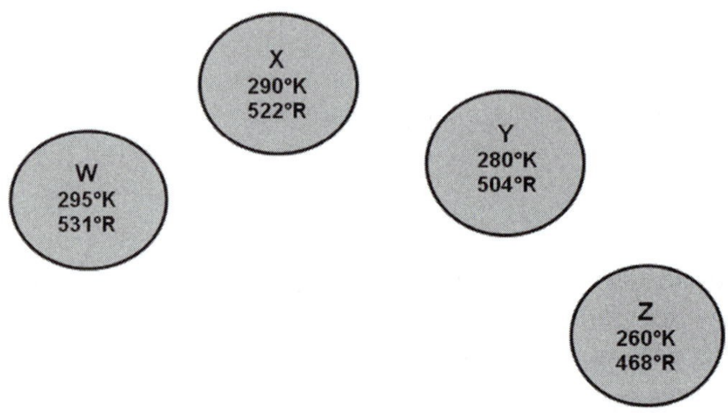

Figure A3-1. Entropy

Solution (a):

In an ***isothermal*** (constant temperature) process, the entropy production, Δs, is a function of the energy transfer rate and the change in the enthalpy, heat transferred and the absolute temperature of the object in question is governed by **Eq. 3-5**:

$$\Delta s = q / T_{abs} \qquad \textbf{Eq. 3-5}$$

$\therefore$ $\Delta s_X = (-1300 \text{ Btu/lbm})/(522°\text{R})$
$= -\ 2.49$ **Btu/lbm.°R** {Due to heat loss by Planet X}

And,

$\Delta s_Y = (+1{,}300 \text{ Btu/lbm})/(504°\text{R})$
$= +2.5794$ **Btu/lbm.°R** {Due to heat gain by Planet Y}

Solution (b):

According to **Eq. 3-4**:

$$\text{Overall } \Delta s_{\text{Planetary System}} = \sum (\Delta s_i)$$

In this case,

$$\Delta s_{\text{Planetary System}} = \Delta s_X + \Delta s_Y + \Delta s_{YZ} + \Delta s_Z$$

Or,

$\Delta s_{\text{Planetary System}}$ = - 2.49 Btu/lbm.°R +2.5794 Btu/lbm.°R – 2.9 Btu/lbm.°R + 3.1 Btu/lbm.°R

∴ **Overall $\Delta s_{\text{Planetary System}}$ = 0.29 Btu/lbm.°R**

3. If the mass of vapor under consideration in problem 1 were tripled to 3 kg, what would be the impact of such a change on the volume?

Answer:

Since the final value for volume is on per unit mass basis, if the mass is tripled, **the volume would increase three (3) fold**, as well, if all other parameters stay constant. This is obvious from inspection of Eq. 3-1a

$$\mathbf{V = (h - u)/p} \qquad \textbf{Eq. 3-1a}$$

4. Would **Eq. 3-2** be suitable for calculation of enthalpy if all available data is in SI (Metric) units?

Answer:

The answer is **NO**. One ostensible reason why Eq. 3-2 cannot be used for calculation of enthalpy in the SI unit realm is the presence of "*J*," the Joules constant. The *J* has a value of 778 ft-lbf/Btu; which is unit conversion factor in the ***US Unit system***. Therefore, Eq. 3-2 pertains to computation of enthalpy in the US unit realm.

CHAPTER 4—SELF ASSESSMENT PROBLEMS & QUESTIONS

1. Using the Mollier diagram, find the entropy of steam at 400°C and 1 Atm.

Solution:

See the Mollier diagram in Figure A4-1. Identify the point of intersection of the 400°C line (or 400°C isotherm) and the constant pressure line (or isobar) of 1 bar. This point of intersection of the two lines is labeled **B** in Figure A4-1.

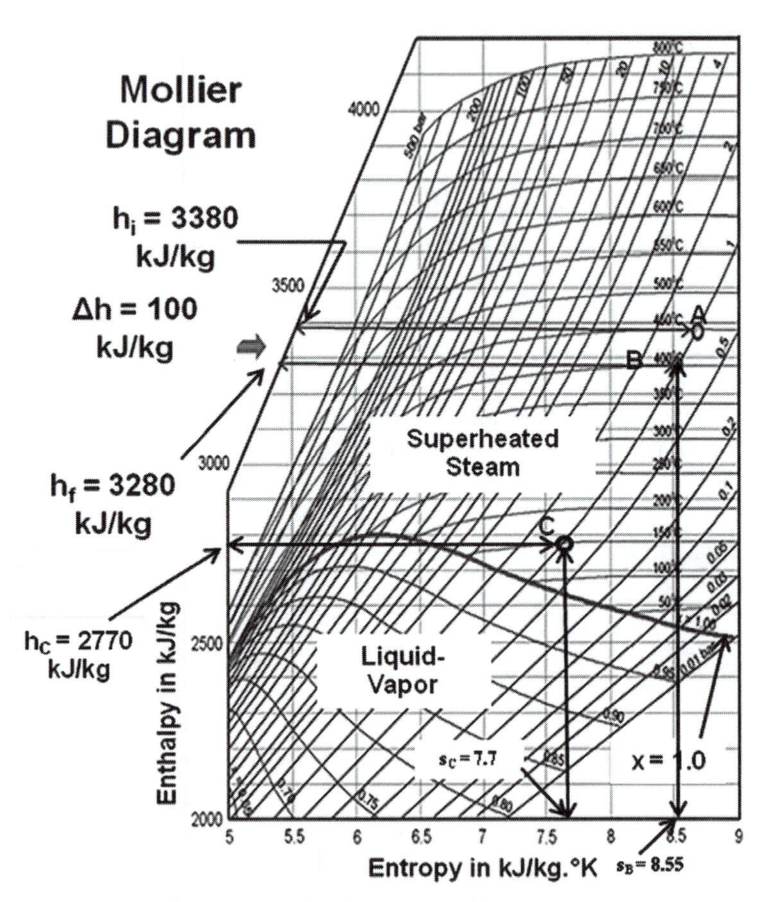

Figure A4-1. Mollier Diagram, SI/Metric Units

Entropy Determination: To determine the entropy at point **B**, draw a straight vertical line from point **B** to the bottom, until it intersects with the entropy line. The vertical line intersects the entropy line at, approximately, 8.55kJ/kg.°K.

Therefore, s_B, or entropy at point B, is 8.55kJ/kg.°K.

2. Heat is removed from a thermodynamic system such that the temperature drops from **450°C, at 1 Atm** to **150°C, at 1 Atm. D**etermine the following:

a) The new, or final, enthalpy

b) The new entropy
c) The state of steam at **150°C** and **1 Atm**?

Solution:

a) Locate the new point, **150°C, 1 Atm**, on the Mollier diagram. This new point is shown as point **C** in Figure A4-1. A horizontal line drawn to the left from point C intersects the enthalpy line at approximately 2770 kJ/kg.

Therefore, the new, or final enthalpy, at 150°C and 1 Atm, is 2770 kJ/kg.

b) A vertical line drawn from point C down to the entropy axis intersects the entropy line at approximately 7.7 kJ/kg.°K.

Therefore, the new or final entropy, at 150°C and 1 Atm, is 7.7 kJ/kg.°K .

c) As shown in Figure A4-1, point **C** lies above the saturation line, in the region labeled superheated steam.

Therefore, at 150°C and 1 Atm, the steam is superheated.

CHAPTER 5—SELF ASSESSMENT PROBLEMS & QUESTIONS

1. Using the saturated liquid enthalpy value for $\mathbf{h_L}$ and the saturated vapor enthalpy value for $\mathbf{h_V}$, at 0.2 MPa and 120.2°C, as listed in the saturated steam tables in Appendix B, calculate the value for $\mathbf{h_{fg}}$.

Solution:

As stated in Eq. 5-1:

$\mathbf{h_{fg} = h_V - h_L}$

As read from Table A5-1:

$\mathbf{h_V = 2706.2\ kJ/kg}$, and

$\mathbf{h_L = 504.68\ kJ/kg}$

$\therefore$ $\mathbf{h_{fg} = h_V - h_L}$
$\mathbf{= 2706.2\ kJ/kg - 504.68\ kJ/kg}$
$\mathbf{= 2201.5\ kJ/kg}$

The value for $\mathbf{h_{fg}}$, at 0.2 MPa and 120.2°C, as listed in Table A5-1, is **2201.9 kJ/kg**.

Table A5-1. Properties of Saturated Steam, by Pressure, SI Units

Properties of Saturated Steam By Pressure Metric/SI Units									
Abs. Press. MPa	Temp. °C	Specific Volume m^3/kg		Enthalpy kJ/kg			Entropy kJ/kg		Abs. Press. MPa
		Sat. Liquid v_L	Sat. Vapor v_V	Sat. Liquid h_L	Evap. h_{fg}	Sat. Vapor h_V	Sat. Liquid s_L	Sat. Vapor s_V	
0.010	45.81	0.0010103	14.671	191.81	2392.8	2583.9	0.6492	8.1489	0.010
0.10	99.61	0.0010431	1.6940	417.44	2258.0	2674.9	1.3026	7.3588	0.10
0.20	120.21	0.0010605	0.88574	504.68	2201.9	2706.2	1.5301	7.1269	0.20
1.00	179.89	0.0011272	0.19435	762.68	2015.3	2777.1	2.1384	6.5850	1.00

2. Determine the enthalpy of 450 psia and 970°F superheated steam.

Solution:

As you examine the superheated steam tables for these parameters, in Appendix B, you realize that exact match for this data is not available in the table. See Table A5-2, for excerpts from the superheated steam tables in Appendix B.

While the given pressure of 450 psia is listed, the stated temperature of 970°F is not listed. Therefore, the enthalpy for 450 psia and 970°F superheated steam must be derived by applying interpolation to the enthalpy data listed in the tables for 900°F and 1,000°F.

The formula for single interpolation, applied between the stated or available enthalpy values for 900°F and 1000°F, at 470 psia, is as follows:

$$\mathbf{h_{970\,°F,\,450\,psia} = ((h_{1000\,°F,\,450\,psia} - h_{900\,°F,\,450\,psia})/(1000°F - 900°F)).\ (970 - 900) + h_{900\,°F,\,450\,psia}}$$

By substituting enthalpy values and other given data from superheated steam table excerpt, shown in Table A5-2:

$$\mathbf{h_{950\,^\circ F,\,450\,psia} = ((1552.4\ Btu/lbm - 1468.6\ Btu/lbm)/(1000^\circ F - 900^\circ F)).(970-900) + 1468.6\ Btu/lbm = 1527.3\ Btu/lbm}$$

Note: The available enthalpy values are circled in Table A5-2.

3. Determine the enthalpy of saturated water at 50°C and 1 Bar.

Table A5-2. Superheated Steam Table Excerpt, US/Imperial Units

Properties of Superheated Steam **US/Imperial Units**					
Abs. Press. psia		**Temp. °F**	**Note: v is in ft³/lbm, *h* is in BTU/lbm and *s* is in BTU/(lbm-°R)**		
(Sat. Temp. °F)		500	700	900	1000
260 (404.45)	*v*	2.062	2.5818	3.0683	3.3065
	h	1262.5	1370.8	1475.2	1527.8
	s	1.5901	1.6928	1.7758	1.8132
360 (434.43)	*v*	1.446	1.8429	2.2028	2.3774
	h	1250.6	1365.2	1471.7	1525
	s	1.5446	1.6533	1.7381	1.7758
450 (456.32)	*v*	1.1232	1.4584	1.7526	1.8942
	h	1238.9	1360	1468.6	1522.4
	s	1.5103	1.6253	1.7117	1.7499
600 (486.25)	*v*			1.3023	1.411
	h			1463.2	1518
	s			1.577	1.7159

Solution:

The saturation temperature at 1 Bar, 1 Atm, or 101 kPa, as stated in the saturated steam tables in Appendix B, is 99.6°C or, approximately, 100°C. The saturated water in this problem is at 50°C; clearly below the saturation temperature. Therefore, the water is in a subcooled state.

In the subcooled state, saturated water's enthalpy is determined by its temperature and not the pressure. Hence, the enthalpy of saturated water at 50°C and 1 Bar must be retrieved from the temperature based saturated steam tables.

From Appendix B, and as circled in Table A5-3, the enthalpy of saturated water at 50°C is **209.34 kJ/kg**[1].

[1]Note: Since the water is referred to as "***saturated <u>water</u>***" and is clearly identified to be subcooled, the enthalpy value selected from the tables is $\mathbf{h_L}$ and not $\mathbf{h_V}$.

4. Determine the enthalpy and specific volume for 14 psia steam with a quality of 65%.

Solution:
Given:

Quality, x = **0.65**
Absolute Pressure = **14 psia**

Table A5-3. Properties of Saturated Steam, by Temperature, SI Units

Properties of Saturated Steam By Temperature, Metric/SI Units									
Temp. °C	Abs. Press. MPa	Specific Volume m^3/kg		Enthalpy kJ/kg			Entropy kJ/kg		Temp. °C
		Sat. Liquid v_L	Sat. Vapor v_V	Sat. Liquid h_L	Evap. h_{fg}	Sat. Vapor h_V	Sat. Liquid s_L	Sat. Vapor s_V	
20	0.002339	0.0010018	57.7610	83.920	2454.1	2537.5	0.2965	8.6661	20
50	0.012351	0.0010121	12.0280	209.34	2382.7	2591.3	0.7038	8.0749	50
100	0.101420	0.0010435	1.6719	419.10	2257.0	2675.6	1.3070	7.3541	100
200	1.554700	0.0011565	0.1272	852.39	1940.7	2792.1	2.3308	6.4303	200

From saturated steam tables in Appendix B, and the excerpt in Table A5-4, the values of enthalpies and specific volumes, at 14 psia, are:

$\mathbf{h_L}$ = **177.68** Btu/lbm
$\mathbf{h_V}$ = **1149.4** Btu/lbm
v_L = **0.01669** ft^3/lbm
v_v = **28.048** ft^3/lbm

Apply equations **5-2** and **5-5**:

$$h_x = (1-x) \cdot h_L + x.\ h_V \qquad \text{Eq. 5-2}$$

$$v_x = (1-x) \cdot v_L + x\ .v_v \qquad \text{Eq. 5-5}$$

Then,

h_x = **(1- 0.65) •= (177.68** Btu/lbm**) + (0.65) • (1149.4** Btu/lbm**)**
h_x = **809.30** Btu/lbm

And,

v_x = **(1- 0.65) • (0.01669) + (0.65) . (28.048** ft^3/lbm)
v_x = **18.24** ft^3/lbm

Table A5-4. Properties of Saturated Steam, by Pressure, US Units

Properties of Saturated Steam By Pressure US/Imperial Units									
Abs. Press. psia	Temp. °F	Specific Volume ft^3/lbm		Enthalpy Btu/lbm			Entropy Btu/(lbm.°R)		Abs. Press. psia
		Sat. Liquid v_L	Sat. Vapor v_V	Sat. Liquid h_L	Evap. h_{fg}	Sat. Vapor h_V	Sat. Liquid s_L	Sat. Vapor s_V	
1.0	101.69	0.016137	333.51	69.728	1036	1105.4	0.1326	1.9776	1.0
4.0	152.91	0.016356	90.628	120.89	1006.4	1126.9	0.2198	1.8621	4.0
14.0	209.52	0.01669	28.048	177.68	972.0	1149.4	0.3084	1.7605	14.0
100	327.82	0.017736	4.4324	298.57	889.2	1187.5	0.4744	1.6032	100

CHAPTER 6—SELF ASSESSMENT PROBLEMS & QUESTIONS

1. A boiler is relocated from sea level to a location that is at an elevation of 10,000 ft MSL. Using the table below and Table A6-2,

determine the temperature at which the water will boil if the boiler is assumed to be open to atmosphere.

Solution/Answer:

In order to determine the boiling point of the water, at an altitude of 10,000 ft, we need to assess the pressure at that altitude. The pressures at a range of altitudes are available through the given table. In US units, the pressure at 10,000 ft is 10.1 psia, or simply, 10 psia.

The next step is to retrieve the saturation temperature at 10 psia, from the saturated steam tables. An excerpt of the Saturated Steam Tables, for the range of pressures surrounding 10 psia, is shown in Table A6-2.

Table A6-1

Altitude With Mean Sea Level as Ref.		Absolute Pressure in Hg Column		Absolute Atmospheric Pressure		
Feet	Meters	Inches Hg Column	mm Hg Column	psia	kg/cm^2	kPa
0	0	29.9	765	14.7	1.03	101
500	152	29.4	751	14.4	1.01	99.5
1000	305	28.9	738	14.2	0.997	97.7
1500	457	28.3	724	13.9	0.979	96
2000	610	27.8	711	13.7	0.961	94.2
2500	762	27.3	698	13.4	0.943	92.5
3000	914	26.8	686	13.2	0.926	90.8
3500	1067	26.3	673	12.9	0.909	89.1
4000	1219	25.8	661	12.7	0.893	87.5
4500	1372	25.4	649	12.5	0.876	85.9
5000	1524	24.9	637	12.2	0.86	84.3
6000	1829	24	613	11.8	0.828	81.2
7000	2134	23.1	590	11.3	0.797	78.2
8000	2438	22.2	568	10.9	0.768	75.3
9000	2743	21.4	547	10.5	0.739	72.4
10000	3048	20.6	526	10.1	0.711	69.7
15000	4572	16.9	432	8.29	0.583	57.2
20000	6096	13.8	352	6.75	0.475	46.6

Table A6-2. Properties of Saturated Steam By Pressure, US Units

Properties of Saturated Steam By Pressure US/Imperial Units										
Abs. Press. psia	Temp in °F	Specific Volume ft^3/lbm		Enthalpy BTU/lbm			Entropy BTU/(lbm.°R)		Abs. Press. psia
		Sat. Liquid v_L	Sat. Vapor v_V	Sat. Liquid h_L	Evap. h_{fg}	Sat. Vapor h_V	Sat. Liquid s_L	Sat. Vapor s_V	
1.0	102	0.0161	333.	69.73	1036	1105	0.133	1.978	1.0
4.0	153	0.0164	90.0	120.8	1006	1127	0.220	1.862	4.0
10.0	193	0.0166	38.42	161.2	986	1143	0.2836	1.788	10.0
14.0	209	0.0167	28.0	177.6	972	1149	0.308	1.761	14.0
100	328	0.0177	4.4	298.5	889	1188	0.474	1.603	100

Reading directly across, to the right, from the 10 psia saturation pressure field, we see the listed saturation temperature of 193°F.

Answer: Therefore, the boiling point for the water at the listed altitude of 10,000 ft would be **193°F.**

2. In problem (1), if the objective is just to heat the water close to the boiling point, will the boiler consume more or less fuel than it did when it was located at the sea level?

Solution/Answer:

In order to assess if it would take more heat or less heat to bring the water to a boil at the higher altitude of 10,000 ft, we need to compare the difference between the enthalpy values of water, in saturated liquid phase, at the two different altitudes and pressures.

In other words, we need to compare the enthalpy, $\mathbf{h_L}$, at 14 psia (pressure at sea level) and 10 psia (pressure at 10,000 ft altitude).

Based on the saturated steam table excerpt in Table A6-2:

$\mathbf{h_{l,\ at\ 10\ psia}}$ = 161.2 Btu/lbm, and
$\mathbf{h_{l,\ at\ 14\ psia}}$ = 177.6 Btu/lbm

Answer: As apparent from the heat content values or enthalpy values above, the water that comes to a boil at the higher altitude of 10,000 ft will do so with **less heat content**, or would require less heat. Conversely, water that comes to a boil at higher pressure would do so at higher enthalpy value, or greater heat content.

3. Answer the following questions for water at a temperature of 193°F and pressure of 10 psia:
   a) Heat content for saturated water.
   b) Specific heat (Btu/lbm) required to evaporate the water.
   c) If the water were evaporated, what would the saturated vapor heat content be?
   d) What state or phase would the water be in at the stated temperature and pressure?
   e) What would the entropy of the water be while it is in saturated liquid phase?
   f) What would the specific volume of the water be while it is in saturated vapor phase?
   g) What would the phase of the water if the pressure is increased to 20 psia while keeping the temperature constant at 193°F?

Solution/Answer:

a) At193°F and pressure of 10 psia, the water is in saturated liquid form. According to saturated steam table excerpt in Table A6-2, the saturated water enthalpy at 193°F and pressure of 10 psia is 161.2 Btu/lbm. This value is listed under column labeled h_L, in Table A6-2, in the row representing temperature of193°F and pressure of 10 psia.

Therefore, the enthalpy or heat content for saturated water, at the given temperature and pressure, is **161.2 Btu/lbm.**

b) The specific heat, in Btu/lbm, required to evaporate the water from saturated liquid phase to saturated vapor phase, is represented by the term $\mathbf{h_{fg}}$. The value of h_{fg}, for saturated water at 193°F and a pressure of 10 psia, as read from Table A6-2, is 986 Btu/lbm. See circled values in Table A6-2.

Answer: $\mathbf{h_{fg\ at193°F\ and\ 10\ psia}}$ **= 986 Btu/lbm**

c) Saturated vapor heat content if the water were evaporated would be the value for $\mathbf{h_{v\ at193°F\ and\ 10\ psia}}$, and from Table A6-2 this value is **1143 Btu/lbm.**

d) The water would be in **saturated liquid phase** at the stated temperature and pressure. All stated saturation temperatures and pressures, in the saturated steam tables, represent the current state of water in saturated liquid phase.

e) The entropy of water at 193°F and a pressure of 10 psia, in saturated liquid phase, as read from Table A6-2 would be s_L= **0.2836 Btu/(lbm.°R).**
Note that the s_L value is retrieved form the table and not the s_V value. This is because the problem statement specifies the ***liquid phase.***

f) The specific volume of water at 193°F and a pressure of 10 psia, in saturated vapor phase, as read from Table A6-2 would be v_V = 38.42 **ft³/lbm**. Note that the v_V value is retrieved form the table and not the v_L value. This is because the problem statement specifies the ***vapor phase.***

g) The phase of the water if the pressure is increased to 20 psia while keeping the temperature constant at 193°F can be determined through the saturated team table excerpt in Table A6-2. As established earlier in part (d), the current phase of the water is **saturated liquid**. If, however, the pressure is doubled to 20 psia, according to Table A6-2, it would take well over 209 °F - and additional heat - to reinstate the water into the saturated water phase. Therefore, the water at 20 psia and 193°F would be in **subcooled phase or state.**

CHAPTER 7—SELF ASSESSMENT PROBLEMS & QUESTIONS

1. Why is the efficiency of this power plant in Case Study 7-1 rather low (17%)?

Solution/Answer:

The efficiency of the power station is low (17%) because of the fact that the working fluid is introduced into the system as -10°C ice. If the working fluid were room temperature water – or return condensate from the discharge side of the turbine, as is the case in Self-Assessment problem number 4—the overall system efficiency would be substantially higher.

2. Using the steam tables in Appendix B and the Double Interpolation Method described in Case Study 7-1, US Unit Version, determine the exact enthalpy of a superheated steam at a pressure of 400 psia and temperature of 950°F.

Solution:

As apparent from the superheated steam tables in Appendix B, the enthalpy value for 400 psia and 950°F is not readily available and, therefore, double interpolation must be conducted between the enthalpy values given for 360 psia, 900°F, and 450 psia, 1000°F, to derive $\mathbf{h}_{\text{400 psia and 950°F}}$. The double interpolation approach, as applied here, will entail three steps.

First step involves determination of $\mathbf{h}_{\text{400 psia and 900°F}}$, the enthalpy value at 400 psia and 900°F. The enthalpy values available and used in this interpolation step are circled in **Table A7-1.** The following formula sums up the mathematical approach to this first step:

$$\begin{aligned}\mathbf{h}_{\text{400 psia and 900°F}} &= ((\mathbf{h}_{\text{360 psia and 900°F}} - \mathbf{h}_{\text{450 psia and 900°F}})/ \\ &\quad (450\text{ psia} - 360\text{ psia})) \cdot (450\text{ psia-}400\text{ psia}) + \mathbf{h}_{\text{450 psia and 900°F}}\end{aligned}$$

Substituting enthalpy values and other given data from superheated steam table excerpt, shown in Table A7-1:

$$\begin{aligned}\mathbf{h}_{\text{400 psia and 900°F}} &= ((1471.7 - 1468.6)/(450\text{ psia} - 360\text{ psia})).(450\text{ psia-}400 \\ &\quad \text{psia}) + 1468.6 \\ &= 1470.32\text{ Btu/lbm}\end{aligned}$$

Second step involves determination of $\mathbf{h}_{\text{400 psia and 1000°F}}$, the enthalpy value at 400 psia and 1000°F. The enthalpy values available and used in this interpolation step are circled in **Table A7-1.** The following formula sums up the mathematical approach to this first step:

$$\begin{aligned}\mathbf{h}_{\text{400 psia and 1000°F}} &= ((\mathbf{h}_{\text{360 psia and 1000°F}} - \mathbf{h}_{\text{450 psia and 1000°F}})/(450\text{ psia} - \\ &\quad 360\text{ psia})).(450\text{ psia-}400\text{ psia}) + \mathbf{h}_{\text{450 psia and 1000°F}}\end{aligned}$$

Substituting enthalpy values and other given data from superheated steam table excerpt, shown in Table A7-1:

Table A7-1. Superheated Steam Table Excerpt, US/Imperial Units

Properties of Superheated Steam US/Imperial Units					
Abs. Press. psia (Sat. T, °F)		Temp. °F	Note: v is in ft³/lbm, *h* is in BTU/lbm and ***s*** is in BTU/(lbm-°R)		
		500	700	900	1000
260 (404.45)	*v*	2.062	2.5818	3.0683	3.3065
	h	1262.5	1370.8	1475.2	1527.8
	s	1.5901	1.6928	1.7758	1.8132
360 (434.43)	*v*	1.446	1.8429	2.2028	2.3774
	h	1250.6	1365.2	1471.7	1525
	s	1.5446	1.6533	1.7381	1.7758
450 (456.32)	*v*	1.1232	1.4584	1.7526	1.8942
	h	1238.9	1360	1468.6	1522.4
	s	1.5103	1.6253	1.7117	1.7499
600 (486.25)	*v*			1.3023	1.411
	h			1463.2	1518
	s			1.577	1.7159

$\mathbf{h_{400\ psia\ and\ 1000°F}}$

= ((1525 Btu/lbm – 1522.4 Btu/lbm)/(450 psia - 360 psia)). (450 psia-400 psia) + 1522.4 Btu/lbm

= 1523.84 Btu/lbm

The final step in the double interpolation process, as applied in this case, involves interpolating between $\mathbf{h_{400\ psia\ and\ 1000°F}}$ and $\mathbf{h_{400\ psia\ and\ 900°F}}$, the enthalpy values derived in the first two steps above, to obtain the desired final enthalpy $\mathbf{h_{400\ psia\ and\ 950°F}}$.

The formula for this final step is as follows:

$$h_{400\ \text{psia and}\ 950°F} = ((h_{400\ \text{psia and}\ 1000°F} - h_{400\ \text{psia and}\ 900°F})/(1000°F - 900°F)) \cdot (950°F - 900°F) + h_{400\ \text{psia and}\ 900°F}$$

Substituting enthalpy values derived in the first two steps above:

$$h_{400\ \text{psia and}\ 950°F} = ((1523.84\ \text{Btu/lbm} - 1470.32\ \text{Btu/lbm})/(1000°F - 900°F)) \cdot (950°F - 900°F) + 1470.32\ \text{Btu/lbm}$$
$$= 1497.08\ \text{Btu/lbm}$$

3. In Case Study 7-1, as an energy engineer you have been retained by Station Zebra to explore or develop an alternative integrated steam turbine and electric power generating system that is capable of generating 10 MW of power with only 6 truck loads, or 54,432 kg, of ice per hour. With all other parameters the same as in the original Case Study 7-1 scenario, determine the total heat flow rate, in kJ/hr, needed to produce 10 MW of electrical power.

Solution:

This problem requires accounting for heat added during each of the five (5) stages of the overall process, with a working fluid mass flow rate of 6 truck loads, or 54,432 kg, per hour. Therefore, the solution is divided into five sub-parts, each involving either a sensible or a latent heat calculation, based on the entry and exit temperature and phase status.

Table 7-2 lists specific heat for water and ice. These heat values will be used in the sensible heat calculations. Table 7-3 lists latent heat values for water. These values will be used to compute the latent heats associated with stages that involve phase transformation.

(i) Calculate the heat required to heat the ice from -10°C to 0°C. Since there is no change in phase involved, the entire heat absorbed by the ice (working substance) in this stage would be sensible heat.

First stage of the overall power generating system is illustrated in Figure A7-4.

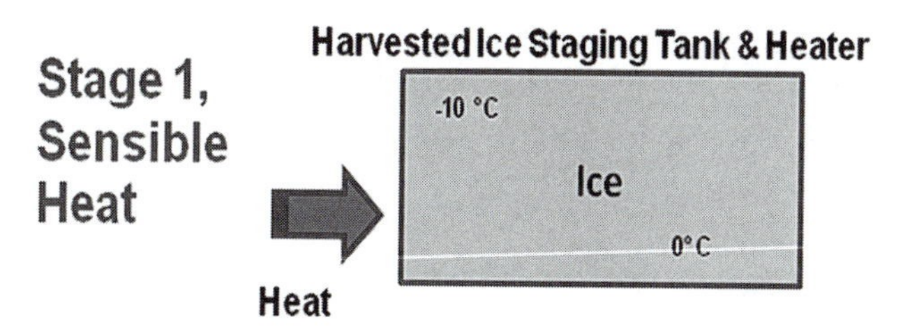

Figure A7-4. Case Study 7-1 Stage 1 Sensible Heat Calculation

Given:

$T_i = -10\ °C$
$T_f = 0\ °C$
$c_{ice} = 2.1\ kJ/kg.\ °K$ {Table 7-3}

Utilizing the given information:

$$\Delta T = T_f - T_i$$
$$\therefore\ \Delta T = 0 - (-10\ °C)$$
$$= +10\ °C$$

Since **ΔT** represents the *change* in temperature and not a specific absolute temperature,

$$\therefore\ \Delta T = +10\ °K$$

Mathematical relationship between sensible heat, mass of the working substance, specific heat of the working substance and change in temperature can be stated as:

$$Q_{s(heat\ ice)} = m \cdot c_{ice} \cdot \Delta T \qquad \text{Eq. 7-14}$$

And,

$$\dot{Q}_{s(heat\ ice)} = \dot{m} \cdot c_{ice} \cdot \Delta T \qquad \text{Eq. 7-15}$$

Where,

$Q_{s(heat\ ice)}$ = Sensible heat required to heat the ice over ΔT
$\dot{Q}_{s(heat\ ice)}$ = Sensible heat flow rate required to heat the ice over ΔT

$\mathbf{m}$ = Mass of ice being heated
c_{ice} = Specific heat of ice = **2.1 kJ/kg. °K**
ΔT = Change in temperature, in **°C or °K**
$\dot{m}$ = Mass flow rate of water/ice
= 60 tons/hr
= (60 tons/hr) • (907.2 kg/ton)
= **54,432 kg/hr**

Then, by application of **Eq. 7-15**:

$$\dot{Q}_{s(heat\ ice)} = \mathbf{(54{,}432\ kg/hr)\ .\ (2.1\ kJ/kg.\ °K)\ .\ (10\ °K)}$$

Or,

$$\dot{Q}_{s(heat\ ice)} = \mathbf{1{,}143{,}072\ kJ/hr}$$

Since there are 1.055 kJ per Btu,

$$\dot{Q}_{s(heat\ ice)} = \mathbf{(1{,}143{,}072\ kJ/hr)/(1.055\ kJ/Btu)}$$

Or,

$$\dot{Q}_{s(heat\ ice)} = \mathbf{1{,}083{,}481\ Btu/hr}$$

(ii) Calculate the heat required to melt the ice at 0°C. Since change in phase *is* involved in this case, the heat absorbed by the ice (working substance) in this stage would be latent heat.

The 2nd stage of the overall power generating system is illustrated in Figure A7-5.

Mathematical relationship between latent heat, mass of the working substance, and the heat of fusion of ice can be stated as:

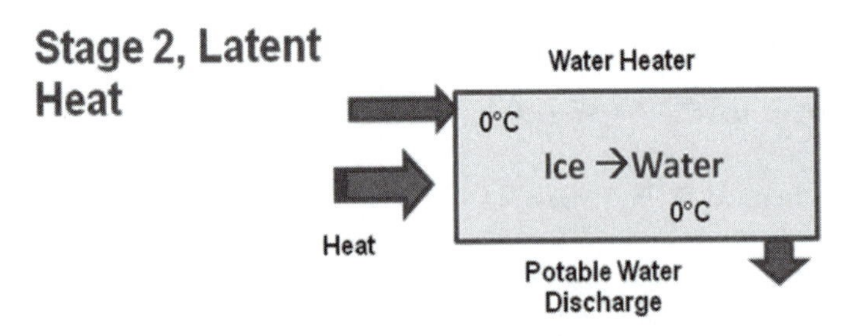

Figure A7-5. Case Study 7-1 Stage 2 Latent Heat Calculation

$$\mathbf{Q_{l(latent\ ice)} = h_{sl\ (ice)} \cdot m} \qquad \textbf{Eq. 7-16}$$

And,

$$\dot{Q}_{l(latent\ ice)} = h_{sl\ (ice)} \cdot \dot{m} \qquad \textbf{Eq. 7-17}$$

Where,

$Q_{l(latent\ ice)}$ = Latent heat required to melt a specific mass of ice, isothermally

$\dot{Q}_{l(latent\ ice)}$ = Latent heat flow rate required to melt a specific mass of ice, isothermally, over a period of time

m = Mass of ice being melted

$\dot{m}$ = Mass flow rate of water/ice
= 60 tons/hr
= (60 tons/hr) • (907.2 kg/ton)
= **54,432 kg/hr,** same as part (a) (i)

$h_{sl\ (ice)}$ = Heat of fusion for Ice
= **333.5 kJ/kg** {Table 7-3}

Then, by application of **Eq. 7-17**:

$$\dot{Q}_{l(latent\ ice)} = h_{sl\ (ice)} \cdot \dot{m}$$

$$\dot{Q}_{l(latent\ ice)} = \mathbf{(333.5\ kJ/kg).\ (54{,}432\ kg/hr)}$$

$$\dot{Q}_{l(latent\ ice)} = \mathbf{18{,}153{,}072\ kJ/hr}$$

Since there are 1.055 kJ per Btu,

$$\dot{Q}_{l(latent\ ice)} = \mathbf{(18{,}153{,}072\ kJ/hr)/(1.055\ kJ/Btu)}$$

Or,

$$\dot{Q}_{l(latent\ ice)} = \mathbf{17{,}206{,}703\ Btu/hr}$$

Note that the specific heat required to melt ice is called heat of fusion because of the fact that the water molecules come closer together as heat is added in the melting process. The water molecules are held apart at specific distances in the crystallographic structure of solid ice. The heat of fusion allows the molecules to overcome the crystallographic forces and "fuse" to form liquid water. This also explains why the density of water is higher than the density of ice.

(iii) Calculate the heat reqd. to heat the water from 0°C to 100°C

The 3rd stage of the overall power generating system is illustrated in Figure A7-6. Since no phase change is involved in this stage, the heat absorbed by the water in this stage would be sensible heat.

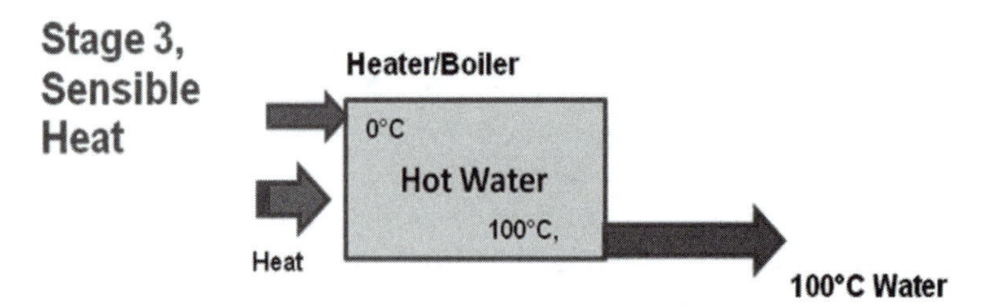

Figure A7-6. Case Study 7-1 Stage 3 Sensible Heat Calculation

Given:

$T_i = 0$ °C

$T_f = 100$ °C

$c_{p\text{-water}} = 4.19$ kJ/kg. °K {Table 7-3}

Utilizing the given information:

$$\Delta T = T_f - T_i$$

$$\therefore \Delta T = 100\ °C - 0°C$$
$$= 100\ °C$$

Since **ΔT** represents the ***change*** in temperature and not a specific absolute temperature,

$$\therefore \Delta T = 100\ °K$$

Mathematical relationship between sensible heat, mass of the working substance, specific heat of water (the working substance), and change in temperature can be stated as:

$$Q_{s(water)} = m \cdot c_{p\text{-water}} \cdot \Delta T \quad \textbf{Eq. 7-18}$$

And,

$$\dot{Q}_{s(water)} = \dot{m} \cdot c_{p\text{-water}} \cdot \Delta T \quad \textbf{Eq. 7-19}$$

Where,

$\mathbf{Q_{s(water)}}$	=	Sensible heat required to heat the water over ΔT
$\dot{Q}_{\mathbf{s(water)}}$	=	Sensible heat ***flow rate*** required to heat the water over ΔT
m	=	Mass of water being heated
$\mathbf{c_{p\text{-}water}}$	=	Specific heat of water = **4.19 kJ/kg. °K**
$\dot{\mathbf{m}}$	=	Mass flow rate of water = **54,432 kg/hr**, as calculated in part (a)
$\mathbf{\Delta T}$	=	Change in temperature, in **°C or °K**

Then, by applying **Eq. 7-19**:

$$\dot{Q}_{s(water)} = \dot{m} \cdot c_{p\text{-}water} \cdot \Delta T$$

$$\dot{Q}_{s(water)} = (54{,}432 \text{ kg/hr}) \cdot (4.19 \text{ kJ/kg. °K}) \,.\, (100 \text{ °K})$$

$$\dot{Q}_{s(water)} = 22{,}807{,}008 \text{ kJ/hr}$$

Since there are 1.055 kJ per Btu,

$$\dot{Q}_{s(water)} = (22{,}807{,}008 \text{ kJ/hr})/(1.055 \text{ kJ/Btu})$$

Or,

$$\dot{Q}_{s(water)} = 21{,}618{,}017 \text{ Btu/hr}$$

(iv) Calculate the heat required to convert 100°C water to 100°C steam

The 4th stage of the overall power generating system is illustrated in Figure A7-7. Since change in phase *is* involved in this case, the heat absorbed by the water in this stage would be latent heat.

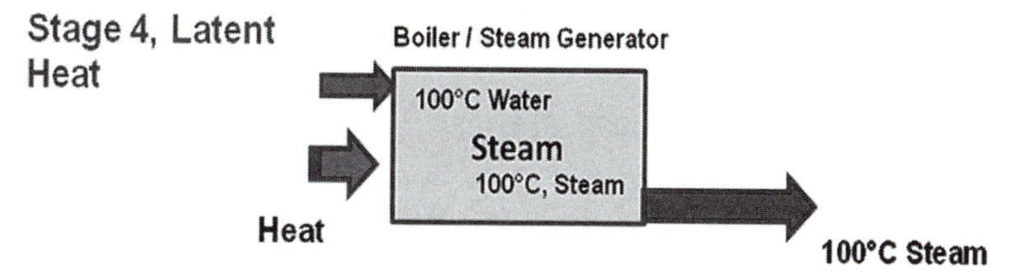

Figure A7-7. Case Study 7-1 Stage 4 Latent Heat Calculation

Mathematical relationship between latent heat of vaporization for water, $\mathbf{h_{fg(water)}}$, mass of the water, and the total heat of vaporization of water, $\mathbf{Q_{l(latent\ water)}}$, can be stated as:

$$\mathbf{Q_{l(latent\ water)} = h_{fg\ (water)} \cdot m} \qquad \textbf{Eq. 7-20}$$

And,

$$\mathbf{\dot{Q}_{l(latent\ water)} = h_{fg\ (water)} \cdot \dot{m}} \qquad \textbf{Eq. 7-21}$$

Where,

$\mathbf{Q_{l(latent\ water)}}$ = Latent heat of vaporization of water required to evaporate a specific mass of water, isothermally

$\mathbf{\dot{Q}_{l(latent\ water)}}$ = Latent heat of vaporization ***flow rate*** required to evaporate a specific mass of water, isothermally, over a given period of time

$\mathbf{m}$ = Mass of water being evaporated

$\mathbf{\dot{m}}$ = Mass flow rate of water
= 60 tons/hr
= (60 tons/hr) • (907.2 kg/ton)
= **54,432 kg/hr,** same as part (a) (i)

$\mathbf{h_{fg\ (water)}}$ = latent heat of vaporization for water
= **2257 kJ/kg** {From the steam tables and Table 7-3}

Then, by application of **Eq. 7-21**:

$$\mathbf{\dot{Q}_{l(latent\ water)} = h_{fg\ (water)} \cdot \dot{m}}$$

$$\mathbf{\dot{Q}_{l(latent\ water)} = (2257\ kJ/kg) \cdot (54{,}432\ kg/hr)}$$

$$\mathbf{\dot{Q}_{l(latent\ water)} = 122{,}853{,}024\ kJ/hr}$$

Since there are 1.055 kJ per Btu,

$$\mathbf{\dot{Q}_{l(latent\ water)} = (122{,}853{,}024\ kJ/hr)/(1.055\ kJ/Btu)}$$

Or,

$$\mathbf{\dot{Q}_{l(latent\ water)} = 116{,}448{,}364\ Btu/hr}$$

(v) Calculate the heat reqd. to heat the steam from 100°C, 1-atm (102 KPa, or 1-bar) to 500°C, 2.5 MPa superheated steam

The 5th stage of the overall power generating system is illustrated in Figure A7-8. Since this stage involves no phase change, the heat absorbed by the steam is sensible heat.

In superheated steam phase, the heat required to raise the temperature and pressure of the steam can be determined using the enthalpy difference between the initial and final conditions.

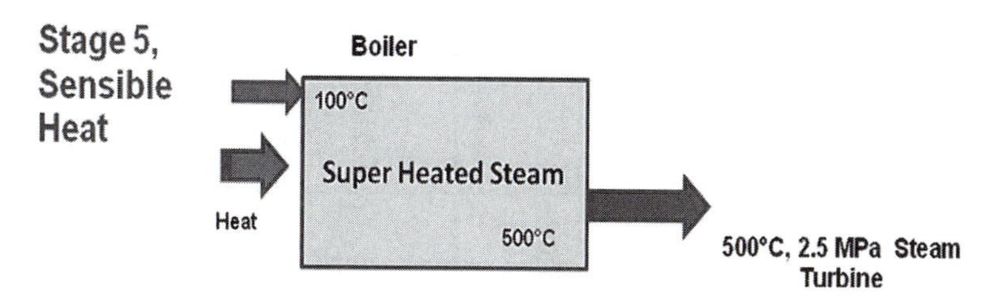

Figure A7-8. Case Study 7-1 Stage 5 Sensible Heat Calculation

Given:

T_i **= 100°C**

P_i **= 1-Atm.** Note: At 100°C, the saturation pressure is 1-Atm, 1-Bar, or 102 kPa

T_f **= 500°C**

P_f **= 2.5 MPa**

For the initial and final temperature and pressure conditions stated above, the enthalpy values, as read from saturated steam table excerpt in Table 7-4a and the superheated steam table excerpt in Table 7-4, are as follows:

h_i **= 2676 kJ/kg at 100 °C, 1-Atm**
h_f **= 3462 kJ/kg at 500 °C, 2.5 MPa**

Equations for determining the heat required to boost the steam from **100°C, 1-Atm** to **500°C, 2.5 MPa** are as follows:

$$\Delta Q_{steam} = (h_f - h_i) \cdot m \quad \textbf{Eq. 7-22}$$

$$\Delta \dot{Q}_{steam} = (h_f - h_i) \cdot \dot{m} \quad \textbf{Eq. 7-23}$$

Table A7-4a. Excerpt, Saturated Steam Table, SI Units

Properties of Saturated Steam By Temperature Metric/SI Units										
Temp. °C	Abs. Press. MPa	Specific Volume m^3/kg		Enthalpy kJ/kg			Entropy kJ/kg		Temp. °C
		Sat. Liquid v_L	Sat. Vapor v_V	Sat. Liquid h_L	Evap. h_{fg}	Sat. Vapor h_V	Sat. Liquid s_L	Sat. Vapor s_V	
20	0.002339	0.0010018	57.7610	83.920	2454.1	2537.5	0.2965	8.6661	20
50	0.012351	0.0010121	12.0280	209.34	2382.7	2591.3	0.7038	8.0749	50
100	0.101420	0.0010435	1.6719	419.10	2257.0	2675.6	1.3070	7.3541	100
200	1.554700	0.0011565	0.1272	852.39	1940.7	2792.1	2.3308	6.4303	200

Where,

ΔQ_{steam} = Addition of heat required for a specific change in enthalpy

$\dot{Q}_{steam}$ = Rate of addition of heat for a specific change in enthalpy

h_i = Initial enthalpy

h_f = Final enthalpy

m = Mass of steam being heated

$\dot{m}$ = Mass flow rate of steam as calculated in part (a) of this case study

= **54,432 kg/hr** {From Part (a)}

Then, by applying **Eq. 7-23**:

$$\dot{Q}_{steam} = (h_f - h_i) \cdot \dot{m}$$

$$\dot{Q}_{steam} = (3462\ kJ/kg - 2676\ kJ/kg) \cdot (54{,}432\ kg/hr)$$

$$\dot{Q}_{steam} = 42{,}783{,}552\ kJ/hr$$

Since there are 1.055 kJ per Btu,

$$\dot{Q}_{s(water)} = (42{,}783{,}552\ kJ/hr)/(1.055\ kJ/Btu)$$

Or,

$$\dot{Q}_{s(water)} = 40{,}553{,}130\ Btu/hr$$

After assessing the heat added, per hour, during each of the five (5) stages of the steam generation process, add all of the heat addition rates to compile the total heat addition rate for the power generating station. The tallying of total heat is performed in Btu's/hr as well as kJ/hr.

Total Heat Addition Rate in kJ/hr:

Total Heat Required to Generate 500°C, 2.5 MPa steam from -10°C Ice, at 54,432 kg/hr

= 1,143,072 kJ/hr + 18,153,072 kJ/hr + 22,807,008 kJ/hr + 122,853,024 kJ/hr + 42,783,552 kJ/hr

= **207,739,728 kJ/hr**

Total Heat Addition Rate in Btu's/hr:

Total Heat Required to Generate 500°C, 2.5 MPa steam from -10°C Ice, at 54,432 kg/hr

= 1,083,481 Btu/hr +17,206,703 Btu/hr +21,618,017 Btu/hr +116,448,364 Btu/hr +40,553,130 Btu/hr

= 196,909,695 Btu/hr

4. If all of the working fluid, or steam, discharged from the turbine in Case Study 7-1 is reclaimed, reheated and returned to the turbine, what would be the overall system efficiency? Assume that the mass flow rate is **58,860 kg/hr**, or **65 tons per hour**.

Solution Strategy:

Energy and process flow pertaining to this special case scenario of Case Study 1 is depicted in Figure A7-9. In order to derive the efficiency for this scenario where all of the steam discharged from the turbine is reclaimed and used as working fluid fed into the boiler in the last, superheating, stage, we need to determine the rate of heat addition required to raise the temperature from 150°C to 500°C and the pressure from 50 kPa to 2.5 MPa.

This rate of addition of heat can be determined using the following formula:

$$\dot{Q}_{steam} = (h_f - h_i) \cdot \dot{m} \qquad \textbf{Eq. 7-23}$$

Once $\dot{Q}_{steam}$ is determined, we can convert it into equivalent power (kW or MW) units for computation of efficiency through the following formula:

Total Station Energy Efficiency, in Percent
= Power Output /Power Input • 100

Or,

Total Station Energy Efficiency, in Percent
= (Power Output in MW /$\dot{Q}_{steam\ in\ MW}$) • 100

Note: With the exception of the provisions stipulated in the problem statement above, all of the pertinent given data from Case Study 1 remains the same, as stated below:

Solution:
Given:

$\mathbf{P_{shaft} = (10\ MW)\ /\eta_g}$
$= (10\ MW)\ /(0.9)$
$= 11.11 \times 10^6\ W$
$= 11.11 \times 10^6\ J/s$

$\dot{\mathbf{m}} = 58{,}860\ kg/hr$

$\mathbf{h_i} = 2780\ kJ/kg = 2780 \times 10^3\ J/kg$ {See Table A7-4}

$\mathbf{h_f} = 3462\ kJ/kg = 3462 \times 10^3\ J/kg$ {See Table A7-4}

Then, by applying Eq. 7-23:

$\dot{Q}_{\ steam} = \mathbf{(h_f - h_i)\ .\ \dot{m}}$
$= (3462\ kJ/kg - 2780\ kJ/kg)\ .\ 58{,}860\ kg/hr$
$= \mathbf{40{,}142{,}520\ kJ/hr}$

Or,

$\dot{Q}_{\ steam\ in\ kW} = (40{,}142{,}520\ kJ/hr)\ /(3600\ sec/hr)$
$= 11{,}151\ kJ/sec$

Or, since 1J/sec = 1W, and 1kJ/sec = 1kW,

$\dot{Q}_{\ steam\ in\ kW} = = 11{,}151\ kW$

Or,

$\dot{Q}_{\ steam\ in\ MW} = (11{,}151\ kW)/(1000kW/MW)$
$= 11.15\ MW$

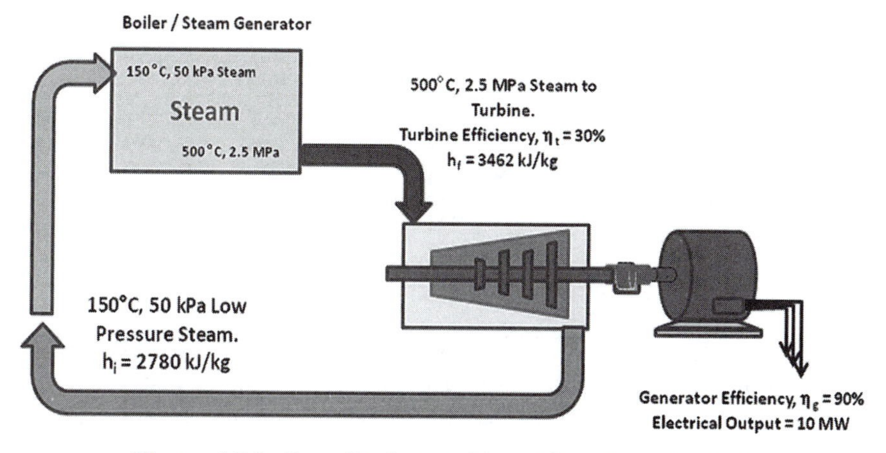

Figure A7-9. Case Study 7-1, Mass Flow Rate Analysis

Then,

Total Station Energy Efficiency, in Percent
= Power Output /Power Input • 100
= 10 MW/11.15MW
= 89.68%

Table A7-4. Excerpt, Superheated Steam Table, SI.

Properties of Superheated Steam Metric/SI Units						
Abs. Press. MPa (Sat. T, °C)		Temp. °C				
		150	300	500	650	800
0.05 (81.33)	n	3.889	5.284	7.134		
	h	2780.1	3075.5	3488.7		
	s	7.9401	8.5373	9.1546		
0.1 (99.61)	n	1.9367	2.6398	3.5656		
	h	2776.6	3074.5	3488.1		
	s	7.6147	8.2171	8.8361		
1.0 (179.89)	n		0.2580	0.3541	0.4245	0.4944
	h		3051.7	3479.0	3810.5	4156.2
	s		7.1247	7.7640	8.1557	8.5024
2.5 (223.99)	n		0.0989	0.13998	0.1623	0.1896
	h		3008.8	3462.1	3799.7	4148.9
	s		6.6438	7.3234	7.7056	8.0559
3.0 (233.86)	n		0.0812	0.1162	0.1405	0.1642
	h		2994.3	3457.0	3797.0	4147.0
	s		6.5412	7.2356	7.6373	7.9885
4.0 (250.36)	n		0.0589	0.0864	0.1049	0.1229
	h		2961.7	3445.8	3790.2	4142.5
	s		6.3638	7.0919	7.4989	7.8523

CHAPTER 8—SOLUTIONS FOR SELF ASSESSMENT PROBLEMS AND QUESTIONS:

1. An ideal heat engine always includes a boiler with superheating function.

Answer: **B. False.**

- Heat engine equipped with superheating function, albeit common, is just one type of heat engine.

2. In the heat engine represented by the enthalpy versus entropy graph in Figure 8.22, the heat is added to working fluid in:

Answer: **(iv) Steam generation stage and the Steam superheating stage**

- This is evident from Figure A8-1. Path 2→3→4.

3. In the heat engine represented by the enthalpy versus entropy graph in Figure 8-22, the energy contained in the superheated working fluid is converted into the rotational kinetic energy in:

Answer: (ii) Process transition from point 4 to 5.

- Figure A8-1 is a duplicate of Figure 8-22, referred to in the problem statement. As evident from Figure A8-1, path 4 to 5 represents the conversion of enthalpy contained in the superheated steam into mechanical kinetic energy of the turbine.

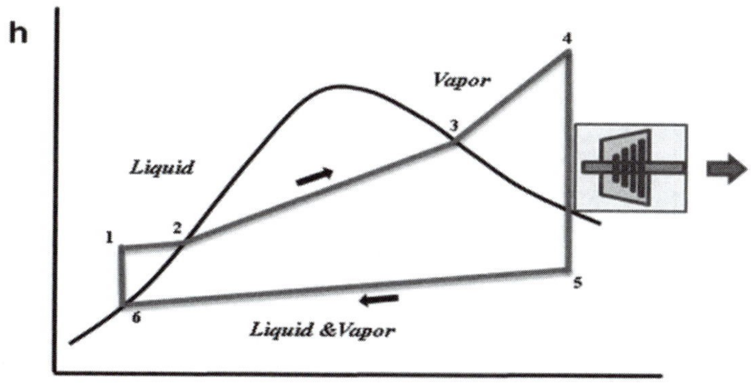

Figure A8-1. Heat Cycle in a Rankin Engine with Superheat, h versus s.

4. A thermodynamic system consists of a Rankin engine with superheat function. The enthalpy versus entropy graph for this system is shown in Figure A8-2.

The mass flow rate of the system or working fluid is 100 lbm/sec. Answer the following questions based on the data provided:

a) If the enthalpy of the fluid is approx. 1850 Btu/lbm and the entropy of the system is approximately 1.8900 Btu/(lbm-°R), what phase or stage is the system in:

Note: Figure 8-33, referred to in the problem statement is duplicated in Figure A8-2, with pertinent annotations, for solution illustration purposes.

Solution/Answer: (i) Work producing turbine stage

- The point representing an enthalpy of approximately 1850 Btu/lbm and the entropy of approximately 1.8900 Btu/(lbm-°R) is labeled as "a" in the Figure A8-2 enthalpy—entropy diagram.

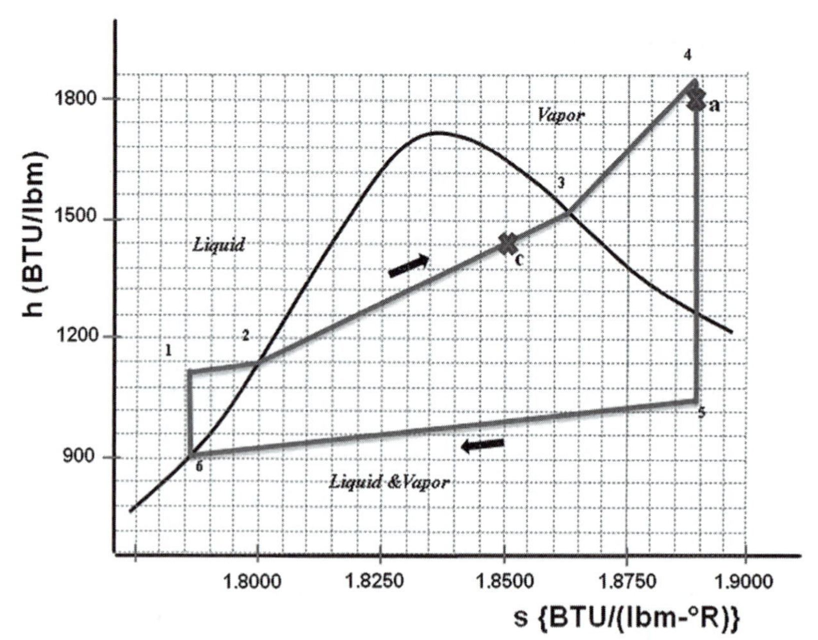

Figure A8-2. Heat Cycle in a Rankin Engine with Superheat, h versus υ.

This point is clearly on thermodynamic transition path 4 to 5. This path represents the work producing turbine stage.

b) If the enthalpy of the subcooled water entering the boiler is **900 Btu/lbm** and the enthalpy of the water at a downstream point in the boiler is **1080 Btu/lbm**, what is approximate amount of heat added in **MMBtu per hour**? Assume that there is no heat loss.

Answer: **(iii) 64.8 MMBtu/hr**

Solution:

Since we are allowed to assume that there is not heat loss, the difference between the enthalpy of the subcooled water entering the boiler is and the enthalpy of the water at a downstream point in the boiler would be equal to the heat energy added on per lbm basis. Also, the mass flow rate, $\dot{m}$, of the working fluid is given.

∴ **Heat energy delivered to the water,** in Btu's/sec

$Q = (h_f - h_i) \cdot (\dot{m})$

Where,

$\dot{m}$ = Mass flow rate of the working fluid or water
= **100 lbm/sec** {Given}

h_i = Enthalpy of steam entering the turbine
= **900 Btu/lbm 1860** {Given}

h_f = Enthalpy of steam discharged
= **1080 Btu/lbm** {Given}

Then, application of Eq. 8-2 would yield:

Heat energy delivered to the water, in Btu's/sec
= **(1080** Btu/lbm **– 900** Btu/lbm**) . (100 lbm/sec)**
= **18,000 Btu/sec**

Heat energy delivered to the water, in Btu's/hr
= **(18,000 Btu/sec). (3600 sec/hr.)**
= **64,800,000 Btu's/hr**

Since there are 1,000,000 Btu's per MMBtu,

Heat energy delivered to the water, in MMBtu's/hr

$$= \frac{64{,}800{,}000 \text{ Btu's/hr}}{1{,}000{,}000 \text{ Btu's}}$$

$$= \mathbf{64.8\ MMBtu/hr}$$

Therefore, the answer is:

(ii) 64.8 MMBtu/hr

Ancillary to part (b):

As apparent upon examination of Figure A8-2, the rise in enthalpy stated in part (b) - from 900 Btu/lbm to1080 Btu/lbm - occurs while the working fluid ***remains in subcooled realm***. If the final enthalpy, $\mathbf{h_f}$, were in the vapor-liquid mixture region, the solution for this part of the problem would have required more steps. This is because the enthalpy of the working fluid at final point would be a sum of the saturated vapor and liquid enthalpies, added in proportion determined by the humidity ratio, ω.

c) If the enthalpy of the working fluid is 1440 Btu/lbm and the entropy is 1.8500 Btu/(lbm-°R), what is the phase of the working fluid?

Answer: (iii) A mixture of vapor and liquid

To address this question, we must locate the point on the graph where **h** = 1440 Btu/lbm and **s** = 1.8500 Btu/(lbm-°R). This point is labeled as point "**c**" on the enthalpy versus entropy graph in Figure A8-2. Point **c** lies in the region that falls between **saturated liquid line** and the **saturated vapor line**. In other words, at point **c**, the working fluid is in a phase that consists of a mixture of liquid and vapor.

CHAPTER 9—SELF ASSESSMENT PROBLEMS & QUESTIONS

1. A nozzle is fed from a superheated steam reservoir. The superheated steam in the reservoir is at 500°C (932°F) and 2.0 MPa (290 psia). The duct or hose connecting the nozzle to the reservoir is short and the frictional head loss in the hose is negligible. Based on these practical assumptions, the velocity of the superheated steam in the

hose can be neglected. The steam exits the nozzle at 1.0 bar (14.5 psia) and 95% quality. Determine the exit velocity of the steam at the nozzle in SI units.

Solution:
Given:

T_o = 500°C
P_o = 2.0 MPa
$v_1 = v_0 = 0$
x = Quality = 95%
P_2 = 1.0 bar or 14.5 psia

SI Unit System:

Apply Eq. 9-4 to calculate the exit velocity of the superheated steam in the SI units:

$$v_2 = \sqrt{2(h_0 - h_2)} \qquad \{\text{SI Unit System}\} \qquad \textbf{Eq. 9-4}$$

From the steam tables in Appendix B, in the SI units:
h_o = 3468 kJ/kg

From Mollier diagram, in **Figure A9-1**:
h_2 = 2550 kJ/kg

Note: The enthalpies are converted into J/kg, from kJ/kg because Eq. 9-4 is premised on Joules and not kilo Joules.

Then, by applying **Eq. 9-4**:

$$v_2 = \sqrt{(2).(3468 - 2550 kJ/kg).(1000 J/kJ)}$$

$$v_2 = 1355 m/s$$

2. Solve Problem 1 in US Units. Use Mollier diagram for all enthalpy identification and compare the resulting steam speed with results from computation conducted in SI units, in Problem 1.

Solution:
Given:

T_o = 932°F
P_o = 290 psia

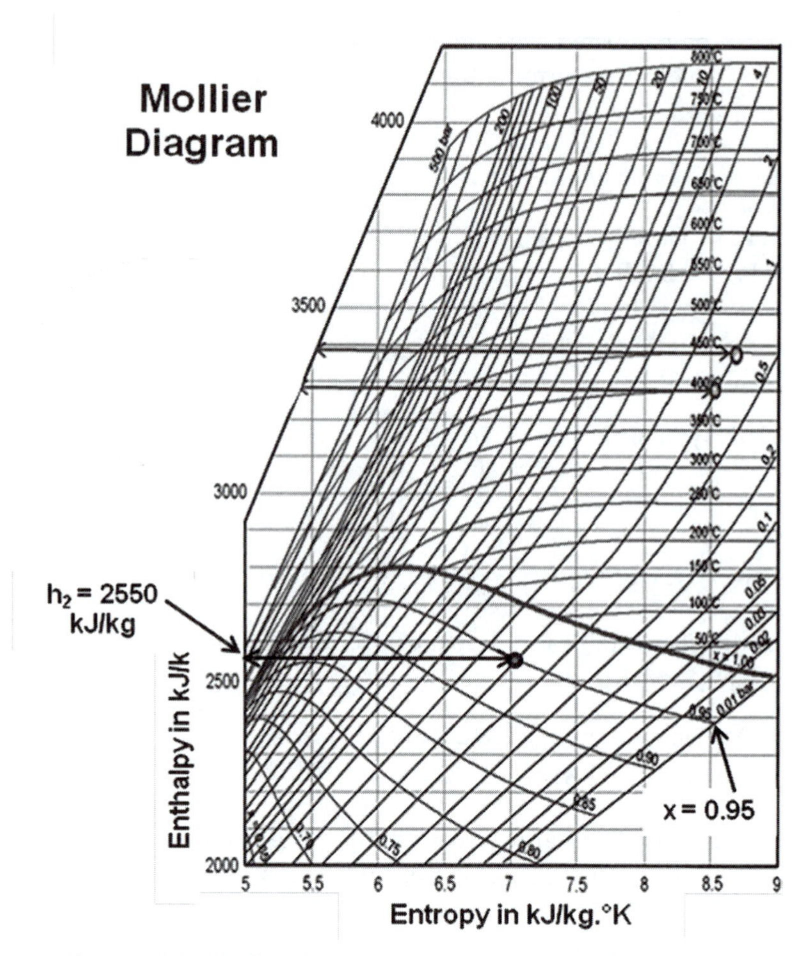

Figure A9-1. Mollier Diagram, Self Assessment Problem 1

$\mathbf{v_1 = v_0 = 0}$
$\mathbf{x}$ = Quality = 95%
$\mathbf{P_2}$ = 14.5 psia

US Unit System:

Apply **Eq. 9-5** do calculate the exit velocity of the superheated steam in the US units:

$$v_2 = \sqrt{2g_c J(h_0 - h_2)} \qquad \{\text{US Unit System}\} \qquad \textbf{Eq. 9-5}$$

From the Mollier Diagram, in Appendix B:

$\mathbf{h_o = 1490\ Btu/lbm}$
$\mathbf{h_2 = 1100Btu/lbm}$

Then, by applying **Eq. 9-5**:

$$v_2 = \sqrt{2.\left(32.2\frac{lbm-ft}{lbf-s^2}\right)\cdot\left(778\frac{ft-lbf}{Btu}\right)\cdot\left(1490-1100\frac{Btu}{lbm}\right)}$$

$$v_2 = 4420 ft/s$$

Comparison with the SI (Metric) $\mathbf{v_2}$ calculation:

$\mathbf{v_2}$ from SI (Metric) Unit calculation is **1355 m/s.** If this Metric speed is converted to US Units, through simple unit conversion, $\mathbf{v_2}$ in US units amounts to: **4442 ft/s**

$\mathbf{v_2}$ from US (Imperial) Unit calculation, as computed above is: **4420 ft/s**

The difference in nozzle speed calculated in the SI realm versus the US realm, is:

= **((4442 ft/s /4420 ft/s) – 1) . (100%)**
= **0.49%**

This verifies that while the US Unit method, utilizing Eq. 9-5, and the SI Unit method, utilizing Eq. 9-4, yields nozzle speed results that are slightly different, the difference is less than a percent.

3. The SFEE Equation 9-2 can be applied to compute the exit speed of gas, in high speed gas applications, under which of the following conditions?

A. When data are available in US units
B. When data are available in SI units
C. When the reservoir is large enough such that v0 = 0, applies.
D. Both B and C.
E. Both A and B.

Answer:

As explained in Chapter 9, SFEE Equation 9-2 can be applied to compute the exit speed of gas, in high speed gas applications, in the

SI Unit System, when the reservoir is large enough such that $\mathbf{v_0 = 0}$, applies. Therefore, the correct answer is "**D**."

4. Which of the following statements is true about shock waves?
   A. Shock waves require superheated steam
   B. Shock waves travel parallel to the direction of the flow of gas.
   C. Shock waves travel perpendicular to the direction of the flow of gas.
   D. Both A and B.

Answer: C.

Shock waves travel perpendicular to the direction of the flow of gas.

CHAPTER 10—SOLUTIONS

Problem 1

1. [*Note: Problem restated for reader's convenience*] In an environment that is estimated to contain, approximately, 400 kg of air, the dry-bulb is measured to be 40°C and the wet-bulb is at 27.3°C. Later, the air is cooled to 20°C and, in the process of lowering the dry-bulb temperature, the relative humidity drops to 47%. As an Energy Engineer, you are to perform the following psychrometric analysis on this system using the psychrometric chart in Figure 10-6:

a) Find the initial humidity ratio, ωf.
b) Find the final humidity ratio, ωf.
c) Find the total amount of heat removed.
d) Find the amount of sensible heat removed.
e) Find the amount of latent heat removed.
f) Find the final wet-bulb temperature.
g) Find the initial dew point.
h) Find the final dew point.
i) Find the amount of moisture condensed/removed.

As explained in Chapter 10 case study solutions, we need to begin the analyses associated with psychrometric problems with the identification or location of initial and final psychrometric points on the psychrometric chart.

Location of the initial psychrometric point in this problem can be established using the following two parameters associated with the point:

- Dry-bulb temperature of 40 °C
- Wet-bulb temperature of 27.3 °C

This point is shown on the psychrometric chart in Figure A10-1, as point A.

Location of the final psychrometric point can be established using the following two pieces of data associated with this point:

- Dry bulb temperature of 20°C
- Relative humidity of 47%.

Relative humidity line, representing an RH of 47% is placed through interpolation between the given 50% and 40% RH lines on the psychrometric chart. The final point would be located at the point of intersection of the 47% RH line and the vertical line representing the dry bulb temperature of 20°C. The final point, thus identified, is shown as point B on the psychrometric chart in Figure A10-1.

a) Find the initial humidity ratio, ω_i.

To determine the initial humidity ratio, draw a horizontal line from the initial point to the vertical humidity ratio scale, on the psychrometric chart as shown in Figure A10-1.

The point of intersection of this horizontal line and the humidity ratio scale represents the humidity ratio ω_i for the initial psychrometric point.

As read from the psychrometric chart in Figure A10-1:

ω_i = **0.018 kg of moisture per kg of dry air**

b) Find the final humidity ratio, ω_f.

Similar to part (a), the humidity ratio for the final psychrometric point can be determined by drawing a horizontal line from the final point to the vertical humidity ratio scale on the psychrometric chart in Figure A10-1.

The point of intersection of this horizontal line and the humidity ratio scale represents the humidity ratio, ω_f, for the final psychrometric point.

As read from the psychrometric chart in Figure A10-1:

ω_f = **0.007** kg of moisture per kg of dry air

c) Find the total amount of heat removed.

The first step is to identify the enthalpies, on the psychrometric chart, at the initial and final points. See Figure A10-1.

At the initial point, the dry-bulb temperature is 40°C, the wet-bulb is 27.3°C, and as shown on the psychrometric chart, h_i = 87 kJ/kg of dry air.

At the final point, dry-bulb is 20°C, with RH at 47%. The enthalpy at this point, h_f = 38 kJ/kg of dry air.

$$\therefore \Delta h = h_f - h_i$$
$$= 38 \text{ kJ/kg} - 87 \text{ kJ/kg}$$
$$= -49 \text{ kJ/kg of dry air.}$$

And,

$$\Delta Q = (\Delta h) \cdot m_{air} \qquad \text{Eq. 10-1}$$
$$= (-49 \text{ kJ/kg}) \cdot m_{air}$$

Where, the mass of *dry* air, m_{air}, needs to be derived through the given combined mass of moisture and air, 400 kg, **(Note: This is the mass of moist air at the initial point)** and the ***initial*** humidity ratio, ω_i. Humidity ratio is defined as:

ω = **mass of moisture (kg) /mass of dry air (kg)**

And, as determined from the psychrometric chart, earlier, in part (a):

ω = **0.018** kg of moisture per kg of dry air, at the initial point

$$\omega = m_{moisture} / m_{dry\ air}$$

Or,

$$\omega = (m_{moist\ air} - m_{dry\ air}) / m_{dry\ air}$$

Through algebraic rearrangement of this equation, we get:

$$\mathbf{(1 + \omega) = m_{moist\ air} / m_{dryair}}$$

Or,

$$\mathbf{m_{dryair} = mass\ of\ dry\ air = m_{moist\ air} / (1 + \omega)}$$

Where the total combined of mass of the moisture and the dry air is given as 400 kg.

$$\therefore m_{dry\ air} = 400\ kg/(1 + 0.018)$$
$$= \mathbf{407\ kg}$$

Then, by applying **Eq. 10-1**:

$$\mathbf{\Delta Q = Total\ Heat\ Removed}$$
$$= \mathbf{(\Delta h)\ .\ m_{dryair}}$$
$$= (-\ 49\ kJ/kg)\ .\ (407\ kg)$$

Or,

$$\mathbf{\Delta Q = Total\ Heat\ Removed = -\ 19{,}943\ kJ}$$

The negative sign, in the answer above, signifies that the heat is extracted or that it exits the system as the air conditioning process transitions from the initial psychrometric point to the final psychrometric point.

d) Find the amount of sensible heat removed.

The first step in determining the amount of sensible heat removed is to identify the SHR, Sensible Heat Ratio, from the psychrometric chart. This process involves drawing a straight line between the initial and final points. This line is shown as a dashed line between the initial and final points. Then draw a line parallel to this dashed line such that it intersects with the SHR Reference Point and the vertical scale representing the SHR. The point of intersection reads, approximately, **0.43**. See Figure A10-1.

Note: For additional discussion on the significance of SHR, refer to Case Study 10-2, part (g) and the self assessment problem 10-2 (g).

SHR, Sensible Heat Ratio, is defined, mathematically, as:

SHR = Sensible Heat /Total Heat

In this case,

SHR = Q_s /Q_t = 0.43

Or,

Q_s = Sensible Heat = (0.43) . (Q_t)

And, since $Q_t = \Delta Q$ = Total Heat Removed = - 19,935 kJ, as calculated earlier,

Q_s = Sensible Heat = (0.43). (-19,943 kJ)

= -8,575 kJ

e) Find the amount of latent heat removed.

The total heat removed consists of sensible and latent heat components.

Or,

$Q_t = Q_s + Q_l$

Q_l = Latent Heat = Q_t - Q_s

= - 19,943 kJ – (- 8,575 kJ)

∴ Q_l = Latent Heat Removed = - 11,368 kJ

f) Find the final wet-bulb temperature.

As explained in the discussion associated with the psychrometric chart interpretation guide in Figure 10-2, wet bulb is read diagonally from the psychrometric point along the wet bulb temperature scale. The wet bulb lines run parallel to the enthalpy lines on the psychrometric chart.

The diagonal line emerging from the final point intersects the wet bulb and dew point scale at approximately 13.5°C as shown on the psychrometric chart in Figure A10-1. Therefore, the wet bulb temperature at the final point is **13.5°C.**

g) Find the initial dew point.

As shown on the psychrometric chart in Figure A10-1, follow the horizontal "dew point" line drawn from the initial point to the left, toward the saturation curve. The point of intersection of the saturation line and the dew point line represents the dew point. This point lies at 23.2°C.

Therefore, the dew point at the initial point, as read off from Figure A10-1, is **23.2°C.**

h) Find the final dew point.

As shown on the psychrometric chart in Figure A10-1, follow the horizontal "dew point" line drawn from the final point to the left, toward the saturation curve. The point of intersection of the saturation line and the dew point line represents the dew point. This point lies at 8.6°C.

Therefore, the dew point at the initial point, as read off from Figure A10-1, is **8.6°C.**

i) Find the amount of moisture condensed/removed.

In order to calculate amount of moisture condensed or removed, we need to find the difference between the humidity ratios for the initial and final points.

Humidity ratio can be defined mathematically as:

Humidity Ratio, ω = mass of moisture (kg) /mass of dry air (kg)

From the psychrometric chart, in Figure A10-1:

ω_i = **0.018** kg of moisture per kg of dry air, at the initial point
ω_f = **0.007** kg of moisture per kg of dry air, at the final point

∴ **$\Delta\omega$** = Change in the Humidity Ratio
= **0.018 - 0.007**
= **0.011** kg of moisture per kg of dry air

The amount of moisture condensed or removed
= ($\Delta\omega$) . (Total mass of Dry Air)
= ($\Delta\omega$) . ($m_{dry\ air}$)

Where, $m_{dry\ air}$ = 407 kg of dry air, as calculated in part (a)

∴ **The amount of moisture condensed or removed**
= (0.011 kg of moist./kg of dry air) . (407 kg of dry air)
= 4.477 kg

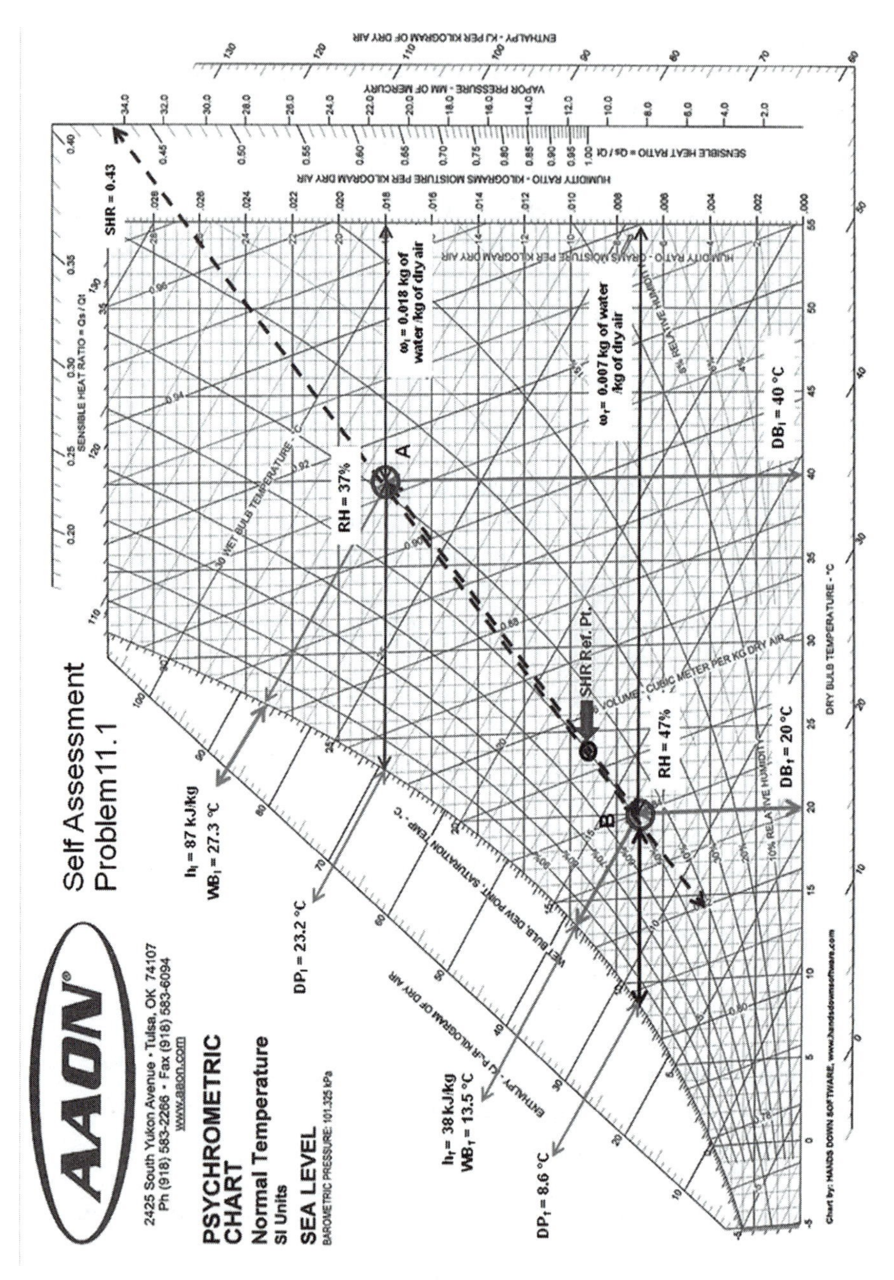

Figure A10-1. Psychrometric Chart

Solution – Problem 2

2. [*Note: Problem restated for reader's convenience*] Psychrometric chart for the initial and final conditions in a commercial warehouse is shown in Figure 10-7. Dry Bulb and Wet Bulb data associated with the initial and final points is labeled on the chart. Assess the disposition and performance of the HVAC system in this building as follows:

a) What is the initial Dew Point?
b) What is the final Dew Point?
c) Based on the results of dew point determination in parts (a) and (b), define the type of thermodynamic process this system undergoes in the transition from initial point to the final point.
d) What is the RH, Relative Humidity, at the initial point?
e) What is the RH, Relative Humidity, at the final point?
f) Determine the SHR for the change in conditions from the initial to the final point.
g) Comment on why the SHR for this scenario is significantly higher than the scenario analyzed in Case Study 10-2?

As explained in the solution segment of the case studies we, typically, begin our analyses of psychrometric problems with the identification or location of the initial and final psychrometric points on the psychrometric chart.

However, in this problem, the initial and final points are already plotted on the psychrometric chart provided. See Figures 10-7 and A10-2.

a) What is the initial Dew Point?

To determine the initial dew point, read the dew point temperature for the initial point on the psychrometric chart in Figure A10-2 by following the horizontal "dew point" line drawn from the initial point to the saturation curve, to the left. The point of intersection of the saturation line and the dew point line represents the dew point.

The dew point at the initial point, as read off from Figure A10-2, is **64.2 °F.**

b) Find the final dew point.

As with the initial dew point determination, to determine the final dew point, read the dew point temperature for the final point on the psychrometric chart in Figure A10-2 by following the horizontal "dew point" line drawn from the final point to the saturation curve, to the left. The point of intersection of the saturation line and the dew point line represents the dew point.

The dew point, at the final point, as read off from See Figure A10-2, is **64.2 °F,** which is the same as the initial dew point.

c) Based on the results of dew point determination in parts (a) and (b), define the type of thermodynamic process this system undergoes in the transition from initial point to the final point.

The initial and final dew points are the same. In other words, the change in conditions from the initial point to the final point represent a true horizontal shift with substantial dry bulb change, some wet bulb change, but no dew point change. This type of transition that represents a substantial dry bulb change but negligible dew point change signifies a definite sensible heat change. This transition is identical to path O-G described in the Psychrometric Transition Process section of Chapter 10. This case involves sensible heat removal in cooling the temperature of the air from **101°F** dry bulb to **87°F** dry bulb.

d) What is the RH, Relative Humidity, at the initial point?

Use the psychrometric chart interpretation guide in Figure 10-2 to locate the relative humidity (RH) lines on the psychrometric chart. The initial point lies directly on the 30% RH line. Therefore, the RH at the initial point is **30%.**

e) What is the RH, Relative Humidity, at the final point?

Use the psychrometric chart interpretation guide in Figure 10-2 to locate the two closest relative humidity (RH) lines on the psychrometric chart. The final point lies between the 40% and the 50% RH lines. Through interpolation, the relative humidity for the final point is estimated to be approximately **47%.**

f) Determine the SHR for the change in conditions from the initial to the final point.

Draw a straight line between the initial and final points as shown in Figure A 10-2. This line is shown as a dashed line between the initial and final points. Then draw a line parallel to this dashed line such that it intersects with the SHR Reference Point and the scale, to the right, representing the SHR. The point of intersection reads, approximately, **1.0**.

g) Comment on why the SHR for this scenario is significantly higher than the scenario analyzed in Case Study 10-2?

The **SHR of 1.0** for the scenario portrayed in this problem is substantially higher than the **SHR of 0.18** for the scenario in Case Study 10-2. The lesson to be taken away from this self assessment problem is that when the dry bulb changes significantly while the dew point is held steady, the sensible heat plays the predominant role. In other words, when the dry bulb changes significantly, at a relatively constant dew point, the sensible heat to total heat ratio approaches 1.0. In this case, the SHR is actually 1.0. An SHR of 1.0 implies that the entire amount of heat removed in this case was sensible heat and no latent heat was involved. The scenario portrayed in Case Study 10-2 was somewhat opposite. In that, the change in dry bulb was minute while the change in dew point was substantial. So, the substantial change in latent heat observed in Case Study 10-2 was indicative of the predominance of latent heat. The SHR of 0.18 in case Study 10-2 implied that only 18% of the heat involved in the process was sensible, while 82% of the heat involved in the process was latent.

See Figure A10-2 on following page.

CHAPTER 11—SOLUTION

Problem 1

(a) Draw the refrigeration cycle on the given diagram:

The process involved in the drawing of the refrigeration cycle is described below:

C-D:

See the pressure-enthalpy diagram in Figure A11-1. Locate the 290 psia and 60 psia points along the pressure (vertical) axis of the chart, name these points "C" and "D," respectively. This is the throttling portion of the refrigeration cycle. Note: HFC-134a, at point C is in, high pressure, saturated liquid phase.

Since the throttling process is adiabatic, and Δ h = 0, draw a straight, vertical, line down from C to D. At point D, HFC-134a would be in liquid-vapor mixture phase.

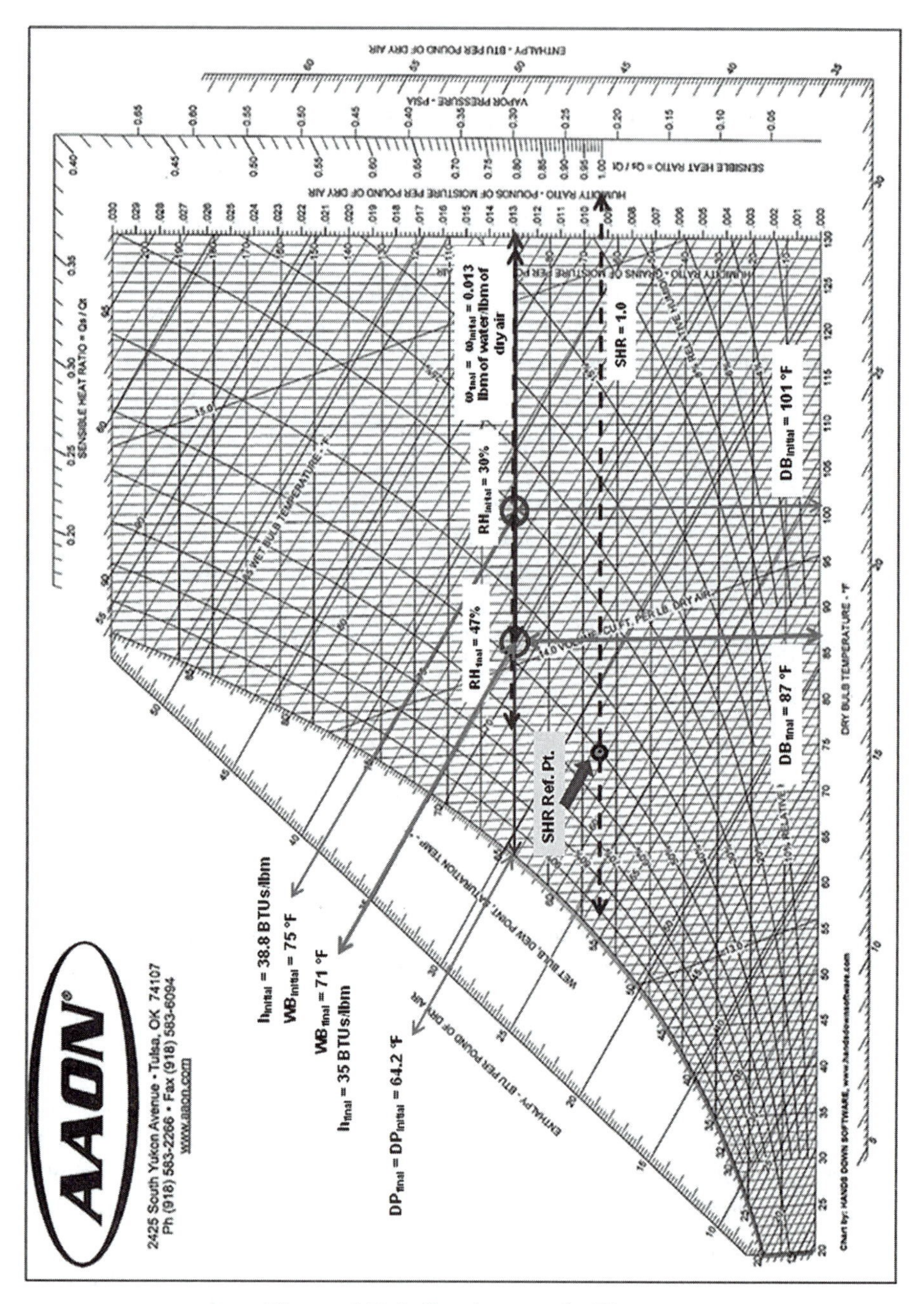

Figure A10-2. Psychrometric Chart

D-A:

See Figure A11-1. The next step involves complete transformation of the refrigerant from low pressure liquid to gaseous phase, through absorption of heat. Therefore, **h ≠ 0** and this is a non-adiabatic, isobaric, process. Draw a straight, horizontal, line from D to A. This step is referred to as the ***evaporator*** segment of the refrigeration cycle. This is where the system (refrigerant) performs cooling of the environment as its phase undergoes latent transformation from liquid to gaseous phase.

A-B:

The next step involves the transformation of HFC-134a from low pressure (60 psia) gaseous phase to high (290 psia) pressure gaseous phase. See Figure A11-1. This is the ***compression*** segment of the refrigeration cycle. This segment of the refrigeration cycle is an isentropic process and **Δs = 0.** Therefore, draw a straight line from point A to B, asymptotic to **s = 0.22 Btu/lbm.F.**

B-C:

The next step involves the transformation of HFC-134a from high (**290 psia**) pressure gaseous phase to high pressure saturated liquid phase. See **Figure A11-1.** This segment constitutes the ***condenser*** segment of the refrigeration cycle. This is an **isobaric** process and **Δ P = 0.** Therefore, draw a straight line from point B to C, along the **P = 290 psia** line.

(b) What is the change in *enthalpy* during the *expansion* process?

Solution/Answer: This is a captious question. In that, as evident from Figure A11-1, during the C-D throttling or expansion process, the enthalpy stays constant, at h = 65 Btu/lbm. Therefore, there is no change in enthalpy during the expansion process, or **Δ h = 0.**

(c) Determine the percentages of liquid and vapor at the end of the throttling segment of the refrigeration cycle.

Solution:

Determine the value of "**x**," the quality, at point "**D**," from the Pressure-Enthalpy diagram. See Figure A11-1. This involves interpo-

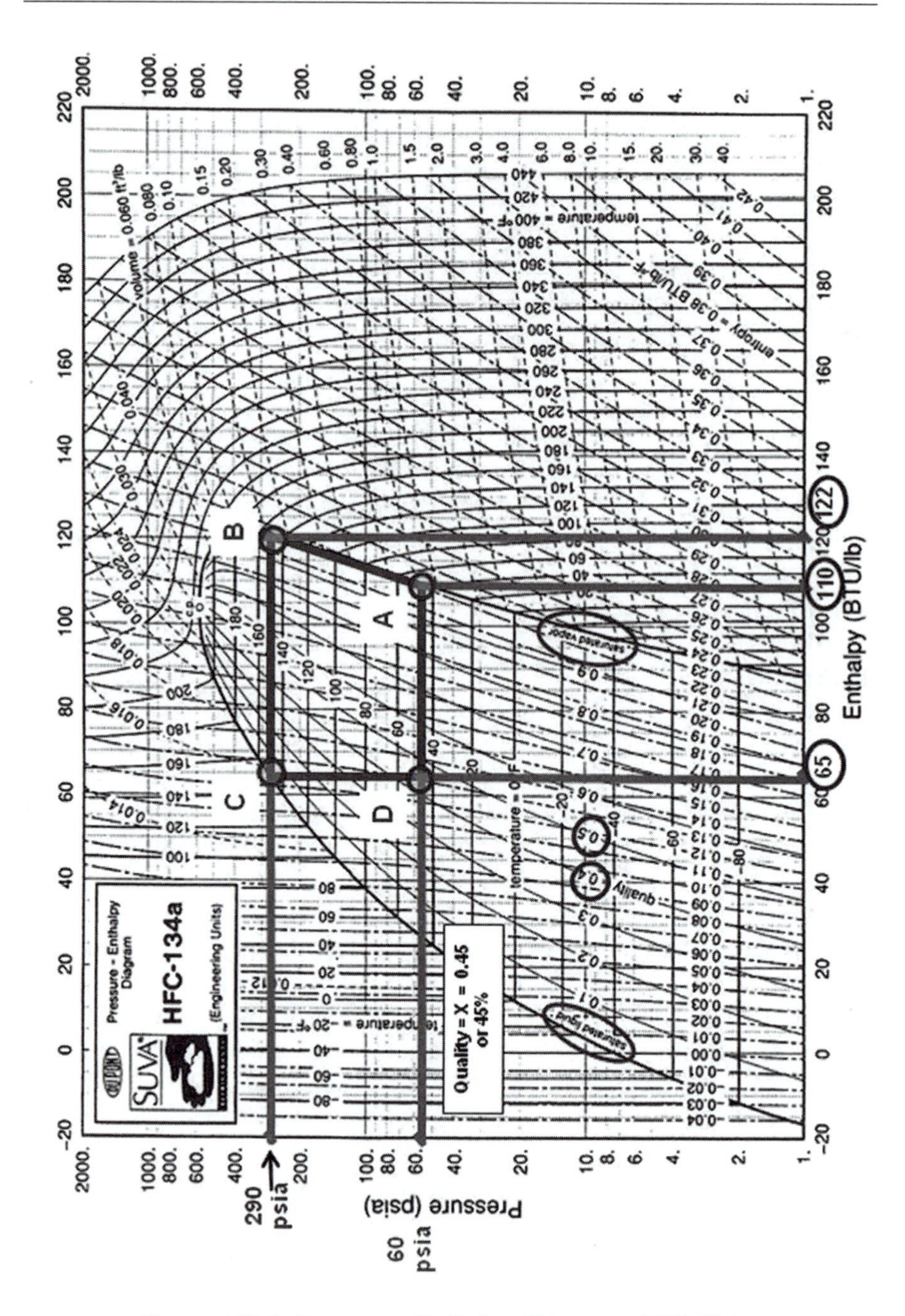

Figure A11-1. Pressure - Enthalpy Diagram, HFC-134a

lation or derivation of a quality curve that would intersect through point D. This derived or interpolated curve would be almost midway in between the x = 0.4 and x = 0.5 curves. Hence, x = 0.45 at point D.

$$\therefore\ \mathbf{x = (m_{vapor}) / (m_{vapor} + m_{liquid}) = 0.45 \text{ or } 45\%}$$

In other words,

$$\mathbf{m_{vapor}\ (\%) = 45}$$

And since

$$\mathbf{(\% m_{vapor} + \% m_{liquid}) = 100\%,}$$
$$\mathbf{m_{liquid}\ (\%) = 100 - 45 = 55\%}$$

(d) How much enthalpy is absorbed by the system (refrigerant) in the evaporation (latent) phase?

Solution:
This involves step **D – A** in Figure A11-1 Pressure-Enthalpy diagram:

$$\mathbf{\Delta h_{\,D-A} = h_A - h_D = 110 - 65 = 45\ Btu/lbm}$$

(e) How much enthalpy is extracted from the system (refrigerant) in the condensation (condenser) phase of the cycle?

Solution:
This involves step **B-C** in Figure A11-1 Pressure-Enthalpy diagram. In this segment of the refrigeration cycle:

$$\mathbf{\Delta h_{\,B-C} = h_C - h_B = 65 - 122 = -57}\ \text{Btu/lbm}$$

Note that $\Delta h_{\,B-C}$ is negative because heat is lost or extracted from the system in this segment of the refrigeration cycle.

(f) Determine the percentages of liquid and vapor at B.

Solution/Answer: As apparent from Figure A11-1, Point B lies directly on the ***saturated vapor line***. And, x = 1 at all points on the saturated vapor line.

According to Eq. 11-1,

$$\mathbf{x = (m_{vapor}) / (m_{vapor} + m_{liquid})}$$

Therefore, at x = 1

$$\mathbf{(m_{vapor}) / (m_{vapor} + m_{liquid}) = 1}$$

Or,

$$\mathbf{(m_{vapor}) = (m_{vapor} + m_{liquid})}$$

This can only hold true if $\mathbf{m_{liquid} = 0\%}$. And if $\mathbf{m_{liquid} = 0\%}$, then $m_{vapor} = 100\%$.

Another way to interpret and dissect **x = 1**, would be through revisiting the definition of quality factor **x.** Since **x** represents the ratio of the mass of vapor to the total combined mass of vapor and liquid, x = 100% implies that vapor constitutes the combined mass in entirety.

Therefore, the answer is:

$$\mathbf{m_{liquid} = 0\% \text{ and } m_{vapor} = 100\%}$$

(g) Assume that the mass flow rate of refrigerant being cycled in this air-conditioning system is 10 lbm/min and the compressor efficiency is 70%. Determine the amount of electrical power demanded by the compressor motor if the compressor motor efficiency is 90%.

Solution:

The amount of electrical power demanded by the compression cycle:

According to Eq. 11-2, net power required to compress the refrigerant:

$$P = \dot{W} = \dot{m}(h_B - h_A)$$

Where, point B is equivalent to point 4 and point A is equivalent to point 3.

And,

Electrical Power Demanded From Utility

$$\mathbf{= P \div (Motor_{efficiency}) \div (Comp._{efficiency})}$$

$$\dot{m} = 10 \text{ lbm/min \{Given\}}$$
$$= (10 \text{ lbm/min})/(60 \text{ sec/min})$$
$$= 0.167 \text{ lbm/sec}$$

$$P = (\dot{m}) \bullet (h_B - h_A)$$
$$= (0.167 \text{ lbm/sec}) \bullet (122 - 110)(\text{Btu/lbm})$$
$$= 2 \text{ Btu/sec}$$

Since 1055 J = 1 Btu,

$$P = (2 \text{ Btu/sec}) \bullet (1055 \text{ J/Btu})$$
$$= 2110 \text{ J/s}$$
$$= 2110 \text{ W or } 2.11 \text{ kW}$$

$\therefore$ **Electrical Power Demanded From Utility**

$$= (2.11 \text{ kW}) \bullet (1/0.7) \bullet (1/0.9)$$

= 4.49 kW

(h) Which leg of the refrigeration cycle would be considered isentropic.

Answer: Path A-B of the HFC-134a refrigeration cycle, as shown in Figure A11-1, adheres to and stays coincident with the entropy line at s = 0.22 Btu/lbm.°F. The entropy stays constant and equal to 0.22 Btu/lbm.°F along path A-B.

Therefore, path A-B represents the isentropic process in the given refrigeration cycle.

Correct answer: **I**

Solution—Problem 2

Provide the mathematical proof for equivalence of 1 tonne (SI/Metric) of refrigeration capacity to 3.86 kW of power:

The amount of heat that must be extracted from 0°C water to freeze it to 0°C ice is equal to the amount of heat that must be added to 0°C ice to melt it to 32°F (0°C) water. This heat of fusion for water, in the metric units is 333.5 kJ/kg

Therefore,

Rate of refrigeration for one tonne of 0°C water

$$= (333.5 \text{ kJ/kg}) \bullet (1000 \text{ kg})/24 \text{ hr}$$
$$= 3.86 \text{ kJ/sec}$$
$$= 3.86 \text{ k.(J/sec)}$$

And, since 1 J/sec = 1 Watt,

Rate of refrigeration for one tonne of 0°C water = 3.86kW

Solution—Problem 3
Given:

SEER Rating: 12 Btu/W-h
Air Conditioning System Rating: 20,000 Btu/hr
Total, Annual, Seasonal Operating Period: 200 days, 10 hours per day.
Average, Combined, Electrical Energy Cost Rate: \$0.20/kWh

Annual Cost of Energy = (\$0.20/kWh).(Total Energy Drawn From The Utility, Annually)

Total Power Demanded From The Utility
= (Air Conditioning System Rating, in Btu/hr)/
(SEER Rating, in Btu /W-hr **Eq. 11-9**

Note: Both Btu values in Eq. 11-9 are outputs, while the W-hr value represents the input energy drawn from the line (utility) side of the power distribution system.

∴**Total Power Demanded From The Utility**
= (20,000 Btu/hr)/(12 Btu /W-hr)
= 1,667 Watts, or 1.67 kW

Then,

Total Energy Drawn From The Utility, Annually
= (1.667 kW).(Total Annual Operating Hours)
= (1.667 kW).(200 Days).(10 Hours/Day)
= 3,334 kWh

∴**Total Annual Cost of Electrical Energy Consumed**
= (3,334 kWh) • (\$0.20/kWh)
= \$666.67

Appendix B

STEAM TABLES

These steam tables, copyright ASME, published with ASME permission, do not include the heat of evaporation value, h_{fg}, values for the saturation temperature and pressures. The saturated steam tables presented in this text are the compact version. However, the h_{fg} values for the saturation temperatures and pressures in these tables can be derived by simply subtracting the available value of h_L from h_v for the respective saturation pressures and temperatures. In other words:

$$\mathbf{h_{fg} = h_v - h_L}$$

Table 1. Properties of Saturated Water and Steam (Temperature)

Temp. °F	Pressure psia	Volume, ft³/lbm v_L	v_V	Enthalpy, Btu/lbm h_L	h_V	Entropy, Btu/(lbm·°R) s_L	s_V	Temp. °F
32	0.08865	0.016022	3302.0	-0.018	1075.2	0.0000	2.1868	32
35	0.09998	0.016020	2945.5	3.004	1076.5	0.0061	2.1762	35
40	0.12173	0.016020	2443.4	8.032	1078.7	0.0162	2.1590	40
45	0.14757	0.016021	2035.6	13.052	1080.9	0.0262	2.1421	45
50	0.17813	0.016024	1702.9	18.066	1083.1	0.0361	2.1257	50
55	0.21414	0.016029	1430.3	23.074	1085.3	0.0459	2.1097	55
60	0.25639	0.016035	1206.1	28.079	1087.4	0.0555	2.0941	60
65	0.30579	0.016043	1020.8	33.080	1089.6	0.0651	2.0788	65
70	0.36334	0.016052	867.19	38.078	1091.8	0.0746	2.0640	70
75	0.43015	0.016062	739.30	43.074	1094.0	0.0840	2.0495	75
80	0.50744	0.016074	632.44	48.069	1096.1	0.0933	2.0353	80
85	0.59656	0.016086	542.84	53.062	1098.3	0.1025	2.0215	85
90	0.69899	0.016100	467.45	58.054	1100.4	0.1116	2.0080	90
95	0.81636	0.016115	403.79	63.046	1102.6	0.1207	1.9948	95
100	0.95044	0.016131	349.87	68.037	1104.7	0.1296	1.9819	100
105	1.1032	0.016148	304.05	73.028	1106.9	0.1385	1.9693	105
110	1.2766	0.016166	264.99	78.019	1109.0	0.1473	1.9570	110
115	1.4730	0.016185	231.60	83.010	1111.1	0.1560	1.9450	115
120	1.6949	0.016205	202.96	88.002	1113.2	0.1647	1.9333	120
125	1.9449	0.016225	178.34	92.994	1115.3	0.1732	1.9218	125
130	2.2258	0.016247	157.10	97.987	1117.4	0.1817	1.9106	130
135	2.5407	0.016269	138.74	102.98	1119.5	0.1902	1.8996	135
140	2.8929	0.016293	122.82	107.98	1121.6	0.1985	1.8888	140
145	3.2858	0.016317	108.99	112.97	1123.7	0.2068	1.8783	145
150	3.7231	0.016342	96.934	117.97	1125.7	0.2151	1.8680	150
155	4.2089	0.016367	86.405	122.97	1127.8	0.2232	1.8580	155
160	4.7472	0.016394	77.186	127.98	1129.8	0.2313	1.8481	160
165	5.3426	0.016421	69.097	132.98	1131.9	0.2394	1.8384	165
170	5.9998	0.016449	61.982	137.99	1133.9	0.2474	1.8290	170
175	6.7237	0.016478	55.710	143.00	1135.9	0.2553	1.8197	175
180	7.5196	0.016507	50.171	148.01	1137.9	0.2631	1.8106	180
185	8.3930	0.016538	45.267	153.03	1139.9	0.2709	1.8017	185
190	9.3497	0.016569	40.918	158.05	1141.8	0.2787	1.7930	190
195	10.396	0.016601	37.053	163.07	1143.8	0.2864	1.7844	195
200	11.538	0.016633	33.611	168.10	1145.7	0.2940	1.7760	200
205	12.782	0.016667	30.540	173.13	1147.6	0.3016	1.7678	205
210	14.136	0.016701	27.796	178.17	1149.5	0.3092	1.7597	210
215	15.606	0.016736	25.339	183.20	1151.4	0.3167	1.7517	215
220	17.201	0.016771	23.135	188.25	1153.3	0.3241	1.7440	220
225	18.928	0.016808	21.155	193.30	1155.1	0.3315	1.7363	225
230	20.795	0.016845	19.373	198.35	1157.0	0.3388	1.7288	230
235	22.811	0.016883	17.766	203.41	1158.8	0.3461	1.7214	235
240	24.985	0.016921	16.316	208.47	1160.5	0.3534	1.7141	240
245	27.326	0.016961	15.004	213.54	1162.3	0.3606	1.7070	245
250	29.843	0.017001	13.816	218.62	1164.0	0.3678	1.7000	250

Table 1. Properties of Saturated Water and Steam (Temperature)

Temp. °F	Pressure psia	Volume, ft³/lbm v_L	v_V	Enthalpy, Btu/lbm h_L	h_V	Entropy, Btu/(lbm·°R) s_L	s_V	Temp. °F
255	32.546	0.017042	12.739	223.70	1165.7	0.3749	1.6930	255
260	35.445	0.017084	11.760	228.79	1167.4	0.3820	1.6862	260
265	38.551	0.017127	10.870	233.88	1169.1	0.3890	1.6796	265
270	41.874	0.017170	10.059	238.99	1170.7	0.3960	1.6730	270
275	45.426	0.017214	9.3196	244.10	1172.3	0.4030	1.6665	275
280	49.218	0.017259	8.6442	249.21	1173.9	0.4099	1.6601	280
285	53.261	0.017305	8.0265	254.34	1175.5	0.4168	1.6538	285
290	57.567	0.017352	7.4610	259.47	1177.0	0.4236	1.6476	290
295	62.150	0.017400	6.9425	264.61	1178.5	0.4305	1.6414	295
300	67.021	0.017449	6.4666	269.76	1180.0	0.4372	1.6354	300
305	72.193	0.017498	6.0293	274.91	1181.4	0.4440	1.6294	305
310	77.680	0.017548	5.6270	280.08	1182.8	0.4507	1.6235	310
315	83.496	0.017600	5.2564	285.26	1184.2	0.4574	1.6177	315
320	89.654	0.017652	4.9148	290.44	1185.5	0.4640	1.6120	320
325	96.168	0.017705	4.5994	295.64	1186.8	0.4706	1.6063	325
330	103.05	0.017760	4.3079	300.85	1188.0	0.4772	1.6007	330
335	110.32	0.017815	4.0384	306.07	1189.3	0.4838	1.5952	335
340	118.00	0.017871	3.7888	311.30	1190.5	0.4903	1.5897	340
345	126.08	0.017929	3.5574	316.54	1191.6	0.4968	1.5843	345
350	134.60	0.017987	3.3428	321.79	1192.7	0.5033	1.5789	350
355	143.57	0.018047	3.1435	327.06	1193.8	0.5097	1.5736	355
360	153.00	0.018108	2.9582	332.34	1194.8	0.5162	1.5684	360
365	162.92	0.018170	2.7859	337.63	1195.8	0.5226	1.5632	365
370	173.33	0.018233	2.6254	342.94	1196.7	0.5289	1.5580	370
375	184.25	0.018297	2.4758	348.26	1197.6	0.5353	1.5529	375
380	195.71	0.018363	2.3363	353.59	1198.5	0.5416	1.5478	380
385	207.72	0.018430	2.2061	358.94	1199.3	0.5479	1.5428	385
390	220.29	0.018498	2.0843	364.31	1200.1	0.5542	1.5378	390
395	233.45	0.018568	1.9705	369.70	1200.8	0.5605	1.5329	395
400	247.22	0.018639	1.8640	375.10	1201.5	0.5667	1.5280	400
405	261.61	0.018711	1.7643	380.52	1202.1	0.5729	1.5231	405
410	276.64	0.018785	1.6708	385.95	1202.6	0.5791	1.5182	410
415	292.34	0.018861	1.5830	391.41	1203.2	0.5853	1.5134	415
420	308.71	0.018938	1.5007	396.89	1203.6	0.5915	1.5086	420
425	325.79	0.019016	1.4234	402.38	1204.0	0.5977	1.5038	425
430	343.59	0.019097	1.3507	407.90	1204.4	0.6038	1.4991	430
435	362.13	0.019179	1.2822	413.44	1204.7	0.6100	1.4943	435
440	381.44	0.019263	1.2179	419.01	1204.9	0.6161	1.4896	440
445	401.53	0.019349	1.1572	424.59	1205.1	0.6222	1.4849	445
450	422.42	0.019437	1.1000	430.20	1205.2	0.6283	1.4802	450
455	444.14	0.019527	1.0461	435.84	1205.2	0.6344	1.4755	455
460	466.71	0.019619	0.9952	441.50	1205.2	0.6405	1.4709	460
465	490.15	0.019713	0.9471	447.19	1205.1	0.6466	1.4662	465
470	514.48	0.019810	0.9016	452.91	1204.9	0.6526	1.4615	470
475	539.73	0.019908	0.8586	458.66	1204.7	0.6587	1.4569	475

Table 1. Properties of Saturated Water and Steam (Temperature)

Temp. °F	Pressure psia	Volume, ft³/lbₘ v_L	v_V	Enthalpy, Btu/lbₘ h_L	h_V	Entropy, Btu/(lbₘ·°R) s_L	s_V	Temp. °F
480	565.92	0.02001	0.8180	464.44	1204.4	0.6648	1.4522	**480**
485	593.07	0.02011	0.7795	470.25	1204.0	0.6708	1.4475	**485**
490	621.20	0.02022	0.7430	476.10	1203.5	0.6769	1.4429	**490**
495	650.35	0.02033	0.7084	481.97	1203.0	0.6829	1.4382	**495**
500	680.53	0.02044	0.6756	487.89	1202.3	0.6890	1.4335	**500**
505	711.77	0.02056	0.6445	493.84	1201.6	0.6951	1.4288	**505**
510	744.09	0.02068	0.6149	499.83	1200.8	0.7011	1.4241	**510**
515	777.52	0.02080	0.5868	505.86	1199.9	0.7072	1.4193	**515**
520	812.08	0.02092	0.5601	511.93	1198.9	0.7133	1.4145	**520**
525	847.81	0.02105	0.5347	518.05	1197.9	0.7194	1.4098	**525**
530	884.73	0.02118	0.5105	524.21	1196.7	0.7255	1.4049	**530**
535	922.85	0.02132	0.4875	530.42	1195.4	0.7316	1.4001	**535**
540	962.23	0.02146	0.4656	536.69	1194.0	0.7377	1.3952	**540**
545	1002.9	0.02161	0.4446	543.00	1192.5	0.7438	1.3903	**545**
550	1044.8	0.02176	0.4247	549.37	1190.8	0.7500	1.3853	**550**
555	1088.1	0.02192	0.4056	555.80	1189.1	0.7562	1.3803	**555**
560	1132.7	0.02208	0.3875	562.29	1187.2	0.7624	1.3752	**560**
565	1178.7	0.02225	0.3701	568.85	1185.2	0.7686	1.3701	**565**
570	1226.2	0.02242	0.3535	575.48	1183.0	0.7749	1.3649	**570**
575	1275.1	0.02260	0.3376	582.18	1180.7	0.7812	1.3596	**575**
580	1325.4	0.02279	0.3223	588.95	1178.2	0.7875	1.3543	**580**
585	1377.3	0.02299	0.3077	595.81	1175.6	0.7939	1.3489	**585**
590	1430.8	0.02319	0.2938	602.75	1172.8	0.8003	1.3433	**590**
595	1485.8	0.02341	0.2803	609.79	1169.8	0.8067	1.3377	**595**
600	1542.5	0.02363	0.2675	616.93	1166.6	0.8133	1.3320	**600**
605	1600.8	0.02387	0.2551	624.17	1163.2	0.8198	1.3261	**605**
610	1660.8	0.02411	0.2432	631.53	1159.6	0.8265	1.3202	**610**
615	1722.6	0.02437	0.2317	639.01	1155.7	0.8332	1.3140	**615**
620	1786.1	0.02465	0.2207	646.62	1151.6	0.8400	1.3077	**620**
625	1851.5	0.02494	0.2101	654.38	1147.2	0.8469	1.3012	**625**
630	1918.8	0.02525	0.1998	662.30	1142.5	0.8539	1.2945	**630**
635	1988.0	0.02558	0.1899	670.40	1137.4	0.8610	1.2876	**635**
640	2059.2	0.02593	0.1802	678.69	1132.0	0.8683	1.2804	**640**
645	2132.4	0.02631	0.1709	687.21	1126.1	0.8757	1.2729	**645**
650	2207.7	0.02672	0.1618	695.99	1119.7	0.8833	1.2651	**650**
655	2285.2	0.02717	0.1530	705.06	1112.8	0.8911	1.2569	**655**
660	2364.8	0.02766	0.1444	714.47	1105.3	0.8991	1.2482	**660**
665	2446.8	0.02821	0.1359	724.30	1097.2	0.9075	1.2390	**665**
670	2531.2	0.02883	0.1276	734.63	1088.1	0.9163	1.2292	**670**
675	2618.0	0.02953	0.1194	745.57	1078.0	0.9255	1.2185	**675**
680	2707.3	0.03035	0.1112	757.30	1066.6	0.9354	1.2068	**680**
685	2799.3	0.03133	0.1030	770.10	1053.5	0.9462	1.1937	**685**
690	2894.2	0.03256	0.09444	784.45	1037.9	0.9582	1.1786	**690**
695	2991.9	0.03422	0.08531	801.35	1018.3	0.9723	1.1602	**695**
700	3092.9	0.03683	0.07466	823.64	990.64	0.9910	1.1350	**700**
705	3197.9	0.04662	0.05338	882.44	913.89	1.0409	1.0679	**705**
T_c	3200.1	0.0497	0.0497	897.48	897.48	1.0538	1.0538	T_c

T_c = 705.1028 °F

Table 2. Properties of Saturated Water and Steam (Pressure)

Pressure psia	Temp. °F	Volume, ft^3/lb_m v_L	v_V	Enthalpy, Btu/lb_m h_L	h_V	Entropy, $Btu/(lb_m{\cdot}°R)$ s_L	s_V	Pressure psia
0.1	35.00	0.016020	2945.0	3.009	1076.5	0.0061	2.1762	**0.1**
0.2	53.13	0.016027	1525.9	21.204	1084.4	0.0422	2.1156	**0.2**
0.3	64.45	0.016042	1039.4	32.532	1089.4	0.0641	2.0805	**0.3**
0.5	79.55	0.016073	641.32	47.618	1095.9	0.0925	2.0366	**0.5**
0.7	90.05	0.016100	466.81	58.100	1100.4	0.1117	2.0079	**0.7**
1.0	101.69	0.016137	333.51	69.728	1105.4	0.1326	1.9776	**1.0**
1.5	115.64	0.016187	227.68	83.650	1111.4	0.1571	1.9435	**1.5**
2.0	126.03	0.016230	173.72	94.019	1115.8	0.1750	1.9195	**2.0**
3.0	141.42	0.016299	118.70	109.39	1122.2	0.2009	1.8858	**3.0**
4.0	152.91	0.016356	90.628	120.89	1126.9	0.2198	1.8621	**4.0**
6	170.00	0.016449	61.979	137.99	1133.9	0.2474	1.8290	**6**
8	182.81	0.016524	47.345	150.83	1139.0	0.2675	1.8056	**8**
10	193.16	0.016589	38.423	161.22	1143.1	0.2836	1.7875	**10**
12	201.91	0.016646	32.398	170.02	1146.4	0.2969	1.7728	**12**
14	209.52	0.016697	28.048	177.68	1149.4	0.3084	1.7605	**14**
16	216.27	0.016745	24.755	184.49	1151.9	0.3186	1.7497	**16**
18	222.36	0.016788	22.173	190.63	1154.2	0.3276	1.7403	**18**
20	227.92	0.016829	20.092	196.25	1156.2	0.3358	1.7319	**20**
25	240.03	0.016922	16.306	208.51	1160.5	0.3534	1.7141	**25**
30	250.30	0.017003	13.748	218.93	1164.1	0.3682	1.6995	**30**
35	259.25	0.017078	11.900	228.03	1167.2	0.3809	1.6873	**35**
40	267.22	0.017146	10.500	236.15	1169.8	0.3921	1.6766	**40**
45	274.42	0.017209	9.4023	243.50	1172.2	0.4022	1.6672	**45**
50	280.99	0.017268	8.5171	250.23	1174.2	0.4113	1.6588	**50**
55	287.06	0.017325	7.7878	256.45	1176.1	0.4196	1.6512	**55**
60	292.69	0.017378	7.1762	262.24	1177.8	0.4273	1.6443	**60**
65	297.96	0.017429	6.6557	267.66	1179.4	0.4345	1.6378	**65**
70	302.92	0.017477	6.2071	272.76	1180.8	0.4412	1.6319	**70**
75	307.59	0.017524	5.8164	277.59	1182.1	0.4475	1.6264	**75**
80	312.03	0.017569	5.4730	282.18	1183.3	0.4534	1.6212	**80**
85	316.25	0.017613	5.1686	286.55	1184.5	0.4590	1.6163	**85**
90	320.27	0.017655	4.8969	290.73	1185.6	0.4644	1.6117	**90**
95	324.12	0.017696	4.6528	294.73	1186.6	0.4695	1.6073	**95**
100	327.82	0.017736	4.4324	298.57	1187.5	0.4744	1.6032	**100**
110	334.78	0.017813	4.0496	305.84	1189.2	0.4835	1.5954	**110**
120	341.26	0.017886	3.7286	312.62	1190.7	0.4920	1.5883	**120**
130	347.33	0.017956	3.4554	318.98	1192.1	0.4998	1.5818	**130**
140	353.04	0.018023	3.2199	324.99	1193.4	0.5072	1.5757	**140**
150	358.43	0.018089	3.0148	330.68	1194.5	0.5141	1.5700	**150**
160	363.55	0.018152	2.8345	336.10	1195.5	0.5207	1.5647	**160**
170	368.43	0.018213	2.6746	341.27	1196.5	0.5269	1.5596	**170**
180	373.08	0.018272	2.5320	346.21	1197.3	0.5328	1.5549	**180**
190	377.54	0.018330	2.4038	350.96	1198.1	0.5385	1.5503	**190**
200	381.81	0.018387	2.2880	355.53	1198.8	0.5439	1.5460	**200**
210	385.92	0.018442	2.1829	359.94	1199.5	0.5491	1.5419	**210**

Table 2. Properties of Saturated Water and Steam (Pressure)

Pressure psia	Temp. °F	Volume, ft^3/lb_m v_L	v_V	Enthalpy, Btu/lb_m h_L	h_V	Entropy, $Btu/(lb_m \cdot °R)$ s_L	s_V	Pressure psia
220	389.89	0.018496	2.0870	364.19	1200.1	0.5541	1.5379	**220**
230	393.71	0.018549	1.9992	368.30	1200.6	0.5588	1.5342	**230**
240	397.41	0.018601	1.9184	372.29	1201.1	0.5635	1.5305	**240**
250	400.98	0.018653	1.8439	376.16	1201.6	0.5679	1.5270	**250**
260	404.45	0.018703	1.7749	379.92	1202.0	0.5723	1.5236	**260**
270	407.82	0.018753	1.7108	383.58	1202.4	0.5764	1.5203	**270**
280	411.09	0.018801	1.6512	387.14	1202.8	0.5805	1.5172	**280**
290	414.27	0.018849	1.5955	390.61	1203.1	0.5844	1.5141	**290**
300	417.37	0.018897	1.5434	394.00	1203.4	0.5883	1.5111	**300**
320	423.33	0.018990	1.4487	400.54	1203.9	0.5956	1.5054	**320**
340	429.01	0.019081	1.3647	406.81	1204.3	0.6026	1.5000	**340**
360	434.43	0.019170	1.2898	412.82	1204.6	0.6093	1.4949	**360**
380	439.63	0.019257	1.2224	418.60	1204.9	0.6156	1.4900	**380**
400	444.63	0.019343	1.1616	424.18	1205.0	0.6217	1.4853	**400**
420	449.43	0.019427	1.1064	429.56	1205.1	0.6276	1.4807	**420**
440	454.06	0.019510	1.0560	434.78	1205.2	0.6333	1.4764	**440**
460	458.53	0.019592	1.0098	439.84	1205.2	0.6387	1.4722	**460**
480	462.86	0.019672	0.9673	444.75	1205.1	0.6440	1.4682	**480**
500	467.05	0.019752	0.9282	449.53	1205.0	0.6490	1.4643	**500**
520	471.11	0.019831	0.8919	454.19	1204.9	0.6540	1.4605	**520**
540	475.05	0.019909	0.8582	458.72	1204.7	0.6588	1.4568	**540**
560	478.89	0.019987	0.8268	463.15	1204.4	0.6634	1.4532	**560**
580	482.62	0.02006	0.7976	467.48	1204.2	0.6679	1.4498	**580**
600	486.25	0.02014	0.7702	471.71	1203.9	0.6723	1.4464	**600**
620	489.79	0.02022	0.7445	475.85	1203.5	0.6766	1.4430	**620**
640	493.24	0.02029	0.7203	479.91	1203.2	0.6808	1.4398	**640**
660	496.62	0.02037	0.6976	483.88	1202.8	0.6849	1.4367	**660**
680	499.91	0.02044	0.6761	487.79	1202.4	0.6889	1.4336	**680**
700	503.14	0.02051	0.6559	491.62	1201.9	0.6928	1.4305	**700**
720	506.29	0.02059	0.6367	495.38	1201.4	0.6966	1.4276	**720**
740	509.38	0.02066	0.6185	499.08	1200.9	0.7004	1.4246	**740**
760	512.40	0.02073	0.6012	502.72	1200.4	0.7040	1.4218	**760**
780	515.36	0.02081	0.5848	506.30	1199.9	0.7076	1.4190	**780**
800	518.27	0.02088	0.5692	509.83	1199.3	0.7112	1.4162	**800**
820	521.12	0.02095	0.5543	513.30	1198.7	0.7146	1.4135	**820**
840	523.92	0.02102	0.5401	516.73	1198.1	0.7181	1.4108	**840**
860	526.67	0.02110	0.5265	520.10	1197.5	0.7214	1.4082	**860**
880	529.37	0.02117	0.5135	523.43	1196.8	0.7247	1.4056	**880**
900	532.02	0.02124	0.5011	526.72	1196.2	0.7279	1.4030	**900**
920	534.63	0.02131	0.4892	529.96	1195.5	0.7311	1.4005	**920**
940	537.20	0.02138	0.4777	533.17	1194.8	0.7343	1.3980	**940**
960	539.72	0.02146	0.4667	536.34	1194.1	0.7374	1.3955	**960**
980	542.21	0.02153	0.4562	539.47	1193.3	0.7404	1.3930	**980**
1000	544.65	0.02160	0.4461	542.56	1192.6	0.7434	1.3906	**1000**
1050	550.61	0.02178	0.4223	550.15	1190.6	0.7507	1.3847	**1050**

Table 2. Properties of Saturated Water and Steam (Pressure)

Pressure psia	Temp. °F	Volume, ft³/lbm v_L	v_V	Enthalpy, Btu/lbm h_L	h_V	Entropy, Btu/(lbm·°R) s_L	s_V	Pressure psia
1100	556.35	0.02196	0.4006	557.55	1188.6	0.7578	1.3789	**1100**
1150	561.90	0.02214	0.3808	564.77	1186.4	0.7647	1.3733	**1150**
1200	567.26	0.02233	0.3625	571.84	1184.2	0.7714	1.3677	**1200**
1250	572.46	0.02251	0.3456	578.76	1181.9	0.7780	1.3623	**1250**
1300	577.50	0.02270	0.3299	585.55	1179.5	0.7843	1.3570	**1300**
1350	582.39	0.02288	0.3153	592.21	1177.0	0.7905	1.3517	**1350**
1400	587.14	0.02307	0.3017	598.77	1174.4	0.7966	1.3465	**1400**
1450	591.76	0.02327	0.2890	605.23	1171.8	0.8025	1.3414	**1450**
1500	596.27	0.02346	0.2770	611.59	1169.0	0.8084	1.3363	**1500**
1550	600.66	0.02366	0.2658	617.87	1166.2	0.8141	1.3312	**1550**
1600	604.93	0.02386	0.2553	624.07	1163.3	0.8197	1.3262	**1600**
1650	609.11	0.02407	0.2453	630.21	1160.3	0.8253	1.3212	**1650**
1700	613.19	0.02428	0.2358	636.28	1157.2	0.8307	1.3163	**1700**
1750	617.18	0.02449	0.2269	642.30	1154.0	0.8361	1.3113	**1750**
1800	621.07	0.02471	0.2184	648.27	1150.7	0.8415	1.3063	**1800**
1850	624.89	0.02493	0.2103	654.20	1147.3	0.8467	1.3014	**1850**
1900	628.62	0.02516	0.2026	660.09	1143.8	0.8519	1.2964	**1900**
1950	632.27	0.02539	0.1952	665.96	1140.2	0.8571	1.2914	**1950**
2000	635.85	0.02563	0.1882	671.80	1136.5	0.8622	1.2864	**2000**
2050	639.36	0.02588	0.1814	677.62	1132.7	0.8673	1.2814	**2050**
2100	642.81	0.02614	0.1750	683.44	1128.7	0.8724	1.2763	**2100**
2150	646.18	0.02640	0.1687	689.26	1124.6	0.8774	1.2711	**2150**
2200	649.50	0.02668	0.1627	695.09	1120.4	0.8825	1.2659	**2200**
2250	652.75	0.02696	0.1569	700.93	1116.0	0.8875	1.2606	**2250**
2300	655.94	0.02726	0.1514	706.80	1111.5	0.8926	1.2553	**2300**
2350	659.08	0.02757	0.1459	712.71	1106.8	0.8976	1.2498	**2350**
2400	662.16	0.02789	0.1407	718.67	1101.9	0.9027	1.2443	**2400**
2450	665.19	0.02823	0.1356	724.69	1096.8	0.9078	1.2387	**2450**
2500	668.17	0.02859	0.1307	730.78	1091.5	0.9130	1.2329	**2500**
2550	671.10	0.02897	0.1258	736.97	1086.0	0.9183	1.2269	**2550**
2600	673.98	0.02938	0.1211	743.27	1080.2	0.9236	1.2208	**2600**
2650	676.81	0.02981	0.1165	749.71	1074.1	0.9290	1.2144	**2650**
2700	679.60	0.03028	0.1119	756.32	1067.6	0.9346	1.2078	**2700**
2750	682.34	0.03078	0.1074	763.13	1060.7	0.9403	1.2009	**2750**
2800	685.03	0.03134	0.1029	770.20	1053.4	0.9462	1.1936	**2800**
2850	687.69	0.03195	0.09843	777.59	1045.5	0.9524	1.1859	**2850**
2900	690.30	0.03264	0.09391	785.39	1036.8	0.9590	1.1776	**2900**
2950	692.88	0.03344	0.08930	793.75	1027.3	0.9660	1.1686	**2950**
3000	695.41	0.03438	0.08453	802.90	1016.5	0.9736	1.1585	**3000**
3050	697.90	0.03554	0.07945	813.22	1003.8	0.9823	1.1469	**3050**
3100	700.35	0.03708	0.07381	825.57	988.14	0.9926	1.1328	**3100**
3150	702.75	0.03947	0.06686	842.34	966.17	1.0068	1.1133	**3150**
3200	705.10	0.04897	0.05052	893.85	901.07	1.0507	1.0569	**3200**
p_c	705.1028	0.0497	0.0497	897.48	897.48	1.0538	1.0538	p_c

p_c = 3200.11 psia

Table 3. Superheated Steam

Pressure psia (Sat. T)		Temperature—Degrees Fahrenheit 200	250	300	350	400	450	500	600	700	800	900	1000	1200
1	*v*	392.53	422.42	452.28	482.11	511.93	541.74	571.55	631.15	690.74	750.32	809.91	869.48	988.64
(101.69)	*h*	1150.1	1172.8	1195.7	1218.6	1241.8	1265.1	1288.6	1336.2	1384.6	1433.9	1484.1	1535.1	1640.0
	s	2.0510	2.0842	2.1152	2.1445	2.1723	2.1986	2.2238	2.2710	2.3146	2.3554	2.3937	2.4299	2.4973
5	*v*	78.155	84.220	90.248	96.254	102.25	108.23	114.21	126.15	138.09	150.02	161.94	173.87	197.71
(162.18)	*h*	1148.5	1171.7	1194.8	1218.0	1241.3	1264.7	1288.2	1335.9	1384.4	1433.7	1483.9	1535.0	1640.0
	s	1.8716	1.9055	1.9370	1.9665	1.9944	2.0209	2.0461	2.0934	2.1371	2.1779	2.2162	2.2525	2.3198
10	*v*	38.851	41.942	44.993	48.022	51.036	54.042	57.042	63.030	69.008	74.980	80.949	86.915	98.841
(193.16)	*h*	1146.4	1170.2	1193.8	1217.2	1240.6	1264.1	1287.8	1335.6	1384.2	1433.5	1483.8	1534.9	1639.9
	s	1.7926	1.8275	1.8595	1.8893	1.9174	1.9440	1.9693	2.0167	2.0605	2.1013	2.1397	2.1760	2.2434
15	*v*		27.846	29.906	31.943	33.966	35.979	37.986	41.988	45.981	49.968	53.950	57.931	65.886
(212.99)	*h*		1168.7	1192.7	1216.3	1239.9	1263.6	1287.3	1335.3	1383.9	1433.3	1483.6	1534.7	1639.8
	s		1.7811	1.8137	1.8438	1.8721	1.8989	1.9243	1.9718	2.0156	2.0565	2.0949	2.1312	2.1986
20	*v*		20.796	22.362	23.903	25.430	26.947	28.458	31.467	34.467	37.461	40.451	43.438	49.408
(227.92)	*h*		1167.2	1191.6	1215.5	1239.3	1263.0	1286.9	1334.9	1383.6	1433.1	1483.4	1534.6	1639.7
	s		1.7477	1.7808	1.8113	1.8398	1.8667	1.8922	1.9398	1.9838	2.0247	2.0631	2.0994	2.1669
25	*v*		16.565	17.835	19.079	20.308	21.528	22.741	25.155	27.559	29.957	32.352	34.743	39.521
(240.03)	*h*		1165.6	1190.4	1214.6	1238.6	1262.5	1286.4	1334.6	1383.4	1432.9	1483.3	1534.5	1639.6
	s		1.7213	1.7551	1.7859	1.8146	1.8417	1.8673	1.9150	1.9590	2.0000	2.0384	2.0748	2.1422
30	*v*			14.816	15.863	16.894	17.915	18.930	20.947	22.954	24.955	26.952	28.946	32.930
(250.30)	*h*			1189.3	1213.8	1237.9	1261.9	1286.0	1334.3	1383.1	1432.7	1483.1	1534.3	1639.5
	s			1.7338	1.7650	1.7939	1.8211	1.8468	1.8947	1.9387	1.9797	2.0182	2.0546	2.1221
35	*v*			12.659	13.565	14.455	15.334	16.207	17.941	19.664	21.381	23.095	24.806	28.222
(259.25)	*h*			1188.1	1212.9	1237.2	1261.4	1285.5	1333.9	1382.9	1432.5	1482.9	1534.2	1639.4
	s			1.7156	1.7472	1.7764	1.8037	1.8295	1.8774	1.9216	1.9626	2.0011	2.0375	2.1050
40	*v*			11.041	11.841	12.625	13.398	14.165	15.686	17.197	18.702	20.202	21.700	24.691
(267.22)	*h*			1186.9	1212.0	1236.5	1260.8	1285.0	1333.6	1382.6	1432.3	1482.7	1534.0	1639.3
	s			1.6996	1.7316	1.7610	1.7885	1.8144	1.8625	1.9067	1.9478	1.9863	2.0227	2.0903
45	*v*			9.7814	10.500	11.202	11.893	12.577	13.933	15.278	16.617	17.952	19.285	21.945
(274.42)	*h*			1185.7	1211.1	1235.9	1260.3	1284.6	1333.2	1382.3	1432.1	1482.6	1533.9	1639.2
	s			1.6854	1.7178	1.7474	1.7750	1.8010	1.8493	1.8935	1.9347	1.9733	2.0097	2.0772
50	*v*			8.7735	9.4273	10.063	10.688	11.306	12.530	13.743	14.950	16.153	17.353	19.748
(280.99)	*h*			1184.5	1210.2	1235.1	1259.7	1284.1	1332.9	1382.1	1431.9	1482.4	1533.8	1639.1
	s			1.6724	1.7053	1.7352	1.7629	1.7891	1.8374	1.8818	1.9229	1.9615	1.9980	2.0656
55	*v*			7.9484	8.5492	9.1315	9.7027	10.267	11.382	12.487	13.585	14.680	15.772	17.951
(287.06)	*h*			1183.2	1209.3	1234.4	1259.1	1283.6	1332.6	1381.8	1431.7	1482.2	1533.6	1639.0
	s			1.6606	1.6939	1.7240	1.7520	1.7782	1.8267	1.8711	1.9123	1.9509	1.9874	2.0550
60	*v*			7.2604	7.8173	8.3549	8.8813	9.4004	10.425	11.440	12.448	13.453	14.454	16.453
(292.69)	*h*			1181.9	1208.4	1233.7	1258.6	1283.2	1332.2	1381.5	1431.4	1482.1	1533.5	1638.9
	s			1.6496	1.6834	1.7138	1.7419	1.7682	1.8168	1.8613	1.9026	1.9413	1.9777	2.0454
65	*v*			6.6776	7.1978	7.6978	8.1862	8.6673	9.6160	10.554	11.486	12.414	13.340	15.185
(297.96)	*h*			1180.5	1207.4	1233.0	1258.0	1282.7	1331.9	1381.3	1431.2	1481.9	1533.3	1638.8
	s			1.6394	1.6737	1.7043	1.7326	1.7590	1.8078	1.8523	1.8937	1.9323	1.9688	2.0365
70	*v*				6.6666	7.1344	7.5904	8.0389	8.9223	9.7951	10.662	11.524	12.384	14.099
(302.92)	*h*				1206.5	1232.3	1257.4	1282.2	1331.5	1381.0	1431.0	1481.7	1533.2	1638.7
	s				1.6646	1.6955	1.7239	1.7505	1.7994	1.8440	1.8854	1.9241	1.9606	2.0283

v = specific volume, ft^3/lb_m *h* = enthalpy, Btu/lb_m *s* = entropy, $Btu/(lb_m \cdot °R)$

Table 3 (continued). Superheated Steam

Pressure psia (Sat. T)		Temperature—Degrees Fahrenheit 350	400	450	500	550	600	700	800	900	1000	1100	1200	1400
80	*v*	5.8030	6.2186	6.6220	7.0176	7.4081	7.7949	8.5614	9.3216	10.078	10.831	11.583	12.333	13.831
(312.03)	*h*	1204.5	1230.8	1256.2	1281.3	1306.1	1330.8	1380.5	1430.6	1481.4	1532.9	1585.3	1638.5	1747.5
	s	1.6480	1.6795	1.7082	1.7350	1.7602	1.7842	1.8289	1.8704	1.9092	1.9457	1.9804	2.0135	2.0755
90	*v*	5.1307	5.5061	5.8686	6.2232	6.5724	6.9180	7.6019	8.2794	8.9529	9.6237	10.293	10.960	12.292
(320.27)	*h*	1202.5	1229.3	1255.1	1280.3	1305.3	1330.1	1380.0	1430.2	1481.0	1532.6	1585.0	1638.3	1747.4
	s	1.6330	1.6651	1.6943	1.7213	1.7466	1.7707	1.8156	1.8572	1.8960	1.9326	1.9673	2.0004	2.0625
100	*v*	4.5923	4.9358	5.2658	5.5875	5.9039	6.2165	6.8342	7.4456	8.0529	8.6576	9.2602	9.8615	11.061
(327.82)	*h*	1200.4	1227.8	1253.9	1279.3	1304.5	1329.5	1379.4	1429.8	1480.7	1532.3	1584.8	1638.1	1747.2
	s	1.6194	1.6521	1.6816	1.7089	1.7344	1.7586	1.8037	1.8453	1.8842	1.9209	1.9556	1.9887	2.0508
110	*v*	4.1513	4.4689	4.7724	5.0674	5.3568	5.6424	6.2061	6.7634	7.3166	7.8671	8.4156	8.9627	10.054
(334.78)	*h*	1198.3	1226.2	1252.6	1278.3	1303.6	1328.8	1378.9	1429.4	1480.4	1532.1	1584.6	1637.9	1747.1
	s	1.6068	1.6402	1.6701	1.6976	1.7233	1.7476	1.7928	1.8345	1.8735	1.9102	1.9450	1.9781	2.0402
120	*v*	3.7832	4.0796	4.3611	4.6339	4.9009	5.1640	5.6827	6.1949	6.7030	7.2083	7.7117	8.2137	9.2148
(341.26)	*h*	1196.1	1224.6	1251.4	1277.3	1302.8	1328.1	1378.4	1428.9	1480.0	1531.8	1584.3	1637.7	1746.9
	s	1.5950	1.6292	1.6595	1.6872	1.7131	1.7375	1.7829	1.8247	1.8637	1.9005	1.9353	1.9684	2.0306
130	*v*	3.4711	3.7500	4.0130	4.2670	4.5151	4.7592	5.2398	5.7138	6.1838	6.6509	7.1162	7.5800	8.5047
(347.33)	*h*	1193.8	1223.0	1250.2	1276.3	1302.0	1327.3	1377.8	1428.5	1479.7	1531.5	1584.1	1637.5	1746.8
	s	1.5839	1.6189	1.6496	1.6776	1.7037	1.7282	1.7737	1.8156	1.8547	1.8915	1.9263	1.9595	2.0217
140	*v*		3.4673	3.7145	3.9524	4.1843	4.4122	4.8602	5.3015	5.7387	6.1732	6.6057	7.0368	7.8960
(353.04)	*h*		1221.4	1248.9	1275.3	1301.1	1326.6	1377.3	1428.1	1479.3	1531.2	1583.8	1637.3	1746.7
	s		1.6092	1.6404	1.6687	1.6949	1.7195	1.7652	1.8072	1.8464	1.8832	1.9180	1.9512	2.0135
150	*v*		3.2220	3.4557	3.6797	3.8976	4.1115	4.5311	4.9441	5.3530	5.7591	6.1632	6.5660	7.3685
(358.43)	*h*		1219.7	1247.6	1274.3	1300.3	1325.9	1376.8	1427.7	1479.0	1530.9	1583.6	1637.1	1746.5
	s		1.6001	1.6317	1.6602	1.6866	1.7114	1.7573	1.7994	1.8386	1.8754	1.9103	1.9435	2.0058
160	*v*		3.0073	3.2291	3.4411	3.6468	3.8483	4.2432	4.6314	5.0155	5.3968	5.7761	6.1540	6.9069
(363.55)	*h*		1218.0	1246.3	1273.3	1299.5	1325.2	1376.2	1427.2	1478.7	1530.7	1583.4	1636.9	1746.4
	s		1.5914	1.6235	1.6523	1.6789	1.7038	1.7498	1.7920	1.8313	1.8682	1.9031	1.9363	1.9986
170	*v*		2.8176	3.0291	3.2304	3.4253	3.6160	3.9892	4.3555	4.7177	5.0771	5.4345	5.7905	6.4996
(368.43)	*h*		1216.3	1245.0	1272.2	1298.6	1324.5	1375.7	1426.8	1478.3	1530.4	1583.1	1636.7	1746.2
	s		1.5831	1.6157	1.6448	1.6716	1.6966	1.7428	1.7851	1.8244	1.8613	1.8963	1.9296	1.9919
180	*v*		2.6487	2.8512	3.0431	3.2285	3.4095	3.7633	4.1103	4.4530	4.7929	5.1309	5.4674	6.1376
(373.08)	*h*		1214.5	1243.7	1271.2	1297.7	1323.8	1375.1	1426.4	1478.0	1530.1	1582.9	1636.5	1746.1
	s		1.5752	1.6082	1.6377	1.6646	1.6898	1.7361	1.7785	1.8179	1.8549	1.8899	1.9232	1.9855
190	*v*		2.4975	2.6920	2.8755	3.0524	3.2248	3.5613	3.8908	4.2161	4.5387	4.8592	5.1784	5.8137
(377.54)	*h*		1212.7	1242.4	1270.1	1296.9	1323.0	1374.6	1426.0	1477.6	1529.8	1582.7	1636.3	1745.9
	s		1.5676	1.6011	1.6309	1.6580	1.6833	1.7298	1.7723	1.8118	1.8488	1.8838	1.9171	1.9795
200	*v*		2.3612	2.5485	2.7246	2.8938	3.0585	3.3794	3.6933	4.0030	4.3098	4.6147	4.9182	5.5222
(381.81)	*h*		1210.9	1241.0	1269.1	1296.0	1322.3	1374.1	1425.5	1477.3	1529.5	1582.4	1636.1	1745.8
	s		1.5602	1.5943	1.6243	1.6517	1.6771	1.7238	1.7664	1.8059	1.8430	1.8780	1.9114	1.9738
220	*v*		2.1252	2.3006	2.4638	2.6198	2.7712	3.0652	3.3521	3.6348	3.9146	4.1924	4.4688	5.0186
(389.89)	*h*		1207.0	1238.3	1266.9	1294.3	1320.8	1373.0	1424.7	1476.6	1529.0	1581.9	1635.7	1745.5
	s		1.5461	1.5814	1.6121	1.6399	1.6656	1.7126	1.7554	1.7950	1.8322	1.8673	1.9007	1.9631
240	*v*		1.9277	2.0936	2.2462	2.3914	2.5317	2.8034	3.0678	3.3279	3.5852	3.8404	4.0943	4.5990
(397.41)	*h*		1203.0	1235.4	1264.7	1292.5	1319.4	1371.9	1423.8	1475.9	1528.4	1581.5	1635.3	1745.2
	s		1.5327	1.5694	1.6007	1.6289	1.6549	1.7023	1.7453	1.7851	1.8223	1.8575	1.8909	1.9534

v = specific volume, ft^3/lb_m *h* = enthalpy, Btu/lb_m *s* = entropy, $Btu/(lb_m \cdot {}^\circ R)$

Table 3 (continued). Superheated Steam

Pressure psia (Sat. T)		Temperature—Degrees Fahrenheit 450	500	550	600	700	800	900	1000	1100	1200	1300	1400	1500
260	*v*	1.9182	2.0620	2.1980	2.3290	2.5818	2.8272	3.0683	3.3065	3.5426	3.7774	4.0111	4.2439	4.4763
(404.45)	*h*	1232.5	1262.5	1290.7	1317.9	1370.8	1423.0	1475.2	1527.8	1581.0	1634.9	1689.5	1744.9	1801.1
	s	1.5580	1.5901	1.6188	1.6450	1.6928	1.7359	1.7758	1.8132	1.8484	1.8819	1.9138	1.9445	1.9739
280	*v*	1.7676	1.9039	2.0322	2.1552	2.3919	2.6210	2.8457	3.0676	3.2874	3.5058	3.7231	3.9396	4.1556
(411.09)	*h*	1229.5	1260.2	1288.8	1316.3	1369.7	1422.1	1474.5	1527.2	1580.5	1634.5	1689.1	1744.6	1800.8
	s	1.5473	1.5801	1.6092	1.6358	1.6839	1.7273	1.7673	1.8047	1.8400	1.8735	1.9055	1.9362	1.9656
300	*v*	1.6367	1.7668	1.8883	2.0045	2.2272	2.4423	2.6529	2.8605	3.0662	3.2704	3.4735	3.6758	3.8776
(417.37)	*h*	1226.4	1257.9	1287.0	1314.8	1368.5	1421.2	1473.8	1526.7	1580.0	1634.0	1688.8	1744.3	1800.6
	s	1.5370	1.5706	1.6002	1.6271	1.6756	1.7191	1.7593	1.7968	1.8322	1.8657	1.8978	1.9284	1.9579
320	*v*	1.5219	1.6467	1.7624	1.8726	2.0831	2.2859	2.4841	2.6793	2.8726	3.0644	3.2551	3.4450	3.6344
(423.33)	*h*	1223.3	1255.5	1285.1	1313.3	1367.4	1420.4	1473.1	1526.1	1579.6	1633.6	1688.4	1744.0	1800.3
	s	1.5271	1.5615	1.5916	1.6189	1.6677	1.7115	1.7518	1.7894	1.8248	1.8584	1.8905	1.9212	1.9507
340	*v*	1.4203	1.5405	1.6512	1.7562	1.9560	2.1478	2.3351	2.5195	2.7018	2.8826	3.0624	3.2414	3.4198
(429.01)	*h*	1220.0	1253.1	1283.2	1311.7	1366.3	1419.5	1472.4	1525.5	1579.1	1633.2	1688.1	1743.7	1800.1
	s	1.5175	1.5529	1.5835	1.6111	1.6603	1.7043	1.7447	1.7824	1.8179	1.8516	1.8837	1.9144	1.9439
360	*v*	1.3297	1.4460	1.5523	1.6527	1.8429	2.0252	2.2028	2.3774	2.5500	2.7211	2.8911	3.0604	3.2291
(434.43)	*h*	1216.6	1250.6	1281.3	1310.1	1365.2	1418.6	1471.7	1525.0	1578.6	1632.8	1687.7	1743.4	1799.8
	s	1.5082	1.5446	1.5757	1.6036	1.6533	1.6975	1.7381	1.7758	1.8114	1.8451	1.8772	1.9080	1.9375
380	*v*	1.2483	1.3613	1.4637	1.5600	1.7418	1.9154	2.0843	2.2502	2.4141	2.5765	2.7379	2.8984	3.0584
(439.63)	*h*	1213.1	1248.1	1279.3	1308.5	1364.0	1417.8	1471.0	1524.4	1578.1	1632.4	1687.4	1743.1	1799.5
	s	1.4991	1.5365	1.5683	1.5965	1.6466	1.6910	1.7317	1.7696	1.8052	1.8389	1.8711	1.9019	1.9314
400	*v*	1.1747	1.2850	1.3839	1.4765	1.6507	1.8166	1.9777	2.1358	2.2919	2.4464	2.6000	2.7527	2.9048
(444.63)	*h*	1209.5	1245.6	1277.3	1306.9	1362.9	1416.9	1470.3	1523.8	1577.7	1632.0	1687.0	1742.8	1799.3
	s	1.4901	1.5288	1.5611	1.5897	1.6402	1.6848	1.7257	1.7636	1.7993	1.8331	1.8653	1.8961	1.9257
450	*v*		1.1232	1.2151	1.3001	1.4584	1.6079	1.7526	1.8942	2.0337	2.1718	2.3088	2.4449	2.5805
(456.32)	*h*		1238.9	1272.2	1302.8	1360.0	1414.7	1468.6	1522.4	1576.5	1631.0	1686.2	1742.0	1798.6
	s		1.5103	1.5442	1.5737	1.6253	1.6705	1.7117	1.7499	1.7857	1.8196	1.8519	1.8828	1.9124
500	*v*		0.9930	1.0797	1.1587	1.3044	1.4409	1.5725	1.7009	1.8272	1.9521	2.0758	2.1987	2.3211
(467.05)	*h*		1231.9	1267.0	1298.6	1357.0	1412.4	1466.8	1520.9	1575.3	1630.0	1685.3	1741.3	1798.0
	s		1.4928	1.5284	1.5591	1.6117	1.6576	1.6991	1.7375	1.7735	1.8076	1.8399	1.8708	1.9005
550	*v*		0.8856	0.9685	1.0428	1.1783	1.3043	1.4251	1.5428	1.6583	1.7723	1.8852	1.9973	2.1088
(476.98)	*h*		1224.5	1261.5	1294.3	1354.0	1410.2	1465.0	1519.5	1574.0	1629.0	1684.4	1740.5	1797.3
	s		1.4760	1.5137	1.5454	1.5993	1.6457	1.6876	1.7263	1.7624	1.7966	1.8290	1.8600	1.8898
600	*v*		0.7953	0.8754	0.9460	1.0732	1.1904	1.3023	1.4110	1.5175	1.6225	1.7264	1.8295	1.9320
(486.25)	*h*		1216.5	1255.8	1289.9	1351.0	1407.9	1463.2	1518.0	1572.8	1628.0	1683.6	1739.8	1796.7
	s		1.4597	1.4996	1.5325	1.5877	1.6348	1.6770	1.7159	1.7523	1.7865	1.8190	1.8501	1.8799
650	*v*		0.7178	0.7962	0.8639	0.9841	1.0940	1.1983	1.2994	1.3983	1.4957	1.5920	1.6874	1.7823
(494.94)	*h*		1208.0	1249.9	1285.3	1347.9	1405.6	1461.4	1516.6	1571.6	1626.9	1682.7	1739.0	1796.0
	s		1.4434	1.4861	1.5203	1.5768	1.6246	1.6672	1.7063	1.7428	1.7772	1.8098	1.8410	1.8708
700	*v*			0.7280	0.7933	0.9077	1.0113	1.1092	1.2038	1.2962	1.3871	1.4768	1.5657	1.6540
(503.14)	*h*			1243.7	1280.6	1344.8	1403.3	1459.6	1515.1	1570.4	1625.9	1681.8	1738.3	1795.4
	s			1.4730	1.5087	1.5666	1.6150	1.6580	1.6974	1.7341	1.7686	1.8013	1.8325	1.8624
750	*v*			0.6684	0.7319	0.8414	0.9396	1.0320	1.1209	1.2077	1.2929	1.3770	1.4602	1.5428
(510.90)	*h*			1237.3	1275.8	1341.6	1401.0	1457.8	1513.6	1569.2	1624.9	1681.0	1737.5	1794.7
	s			1.4602	1.4975	1.5569	1.6060	1.6494	1.6890	1.7259	1.7605	1.7933	1.8245	1.8545

v = specific volume, ft^3/lb_m h = enthalpy, Btu/lb_m s = entropy, $Btu/(lb_m \cdot {}^\circ R)$

Table 3 (continued). Superheated Steam

Pressure psia (Sat. T)		Temperature—Degrees Fahrenheit 550	600	650	700	750	800	900	1000	1100	1200	1300	1400	1500
800	*v*	0.6159	0.6780	0.7328	0.7834	0.8311	0.8768	0.9643	1.0484	1.1302	1.2105	1.2896	1.3679	1.4456
(518.27)	*h*	1230.5	1270.8	1306.0	1338.4	1369.0	1398.6	1456.0	1512.1	1568.0	1623.9	1680.1	1736.8	1794.0
	s	1.4476	1.4866	1.5191	1.5476	1.5735	1.5975	1.6413	1.6812	1.7182	1.7529	1.7858	1.8171	1.8471
850	*v*	0.5691	0.6302	0.6834	0.7320	0.7777	0.8214	0.9047	0.9844	1.0619	1.1378	1.2125	1.2864	1.3597
(525.30)	*h*	1223.4	1265.7	1302.0	1335.1	1366.3	1396.2	1454.1	1510.7	1566.7	1622.8	1679.2	1736.0	1793.4
	s	1.4351	1.4761	1.5096	1.5388	1.5651	1.5894	1.6336	1.6737	1.7109	1.7457	1.7787	1.8101	1.8401
900	*v*	0.5269	0.5875	0.6394	0.6864	0.7303	0.7721	0.8516	0.9275	1.0011	1.0731	1.1440	1.2140	1.2834
(532.02)	*h*	1215.8	1260.4	1297.9	1331.8	1363.5	1393.8	1452.3	1509.2	1565.5	1621.8	1678.3	1735.3	1792.7
	s	1.4226	1.4658	1.5004	1.5302	1.5570	1.5816	1.6262	1.6666	1.7040	1.7389	1.7720	1.8035	1.8336
950	*v*	0.4887	0.5491	0.5998	0.6454	0.6878	0.7280	0.8041	0.8766	0.9467	1.0153	1.0827	1.1492	1.2152
(538.46)	*h*	1207.8	1254.9	1293.7	1328.4	1360.6	1391.4	1450.4	1507.7	1564.3	1620.8	1677.4	1734.5	1792.1
	s	1.4100	1.4557	1.4914	1.5220	1.5492	1.5742	1.6192	1.6599	1.6974	1.7325	1.7657	1.7972	1.8273
1000	*v*	0.4538	0.5143	0.5641	0.6085	0.6495	0.6883	0.7614	0.8308	0.8978	0.9632	1.0275	1.0909	1.1537
(544.65)	*h*	1199.1	1249.3	1289.4	1324.9	1357.8	1389.0	1448.5	1506.2	1563.0	1619.7	1676.6	1733.7	1791.4
	s	1.3971	1.4457	1.4827	1.5140	1.5418	1.5670	1.6125	1.6535	1.6911	1.7264	1.7596	1.7912	1.8214
1100	*v*		0.4536	0.5022	0.5446	0.5833	0.6196	0.6875	0.7516	0.8133	0.8733	0.9322	0.9902	1.0476
(556.35)	*h*		1237.2	1280.5	1317.9	1351.9	1384.0	1444.7	1503.2	1560.6	1617.7	1674.8	1732.2	1790.1
	s		1.4259	1.4658	1.4987	1.5275	1.5535	1.5999	1.6414	1.6794	1.7149	1.7483	1.7801	1.8104
1200	*v*		0.4020	0.4501	0.4910	0.5279	0.5622	0.6259	0.6856	0.7428	0.7984	0.8528	0.9063	0.9592
(567.26)	*h*		1224.1	1271.1	1310.5	1345.9	1378.9	1440.9	1500.1	1558.1	1615.6	1673.0	1730.7	1788.8
	s		1.4061	1.4494	1.4842	1.5141	1.5408	1.5882	1.6302	1.6686	1.7043	1.7379	1.7698	1.8002
1300	*v*		0.3574	0.4057	0.4455	0.4809	0.5136	0.5738	0.6298	0.6832	0.7350	0.7856	0.8353	0.8843
(577.50)	*h*		1209.8	1261.3	1302.9	1339.7	1373.7	1437.0	1497.0	1555.6	1613.5	1671.3	1729.2	1787.4
	s		1.3859	1.4334	1.4701	1.5012	1.5288	1.5772	1.6198	1.6585	1.6945	1.7283	1.7604	1.7909
1400	*v*		0.3178	0.3671	0.4063	0.4405	0.4718	0.5290	0.5819	0.6321	0.6806	0.7280	0.7744	0.8202
(587.14)	*h*		1193.7	1250.8	1295.0	1333.4	1368.5	1433.1	1493.9	1553.1	1611.4	1669.5	1727.7	1786.1
	s		1.3649	1.4175	1.4566	1.4890	1.5174	1.5668	1.6100	1.6491	1.6854	1.7194	1.7515	1.7821
1500	*v*		0.2819	0.3331	0.3720	0.4054	0.4356	0.4902	0.5403	0.5878	0.6335	0.6780	0.7217	0.7646
(596.27)	*h*		1175.4	1239.6	1286.8	1326.9	1363.1	1429.1	1490.8	1550.5	1609.3	1667.7	1726.1	1784.8
	s		1.3423	1.4016	1.4433	1.4771	1.5064	1.5569	1.6007	1.6403	1.6768	1.7110	1.7433	1.7740
1600	*v*			0.3029	0.3418	0.3745	0.4037	0.4562	0.5040	0.5490	0.5923	0.6343	0.6755	0.7160
(604.93)	*h*			1227.7	1278.3	1320.2	1357.6	1425.1	1487.7	1548.0	1607.2	1665.9	1724.6	1783.5
	s			1.3855	1.4302	1.4656	1.4959	1.5475	1.5920	1.6319	1.6687	1.7031	1.7355	1.7663
1700	*v*			0.2757	0.3149	0.3471	0.3756	0.4262	0.4719	0.5148	0.5559	0.5958	0.6348	0.6731
(613.19)	*h*			1214.7	1269.3	1313.3	1351.9	1421.0	1484.5	1545.4	1605.0	1664.1	1723.0	1782.1
	s			1.3691	1.4172	1.4544	1.4857	1.5385	1.5836	1.6240	1.6610	1.6956	1.7282	1.7591
1800	*v*			0.2507	0.2908	0.3227	0.3505	0.3994	0.4433	0.4844	0.5236	0.5615	0.5986	0.6349
(621.07)	*h*			1200.6	1259.9	1306.2	1346.2	1416.9	1481.3	1542.8	1602.9	1662.3	1721.5	1780.8
	s			1.3520	1.4043	1.4434	1.4758	1.5299	1.5756	1.6164	1.6537	1.6885	1.7212	1.7522
1900	*v*			0.2277	0.2689	0.3007	0.3280	0.3755	0.4178	0.4572	0.4947	0.5309	0.5662	0.6008
(628.62)	*h*			1185.1	1250.1	1298.8	1340.3	1412.7	1478.1	1540.2	1600.8	1660.5	1720.0	1779.5
	s			1.3340	1.3914	1.4325	1.4662	1.5215	1.5680	1.6091	1.6468	1.6817	1.7146	1.7457
2000	*v*			0.2059	0.2489	0.2807	0.3076	0.3539	0.3948	0.4327	0.4686	0.5033	0.5370	0.5701
(635.85)	*h*			1167.5	1239.7	1291.2	1334.3	1408.5	1474.9	1537.6	1598.6	1658.7	1718.4	1778.1
	s			1.3146	1.3783	1.4218	1.4567	1.5134	1.5606	1.6022	1.6401	1.6752	1.7083	1.7395

v = specific volume, ft^3/lb_m h = enthalpy, Btu/lb_m s = entropy, $Btu/(lb_m \cdot {}^\circ R)$

Table 3 (continued). Superheated Steam

Pressure psia (Sat. T)		Temperature—Degrees Fahrenheit 650	700	750	800	850	900	950	1000	1100	1200	1300	1400	1500
2200	v	0.1635	0.2136	0.2459	0.2723	0.2955	0.3166	0.3363	0.3550	0.3903	0.4236	0.4556	0.4867	0.5171
(649.50)	h	1122.0	1217.1	1275.2	1321.8	1362.6	1399.8	1434.8	1468.3	1532.4	1594.3	1655.1	1715.3	1775.5
	s	1.2673	1.3514	1.4006	1.4383	1.4700	1.4980	1.5232	1.5466	1.5891	1.6276	1.6631	1.6964	1.7279
2400	v		0.1827	0.2164	0.2426	0.2651	0.2854	0.3042	0.3219	0.3550	0.3861	0.4159	0.4447	0.4729
(662.16)	h		1191.0	1258.0	1308.8	1352.0	1390.9	1427.2	1461.6	1527.1	1589.9	1651.4	1712.2	1772.8
	s		1.3226	1.3793	1.4204	1.4541	1.4833	1.5094	1.5335	1.5769	1.6159	1.6519	1.6855	1.7172
2600	v		0.1548	0.1909	0.2173	0.2393	0.2589	0.2769	0.2938	0.3251	0.3544	0.3823	0.4092	0.4355
(673.98)	h		1160.0	1239.4	1295.0	1341.0	1381.8	1419.4	1454.8	1521.7	1585.6	1647.8	1709.1	1770.1
	s		1.2905	1.3577	1.4027	1.4386	1.4692	1.4963	1.5210	1.5654	1.6050	1.6414	1.6753	1.7073
2800	v		0.1280	0.1685	0.1953	0.2171	0.2362	0.2535	0.2697	0.2995	0.3272	0.3535	0.3788	0.4034
(685.03)	h		1120.6	1219.0	1280.5	1329.7	1372.5	1411.5	1447.9	1516.3	1581.2	1644.1	1706.0	1767.4
	s		1.2520	1.3353	1.3852	1.4235	1.4555	1.4837	1.5092	1.5545	1.5948	1.6316	1.6658	1.6980
3000	v		0.0984	0.1484	0.1760	0.1977	0.2164	0.2332	0.2487	0.2773	0.3037	0.3286	0.3525	0.3757
(695.41)	h		1059.8	1196.4	1265.2	1317.9	1362.9	1403.4	1441.0	1510.9	1576.7	1640.4	1702.8	1764.7
	s		1.1959	1.3118	1.3675	1.4086	1.4423	1.4716	1.4978	1.5442	1.5851	1.6223	1.6569	1.6893
3200	v			0.1300	0.1589	0.1806	0.1990	0.2154	0.2304	0.2579	0.2830	0.3067	0.3294	0.3514
(705.10)	h			1171.0	1248.9	1305.7	1353.0	1395.1	1433.9	1505.4	1572.3	1636.7	1699.7	1762.0
	s			1.2866	1.3497	1.3939	1.4294	1.4598	1.4869	1.5343	1.5759	1.6136	1.6484	1.6810
3400	v			0.1129	0.1435	0.1654	0.1836	0.1996	0.2143	0.2407	0.2648	0.2875	0.3091	0.3299
	h			1141.8	1231.6	1292.9	1342.9	1386.7	1426.6	1499.8	1567.8	1633.0	1696.6	1759.3
	s			1.2587	1.3316	1.3793	1.4168	1.4484	1.4763	1.5248	1.5671	1.6052	1.6403	1.6732
3600	v			0.0964	0.1296	0.1518	0.1699	0.1856	0.1999	0.2255	0.2487	0.2704	0.2910	0.3109
	h			1107.2	1213.2	1279.7	1332.5	1378.1	1419.3	1494.2	1563.3	1629.2	1693.4	1756.6
	s			1.2269	1.3129	1.3648	1.4043	1.4373	1.4660	1.5157	1.5586	1.5972	1.6327	1.6658
3800	v			0.0802	0.1169	0.1396	0.1576	0.1731	0.1870	0.2118	0.2342	0.2551	0.2748	0.2939
	h			1064.4	1193.4	1266.0	1321.9	1369.4	1411.9	1488.5	1558.8	1625.5	1690.3	1753.9
	s			1.1888	1.2936	1.3502	1.3921	1.4264	1.4561	1.5069	1.5505	1.5896	1.6254	1.6587
4000	v			0.0637	0.1052	0.1285	0.1465	0.1617	0.1754	0.1996	0.2212	0.2413	0.2603	0.2785
	h			1009.2	1172.1	1251.7	1310.9	1360.5	1404.4	1482.8	1554.2	1621.7	1687.1	1751.2
	s			1.1409	1.2734	1.3355	1.3799	1.4157	1.4463	1.4983	1.5427	1.5822	1.6183	1.6519
4500	v			0.0393	0.0796	0.1047	0.1229	0.1378	0.1509	0.1737	0.1938	0.2122	0.2296	0.2462
	h			891.0	1111.1	1213.4	1282.3	1337.5	1385.3	1468.4	1542.7	1612.3	1679.2	1744.5
	s			1.0395	1.2183	1.2980	1.3497	1.3896	1.4229	1.4780	1.5242	1.5650	1.6019	1.6361
5000	v			0.0337	0.0594	0.0855	0.1039	0.1186	0.1313	0.1530	0.1719	0.1890	0.2051	0.2204
	h			853.0	1041.9	1171.5	1252.1	1313.7	1365.5	1453.8	1531.2	1602.9	1671.3	1737.7
	s			1.0053	1.1582	1.2593	1.3198	1.3643	1.4005	1.4590	1.5071	1.5491	1.5869	1.6217
5500	v			0.0313	0.0463	0.0701	0.0885	0.1030	0.1153	0.1361	0.1540	0.1701	0.1851	0.1993
	h			834.1	980.9	1126.9	1220.4	1289.1	1345.4	1439.0	1519.6	1593.4	1663.4	1731.0
	s			0.9872	1.1060	1.2198	1.2899	1.3396	1.3788	1.4409	1.4910	1.5343	1.5729	1.6084
6000	v			0.0298	0.0395	0.0582	0.0759	0.0901	0.1021	0.1221	0.1391	0.1544	0.1684	0.1817
	h			821.7	940.8	1083.1	1187.7	1263.8	1324.8	1424.0	1507.9	1583.9	1655.5	1724.3
	s			0.9747	1.0710	1.1818	1.2604	1.3154	1.3579	1.4237	1.4759	1.5204	1.5599	1.5960
7000	v			0.0279	0.0334	0.0438	0.0576	0.0705	0.0817	0.1004	0.1160	0.1298	0.1424	0.1542
	h			805.6	898.4	1013.3	1124.8	1213.1	1283.4	1394.0	1484.6	1565.1	1639.8	1711.0
	s			0.9570	1.0321	1.1215	1.2051	1.2689	1.3179	1.3913	1.4476	1.4948	1.5361	1.5734

v = specific volume, ft^3/lb_m h = enthalpy, Btu/lb_m s = entropy, $Btu/(lb_m \cdot {}^\circ R)$

Table 3 (continued). Superheated Steam

Pressure psia (Sat. T)		Temperature—Degrees Fahrenheit 750	800	850	900	950	1000	1050	1100	1150	1200	1300	1400	1500
8000	*v*	0.0267	0.0306	0.0371	0.0465	0.0571	0.0672	0.0763	0.0844	0.0919	0.0988	0.1115	0.1230	0.1337
	h	795.1	876.0	971.0	1073.2	1165.6	1243.0	1307.9	1364.3	1414.9	1461.5	1546.4	1624.3	1697.9
	s	0.9441	1.0096	1.0836	1.1601	1.2269	1.2808	1.3246	1.3614	1.3933	1.4218	1.4715	1.5146	1.5532
9000	*v*	0.0258	0.0289	0.0335	0.0401	0.0483	0.0568	0.0650	0.0725	0.0794	0.0858	0.0975	0.1081	0.1179
	h	787.5	861.6	945.1	1036.2	1125.7	1205.8	1275.1	1335.6	1389.6	1438.9	1528.1	1609.1	1685.1
	s	0.9338	0.9938	1.0588	1.1270	1.1917	1.2475	1.2942	1.3337	1.3678	1.3980	1.4502	1.4949	1.5348
10000	*v*	0.0251	0.0276	0.0312	0.0362	0.0425	0.0495	0.0566	0.0633	0.0697	0.0756	0.0864	0.0962	0.1053
	h	781.8	851.3	927.6	1010.1	1094.2	1173.6	1245.0	1308.5	1365.4	1417.2	1510.4	1594.3	1672.6
	s	0.9252	0.9815	1.0409	1.1027	1.1634	1.2188	1.2669	1.3083	1.3442	1.3759	1.4304	1.4768	1.5179
11000	*v*	0.0245	0.0267	0.0296	0.0336	0.0385	0.0442	0.0503	0.0563	0.0621	0.0675	0.0776	0.0867	0.0952
	h	777.3	843.6	915.0	991.3	1070.0	1146.8	1218.5	1283.5	1342.6	1396.5	1493.3	1579.9	1660.4
	s	0.9177	0.9714	1.0269	1.0841	1.1409	1.1945	1.2428	1.2852	1.3225	1.3555	1.4121	1.4600	1.5022
12000	*v*	0.0240	0.0260	0.0285	0.0317	0.0357	0.0404	0.0456	0.0509	0.0561	0.0611	0.0704	0.0789	0.0868
	h	773.8	837.5	905.3	977.1	1051.3	1125.1	1195.6	1261.2	1321.6	1377.1	1477.0	1566.1	1648.6
	s	0.9111	0.9627	1.0155	1.0692	1.1228	1.1743	1.2218	1.2646	1.3027	1.3366	1.3951	1.4444	1.4876
13000	*v*	0.0236	0.0253	0.0275	0.0303	0.0336	0.0376	0.0420	0.0467	0.0513	0.0559	0.0645	0.0724	0.0798
	h	771.0	832.7	897.8	966.1	1036.6	1107.5	1176.3	1241.6	1302.6	1359.2	1461.5	1552.9	1637.2
	s	0.9051	0.9551	1.0057	1.0569	1.1079	1.1573	1.2036	1.2462	1.2847	1.3193	1.3792	1.4297	1.4739
14000	*v*	0.0232	0.0248	0.0267	0.0291	0.0320	0.0354	0.0392	0.0433	0.0475	0.0516	0.0596	0.0670	0.0739
	h	768.7	828.8	891.7	957.3	1024.9	1093.1	1160.1	1224.6	1285.7	1342.8	1446.9	1540.3	1626.3
	s	0.8996	0.9483	0.9973	1.0464	1.0952	1.1428	1.1879	1.2299	1.2685	1.3035	1.3644	1.4161	1.4611
15000	*v*	0.0229	0.0243	0.0261	0.0282	0.0308	0.0337	0.0370	0.0406	0.0443	0.0481	0.0554	0.0624	0.0689
	h	766.9	825.6	886.7	950.1	1015.3	1081.2	1146.5	1209.8	1270.6	1327.9	1433.3	1528.4	1615.8
	s	0.8946	0.9422	0.9897	1.0372	1.0843	1.1303	1.1742	1.2155	1.2539	1.2890	1.3506	1.4032	1.4490

v = specific volume, ft^3/lb_m *h* = enthalpy, Btu/lb_m *s* = entropy, $Btu/(lb_m \cdot {}^\circ R)$

Table 4. Properties of Saturated Water and Steam (Temperature)

Temp. °C	Pressure MPa	Volume, m³/kg v_L	v_V	Enthalpy, kJ/kg h_L	h_V	Entropy, kJ/(kg·K) s_L	s_V	Temp. °C
0.01	0.0006117	0.0010002	206.00	0.001	2500.9	0.0000	9.1555	**0.01**
5	0.0008726	0.0010001	147.02	21.019	2510.1	0.0763	9.0249	**5**
10	0.001228	0.0010003	106.31	42.021	2519.2	0.1511	8.8998	**10**
15	0.001706	0.0010009	77.881	62.984	2528.4	0.2245	8.7804	**15**
20	0.002339	0.0010018	57.761	83.920	2537.5	0.2965	8.6661	**20**
25	0.003170	0.0010030	43.341	104.84	2546.5	0.3673	8.5568	**25**
30	0.004247	0.0010044	32.882	125.75	2555.6	0.4368	8.4521	**30**
35	0.005629	0.0010060	25.208	146.64	2564.6	0.5052	8.3518	**35**
40	0.007384	0.0010079	19.517	167.54	2573.5	0.5724	8.2557	**40**
45	0.009594	0.0010099	15.253	188.44	2582.5	0.6386	8.1634	**45**
50	0.012351	0.0010121	12.028	209.34	2591.3	0.7038	8.0749	**50**
55	0.015761	0.0010145	9.5649	230.24	2600.1	0.7680	7.9899	**55**
60	0.019946	0.0010171	7.6677	251.15	2608.8	0.8312	7.9082	**60**
65	0.025041	0.0010199	6.1938	272.08	2617.5	0.8935	7.8296	**65**
70	0.031201	0.0010228	5.0397	293.02	2626.1	0.9550	7.7540	**70**
75	0.038595	0.0010258	4.1291	313.97	2634.6	1.0156	7.6812	**75**
80	0.047415	0.0010290	3.4053	334.95	2643.0	1.0754	7.6110	**80**
85	0.057867	0.0010324	2.8259	355.95	2651.3	1.1344	7.5434	**85**
90	0.070182	0.0010359	2.3591	376.97	2659.5	1.1927	7.4781	**90**
95	0.084609	0.0010396	1.9806	398.02	2667.6	1.2502	7.4150	**95**
100	0.10142	0.0010435	1.6719	419.10	2675.6	1.3070	7.3541	**100**
105	0.12090	0.0010474	1.4185	440.21	2683.4	1.3632	7.2951	**105**
110	0.14338	0.0010516	1.2094	461.36	2691.1	1.4187	7.2380	**110**
115	0.16918	0.0010559	1.0359	482.55	2698.6	1.4735	7.1827	**115**
120	0.19867	0.0010603	0.89130	503.78	2705.9	1.5278	7.1291	**120**
125	0.23222	0.0010649	0.77011	525.06	2713.1	1.5815	7.0770	**125**
130	0.27026	0.0010697	0.66808	546.39	2720.1	1.6346	7.0264	**130**
135	0.31320	0.0010747	0.58180	567.77	2726.9	1.6872	6.9772	**135**
140	0.36150	0.0010798	0.50852	589.20	2733.4	1.7393	6.9293	**140**
145	0.41563	0.0010850	0.44602	610.69	2739.8	1.7909	6.8826	**145**
150	0.47610	0.0010905	0.39250	632.25	2745.9	1.8420	6.8370	**150**
155	0.54342	0.0010962	0.34650	653.88	2751.8	1.8926	6.7926	**155**
160	0.61814	0.0011020	0.30682	675.57	2757.4	1.9428	6.7491	**160**
165	0.70082	0.0011080	0.27246	697.35	2762.8	1.9926	6.7066	**165**
170	0.79205	0.0011143	0.24262	719.21	2767.9	2.0419	6.6649	**170**
175	0.89245	0.0011207	0.21660	741.15	2772.7	2.0909	6.6241	**175**
180	1.0026	0.0011274	0.19386	763.19	2777.2	2.1395	6.5841	**180**
185	1.1233	0.0011343	0.17392	785.32	2781.4	2.1878	6.5447	**185**
190	1.2550	0.0011414	0.15638	807.57	2785.3	2.2358	6.5060	**190**
195	1.3986	0.0011488	0.14091	829.92	2788.9	2.2834	6.4679	**195**
200	1.5547	0.0011565	0.12722	852.39	2792.1	2.3308	6.4303	**200**
205	1.7240	0.0011645	0.11509	874.99	2794.9	2.3779	6.3932	**205**
210	1.9074	0.0011727	0.10430	897.73	2797.4	2.4248	6.3565	**210**
215	2.1055	0.0011813	0.094689	920.61	2799.4	2.4714	6.3202	**215**
220	2.3193	0.0011902	0.086101	943.64	2801.1	2.5178	6.2842	**220**

Table 4. Properties of Saturated Water and Steam (Temperature)

Temp. °C	Pressure MPa	Volume, m³/kg v_L	v_V	Enthalpy, kJ/kg h_L	h_V	Entropy, kJ/(kg·K) s_L	s_V	Temp. °C
225	2.5494	0.001199	0.078411	966.84	2802.3	2.5641	6.2485	**225**
230	2.7968	0.001209	0.071510	990.21	2803.0	2.6102	6.2131	**230**
235	3.0622	0.001219	0.065304	1013.8	2803.3	2.6561	6.1777	**235**
240	3.3467	0.001229	0.059710	1037.5	2803.1	2.7019	6.1425	**240**
245	3.6509	0.001240	0.054658	1061.5	2802.3	2.7477	6.1074	**245**
250	3.9759	0.001252	0.050087	1085.7	2801.0	2.7934	6.0722	**250**
255	4.3227	0.001264	0.045941	1110.1	2799.1	2.8391	6.0370	**255**
260	4.6921	0.001276	0.042175	1134.8	2796.6	2.8847	6.0017	**260**
265	5.0851	0.001289	0.038748	1159.8	2793.5	2.9304	5.9662	**265**
270	5.5028	0.001303	0.035622	1185.1	2789.7	2.9762	5.9304	**270**
275	5.9463	0.001318	0.032767	1210.7	2785.1	3.0221	5.8943	**275**
280	6.4165	0.001333	0.030154	1236.7	2779.8	3.0681	5.8578	**280**
285	6.9145	0.001349	0.027758	1263.0	2773.7	3.1143	5.8208	**285**
290	7.4416	0.001366	0.025557	1289.8	2766.6	3.1608	5.7832	**290**
295	7.9990	0.001385	0.023531	1317.0	2758.6	3.2076	5.7449	**295**
300	8.5877	0.001404	0.021663	1344.8	2749.6	3.2547	5.7058	**300**
305	9.2092	0.001425	0.019937	1373.1	2739.4	3.3024	5.6656	**305**
310	9.8647	0.001448	0.018339	1402.0	2727.9	3.3506	5.6243	**310**
315	10.556	0.001472	0.016856	1431.6	2715.1	3.3994	5.5816	**315**
320	11.284	0.001499	0.015476	1462.1	2700.7	3.4491	5.5373	**320**
325	12.051	0.001528	0.014189	1493.4	2684.5	3.4997	5.4911	**325**
330	12.858	0.001561	0.012984	1525.7	2666.2	3.5516	5.4425	**330**
335	13.707	0.001597	0.011852	1559.3	2645.6	3.6048	5.3910	**335**
340	14.600	0.001638	0.010784	1594.4	2622.1	3.6599	5.3359	**340**
345	15.540	0.001685	0.009770	1631.4	2595.0	3.7175	5.2763	**345**
350	16.529	0.001740	0.008801	1670.9	2563.6	3.7783	5.2109	**350**
355	17.570	0.001808	0.007866	1713.7	2526.4	3.8438	5.1377	**355**
360	18.666	0.001895	0.006945	1761.5	2481.0	3.9164	5.0527	**360**
365	19.822	0.002016	0.006004	1817.6	2422.0	4.0011	4.9482	**365**
370	21.043	0.002222	0.004946	1892.6	2333.5	4.1142	4.7996	**370**
T_c	22.064	0.00311	0.00311	2087.5	2087.5	4.4120	4.4120	T_c

T_c = 373.946 °C

Table 5. Properties of Saturated Water and Steam (Pressure)

Press. MPa	Temp. °C	Volume, m³/kg v_L	v_V	Enthalpy, kJ/kg h_L	h_V	Entropy, kJ/(kg·K) s_L	s_V	Press. MPa
0.001	6.97	0.0010001	129.18	29.298	2513.7	0.1059	8.9749	**0.001**
0.002	17.50	0.0010014	66.990	73.435	2532.9	0.2606	8.7227	**0.002**
0.003	24.08	0.0010028	45.655	100.99	2544.9	0.3543	8.5766	**0.003**
0.004	28.96	0.0010041	34.792	121.40	2553.7	0.4224	8.4735	**0.004**
0.005	32.88	0.0010053	28.186	137.77	2560.8	0.4763	8.3939	**0.005**
0.006	36.16	0.0010064	23.734	151.49	2566.7	0.5209	8.3291	**0.006**
0.007	39.00	0.0010075	20.525	163.37	2571.8	0.5591	8.2746	**0.007**
0.008	41.51	0.0010085	18.099	173.85	2576.2	0.5925	8.2274	**0.008**
0.009	43.76	0.0010094	16.200	183.26	2580.3	0.6223	8.1859	**0.009**
0.010	45.81	0.0010103	14.671	191.81	2583.9	0.6492	8.1489	**0.010**
0.012	49.42	0.0010119	12.359	206.91	2590.3	0.6963	8.0850	**0.012**
0.014	52.55	0.0010133	10.691	219.99	2595.8	0.7366	8.0312	**0.014**
0.016	55.31	0.0010147	9.4309	231.55	2600.7	0.7720	7.9847	**0.016**
0.018	57.80	0.0010160	8.4433	241.95	2605.0	0.8035	7.9437	**0.018**
0.020	60.06	0.0010171	7.6482	251.40	2608.9	0.8320	7.9072	**0.020**
0.025	64.96	0.0010198	6.2034	271.93	2617.4	0.8931	7.8302	**0.025**
0.030	69.10	0.0010222	5.2286	289.23	2624.6	0.9439	7.7675	**0.030**
0.035	72.68	0.0010244	4.5252	304.25	2630.7	0.9876	7.7146	**0.035**
0.040	75.86	0.0010264	3.9931	317.57	2636.1	1.0259	7.6690	**0.040**
0.045	78.71	0.0010282	3.5761	329.55	2640.9	1.0601	7.6288	**0.045**
0.05	81.32	0.0010299	3.2401	340.48	2645.2	1.0910	7.5930	**0.05**
0.06	85.93	0.0010331	2.7318	359.84	2652.9	1.1452	7.5311	**0.06**
0.07	89.93	0.0010359	2.3649	376.68	2659.4	1.1919	7.4790	**0.07**
0.08	93.49	0.0010385	2.0872	391.64	2665.2	1.2328	7.4339	**0.08**
0.09	96.69	0.0010409	1.8695	405.13	2670.3	1.2694	7.3942	**0.09**
0.10	99.61	0.0010431	1.6940	417.44	2674.9	1.3026	7.3588	**0.10**
0.12	104.78	0.0010473	1.4284	439.30	2683.1	1.3608	7.2976	**0.12**
0.14	109.29	0.0010510	1.2366	458.37	2690.0	1.4109	7.2460	**0.14**
0.16	113.30	0.0010544	1.0914	475.34	2696.0	1.4549	7.2014	**0.16**
0.18	116.91	0.0010576	0.97753	490.67	2701.4	1.4944	7.1620	**0.18**
0.20	120.21	0.0010605	0.88574	504.68	2706.2	1.5301	7.1269	**0.20**
0.25	127.41	0.0010672	0.71870	535.35	2716.5	1.6072	7.0524	**0.25**
0.30	133.53	0.0010732	0.60579	561.46	2724.9	1.6718	6.9916	**0.30**
0.35	138.86	0.0010786	0.52420	584.31	2732.0	1.7275	6.9401	**0.35**
0.40	143.61	0.0010836	0.46239	604.72	2738.1	1.7766	6.8954	**0.40**
0.45	147.91	0.0010882	0.41390	623.22	2743.4	1.8206	6.8560	**0.45**
0.50	151.84	0.0010926	0.37480	640.19	2748.1	1.8606	6.8206	**0.50**
0.55	155.46	0.0010967	0.34259	655.88	2752.3	1.8972	6.7885	**0.55**
0.60	158.83	0.0011006	0.31558	670.50	2756.1	1.9311	6.7592	**0.60**
0.65	161.99	0.0011044	0.29258	684.22	2759.6	1.9626	6.7321	**0.65**
0.70	164.95	0.0011080	0.27276	697.14	2762.7	1.9921	6.7070	**0.70**
0.80	170.41	0.0011148	0.24033	721.02	2768.3	2.0460	6.6615	**0.80**
0.90	175.36	0.0011212	0.21487	742.72	2773.0	2.0944	6.6212	**0.90**
1.00	179.89	0.0011272	0.19435	762.68	2777.1	2.1384	6.5850	**1.00**
1.10	184.07	0.0011330	0.17744	781.20	2780.7	2.1789	6.5520	**1.10**

Table 5. Properties of Saturated Water and Steam (Pressure)

Press.	Temp.	Volume, m^3/kg		Enthalpy, kJ/kg		Entropy, kJ/(kg·K)		**Press.**
MPa	°C	v_L	v_V	h_L	h_V	s_L	s_V	**MPa**
1.2	187.96	0.001139	0.16325	798.50	2783.8	2.2163	6.5217	**1.2**
1.3	191.61	0.001144	0.15117	814.76	2786.5	2.2512	6.4936	**1.3**
1.4	195.05	0.001149	0.14077	830.13	2788.9	2.2839	6.4675	**1.4**
1.5	198.30	0.001154	0.13170	844.72	2791.0	2.3147	6.4431	**1.5**
1.6	201.38	0.001159	0.12373	858.61	2792.9	2.3438	6.4200	**1.6**
1.8	207.12	0.001168	0.11036	884.61	2796.0	2.3978	6.3776	**1.8**
2.0	212.38	0.001177	0.099581	908.62	2798.4	2.4470	6.3392	**2.0**
2.2	217.26	0.001185	0.090695	930.98	2800.2	2.4924	6.3040	**2.2**
2.4	221,80	0.001193	0.083242	951.95	2801.5	2.5344	6.2714	**2.4**
2.6	226.05	0.001201	0.076897	971.74	2802.5	2.5738	6.2411	**2.6**
2.8	230.06	0.001209	0.071428	990.50	2803.0	2.6107	6.2126	**2.8**
3.0	233.86	0.001217	0.066664	1008.4	2803.3	2.6456	6.1858	**3.0**
3.2	237.46	0.001224	0.062475	1025.5	2803.2	2.6787	6.1604	**3.2**
3.4	240.90	0.001231	0.058761	1041.8	2803.0	2.7102	6.1362	**3.4**
3.6	244.19	0.001239	0.055446	1057.6	2802.5	2.7403	6.1131	**3.6**
3.8	247.33	0.001246	0.052468	1072.8	2801.8	2.7690	6.0910	**3.8**
4.0	250.36	0.001253	0.049777	1087.4	2800.9	2.7967	6.0697	**4.0**
4.2	253.27	0.001259	0.047333	1101.6	2799.9	2.8232	6.0492	**4.2**
4.4	256.07	0.001266	0.045103	1115.4	2798.7	2.8488	6.0294	**4.4**
4.6	258.78	0.001273	0.043060	1128.8	2797.3	2.8736	6.0103	**4.6**
4.8	261.40	0.001280	0.041181	1141.8	2795.8	2.8975	5.9917	**4.8**
5.0	263.94	0.001286	0.039446	1154.5	2794.2	2.9207	5.9737	**5.0**
5.5	269.97	0.001303	0.035642	1184.9	2789.7	2.9759	5.9307	**5.5**
6.0	275.59	0.001319	0.032449	1213.7	2784.6	3.0274	5.8901	**6.0**
6.5	280.86	0.001336	0.029728	1241.2	2778.8	3.0760	5.8515	**6.5**
7.0	285.83	0.001352	0.027380	1267.4	2772.6	3.1220	5.8146	**7.0**
7.5	290.54	0.001368	0.025331	1292.7	2765.8	3.1658	5.7792	**7.5**
8.0	295.01	0.001385	0.023528	1317.1	2758.6	3.2077	5.7448	**8.0**
8.5	299.27	0.001401	0.021926	1340.7	2751.0	3.2478	5.7115	**8.5**
9.0	303.35	0.001418	0.020493	1363.7	2742.9	3.2866	5.6790	**9.0**
9.5	307.25	0.001435	0.019203	1386.0	2734.4	3.3240	5.6472	**9.5**
10.0	311.00	0.001453	0.018034	1407.9	2725.5	3.3603	5.6159	**10.0**
11.0	318.08	0.001489	0.015994	1450.3	2706.4	3.4300	5.5545	**11.0**
12.0	324.68	0.001526	0.014269	1491.3	2685.6	3.4965	5.4941	**12.0**
13.0	330.86	0.001566	0.012785	1531.4	2662.9	3.5606	5.4339	**13.0**
14.0	336.67	0.001610	0.011489	1570.9	2638.1	3.6230	5.3730	**14.0**
15.0	342.16	0.001657	0.010340	1610.2	2610.9	3.6844	5.3108	**15.0**
16.0	347.36	0.001710	0.009308	1649.7	2580.8	3.7457	5.2463	**16.0**
17.0	352.29	0.001769	0.008369	1690.0	2547.4	3.8077	5.1785	**17.0**
18.0	356.99	0.001839	0.007499	1732.0	2509.5	3.8717	5.1055	**18.0**
19.0	361.47	0.001925	0.006673	1776.9	2465.4	3.9396	5.0246	**19.0**
20.0	365.75	0.002039	0.005858	1827.1	2411.4	4.0154	4.9299	**20.0**
21.0	369.83	0.002212	0.004988	1889.4	2337.5	4.1093	4.8062	**21.0**
22.0	373.71	0.002750	0.003577	2021.9	2164.2	4.3109	4.5308	**22.0**
p_c	373.946	0.00311	0.00311	2087.5	2087.5	4.4120	4.4120	p_c

$p_c = 22.064$ MPa

Table 6. Superheated Steam – SI Units

Pressure MPa (Sat. T)		Temperature—Degrees Celsius 50	100	150	200	250	300	350	400	450	500	550	600	700
0.005	*v*	29.782	34.419	39.043	43.663	48.281	52.898	57.515	62.131	66.747	71.363	75.979	80.594	89.826
(32.88)	*h*	2593.4	2688.0	2783.4	2879.8	2977.6	3076.9	3177.6	3280.0	3384.0	3489.7	3597.1	3706.3	3929.9
	s	8.4976	8.7700	9.0097	9.2251	9.4216	9.6027	9.7713	9.9293	10.078	10.220	10.354	10.483	10.725
0.01	*v*	14.867	17.197	19.514	21.826	24.136	26.446	28.755	31.064	33.372	35.680	37.988	40.296	44.912
(45.81)	*h*	2592.0	2687.4	2783.0	2879.6	2977.4	3076.7	3177.5	3279.9	3384.0	3489.7	3597.1	3706.3	3929.9
	s	8.1741	8.4488	8.6892	8.9048	9.1014	9.2827	9.4513	9.6093	9.7584	9.8997	10.034	10.163	10.405
0.02	*v*		8.5857	9.7488	10.907	12.064	13.220	14.375	15.530	16.684	17.839	18.993	20.147	22.455
(60.06)	*h*		2686.2	2782.3	2879.1	2977.1	3076.5	3177.4	3279.8	3383.8	3489.6	3597.0	3706.2	3929.8
	s		8.1262	8.3680	8.5842	8.7811	8.9624	9.1311	9.2892	9.4383	9.5797	9.7143	9.8431	10.086
0.05	*v*		3.4188	3.8899	4.3563	4.8207	5.2841	5.7470	6.2095	6.6718	7.1339	7.5959	8.0578	8.9814
(81.32)	*h*		2682.4	2780.2	2877.8	2976.2	3075.8	3176.8	3279.3	3383.5	3489.2	3596.7	3706.0	3929.7
	s		7.6952	7.9412	8.1591	8.3568	8.5386	8.7076	8.8658	9.0150	9.1565	9.2912	9.4200	9.6625
0.10	*v*		1.6960	1.9367	2.1725	2.4062	2.6389	2.8710	3.1027	3.3342	3.5656	3.7968	4.0279	4.4900
(99.61)	*h*		2675.8	2776.6	2875.5	2974.5	3074.5	3175.8	3278.5	3382.8	3488.7	3596.3	3705.6	3929.4
	s		7.3610	7.6147	7.8356	8.0346	8.2171	8.3865	8.5451	8.6945	8.8361	8.9709	9.0998	9.3424
0.15	*v*			1.2856	1.4445	1.6013	1.7571	1.9123	2.0671	2.2217	2.3762	2.5305	2.6847	2.9929
(111.35)	*h*			2772.9	2873.1	2972.9	3073.3	3174.9	3277.8	3382.2	3488.2	3595.8	3705.2	3929.1
	s			7.4207	7.6447	7.8451	8.0284	8.1983	8.3571	8.5067	8.6484	8.7833	8.9123	9.1550
0.20	*v*			0.9599	1.0805	1.1989	1.3162	1.4330	1.5493	1.6655	1.7814	1.8973	2.0130	2.2444
(120.21)	*h*			2769.1	2870.8	2971.3	3072.1	3173.9	3277.0	3381.5	3487.6	3595.4	3704.8	3928.8
	s			7.2809	7.5081	7.7100	7.8940	8.0643	8.2235	8.3733	8.5151	8.6501	8.7792	9.0220
0.25	*v*			0.7644	0.8621	0.9574	1.0517	1.1454	1.2387	1.3317	1.4246	1.5174	1.6101	1.7952
(127.41)	*h*			2765.2	2868.4	2969.6	3070.8	3172.9	3276.2	3380.9	3487.1	3594.9	3704.4	3928.5
	s			7.1707	7.4013	7.6046	7.7895	7.9602	8.1196	8.2696	8.4116	8.5467	8.6759	8.9188
0.30	*v*			0.6340	0.7164	0.7965	0.8753	0.9536	1.0315	1.1092	1.1867	1.2641	1.3414	1.4958
(133.53)	*h*			2761.2	2866.0	2967.9	3069.6	3172.0	3275.4	3380.2	3486.6	3594.5	3704.0	3928.2
	s			7.0791	7.3132	7.5181	7.7037	7.8749	8.0346	8.1848	8.3269	8.4622	8.5914	8.8344
0.35	*v*			0.5408	0.6124	0.6815	0.7494	0.8167	0.8836	0.9503	1.0168	1.0832	1.1495	1.2819
(138.86)	*h*			2757.1	2863.5	2966.3	3068.4	3171.0	3274.6	3379.6	3486.0	3594.0	3703.6	3927.9
	s			7.0002	7.2381	7.4445	7.6310	7.8026	7.9626	8.1130	8.2553	8.3906	8.5199	8.7630
0.40	*v*			0.4709	0.5343	0.5952	0.6549	0.7139	0.7726	0.8311	0.8894	0.9475	1.0056	1.1215
(143.61)	*h*			2752.8	2861.0	2964.6	3067.1	3170.0	3273.9	3379.0	3485.5	3593.6	3703.2	3927.6
	s			6.9305	7.1724	7.3805	7.5677	7.7398	7.9001	8.0507	8.1931	8.3286	8.4579	8.7012
0.45	*v*			0.4164	0.4736	0.5281	0.5814	0.6341	0.6863	0.7384	0.7902	0.8420	0.8936	0.9968
(147.91)	*h*			2748.3	2858.5	2962.8	3065.9	3169.0	3273.1	3378.3	3484.9	3593.1	3702.8	3927.3
	s			6.8677	7.1139	7.3237	7.5117	7.6843	7.8449	7.9957	8.1383	8.2738	8.4032	8.6466
0.50	*v*				0.4250	0.4744	0.5226	0.5701	0.6173	0.6642	0.7109	0.7576	0.8041	0.8970
(151.84)	*h*				2855.9	2961.1	3064.6	3168.1	3272.3	3377.7	3484.4	3592.6	3702.5	3927.0
	s				7.0611	7.2726	7.4614	7.6345	7.7954	7.9464	8.0891	8.2247	8.3543	8.5977
0.55	*v*				0.3853	0.4305	0.4745	0.5178	0.5608	0.6035	0.6461	0.6885	0.7308	0.8153
(155.46)	*h*				2853.3	2959.4	3063.3	3167.1	3271.5	3377.0	3483.9	3592.2	3702.1	3926.8
	s				7.0128	7.2261	7.4158	7.5894	7.7505	7.9017	8.0446	8.1803	8.3099	8.5535
0.60	*v*				0.3521	0.3939	0.4344	0.4743	0.5137	0.5530	0.5920	0.6309	0.6698	0.7473
(158.83)	*h*				2850.7	2957.7	3062.1	3166.1	3270.7	3376.4	3483.3	3591.7	3701.7	3926.5
	s				6.9684	7.1834	7.3740	7.5480	7.7095	7.8609	8.0039	8.1398	8.2694	8.5131

v = specific volume, m^3/kg *h* = enthalpy, kJ/kg *s* = entropy, kJ/(kg·K)

Table 6 (continued). Superheated Steam – SI Units

Pressure MPa (Sat. T)		Temperature—Degrees Celsius 200	250	300	350	400	450	500	550	600	650	700	750	800
0.65	*v*	0.3241	0.3629	0.4005	0.4374	0.4739	0.5102	0.5463	0.5822	0.6181	0.6539	0.6897	0.7254	0.7611
(161.99)	*h*	2848.0	2955.9	3060.8	3165.1	3269.9	3375.7	3482.8	3591.3	3701.3	3812.9	3926.2	4041.1	4157.7
	s	6.9270	7.1439	7.3354	7.5099	7.6717	7.8233	7.9665	8.1024	8.2321	8.3564	8.4759	8.5911	8.7024
0.70	*v*	0.3000	0.3364	0.3714	0.4058	0.4398	0.4735	0.5070	0.5405	0.5738	0.6071	0.6403	0.6735	0.7067
(164.95)	*h*	2845.3	2954.1	3059.5	3164.1	3269.1	3375.1	3482.3	3590.8	3700.9	3812.6	3925.9	4040.8	4157.5
	s	6.8884	7.1071	7.2995	7.4745	7.6366	7.7884	7.9317	8.0678	8.1976	8.3220	8.4415	8.5567	8.6680
0.75	*v*	0.2791	0.3133	0.3462	0.3784	0.4102	0.4417	0.4731	0.5043	0.5354	0.5665	0.5975	0.6285	0.6595
(167.76)	*h*	2842.5	2952.3	3058.2	3163.1	3268.4	3374.4	3481.7	3590.4	3700.5	3812.2	3925.6	4040.6	4157.3
	s	6.8520	7.0727	7.2660	7.4415	7.6039	7.7559	7.8994	8.0355	8.1654	8.2898	8.4094	8.5246	8.6360
0.80	*v*	0.2609	0.2932	0.3242	0.3544	0.3843	0.4139	0.4433	0.4726	0.5019	0.5310	0.5601	0.5892	0.6182
(170.41)	*h*	2839.8	2950.5	3056.9	3162.2	3267.6	3373.8	3481.2	3589.9	3700.1	3811.9	3925.3	4040.3	4157.1
	s	6.8176	7.0403	7.2345	7.4106	7.5733	7.7255	7.8690	8.0053	8.1353	8.2598	8.3794	8.4947	8.6060
0.90	*v*	0.2304	0.2596	0.2874	0.3145	0.3411	0.3675	0.3938	0.4199	0.4459	0.4718	0.4977	0.5236	0.5494
(175.36)	*h*	2834.1	2946.9	3054.3	3160.2	3266.0	3372.5	3480.1	3589.0	3699.3	3811.2	3924.7	4039.8	4156.6
	s	6.7538	6.9806	7.1768	7.3538	7.5172	7.6698	7.8136	7.9501	8.0803	8.2049	8.3246	8.4399	8.5513
1.0	*v*	0.2060	0.2327	0.2580	0.2825	0.3066	0.3304	0.3541	0.3777	0.4011	0.4245	0.4478	0.4711	0.4944
(179.89)	*h*	2828.3	2943.2	3051.7	3158.2	3264.4	3371.2	3479.0	3588.1	3698.6	3810.5	3924.1	4039.3	4156.2
	s	6.6955	6.9266	7.1247	7.3028	7.4668	7.6198	7.7640	7.9007	8.0309	8.1557	8.2755	8.3909	8.5024
1.1	*v*	0.1860	0.2107	0.2339	0.2563	0.2783	0.3001	0.3217	0.3431	0.3645	0.3858	0.4070	0.4282	0.4494
(184.07)	*h*	2822.3	2939.5	3049.1	3156.2	3262.8	3369.9	3477.9	3587.2	3697.8	3809.9	3923.5	4038.8	4155.7
	s	6.6414	6.8772	7.0773	7.2564	7.4210	7.5745	7.7189	7.8558	7.9863	8.1111	8.2310	8.3465	8.4580
1.2	*v*	0.1693	0.1924	0.2139	0.2345	0.2548	0.2748	0.2946	0.3143	0.3339	0.3535	0.3730	0.3924	0.4118
(187.96)	*h*	2816.1	2935.7	3046.4	3154.1	3261.2	3368.6	3476.8	3586.2	3697.0	3809.2	3922.9	4038.3	4155.2
	s	6.5908	6.8314	7.0336	7.2138	7.3791	7.5330	7.6777	7.8148	7.9454	8.0704	8.1904	8.3059	8.4175
1.3	*v*	0.1552	0.1769	0.1969	0.2161	0.2349	0.2534	0.2718	0.2900	0.3081	0.3262	0.3442	0.3621	0.3801
(191.61)	*h*	2809.6	2931.8	3043.7	3152.1	3259.6	3367.3	3475.7	3585.3	3696.2	3808.5	3922.4	4037.8	4154.8
	s	6.5430	6.7888	6.9931	7.1745	7.3404	7.4947	7.6397	7.7770	7.9078	8.0329	8.1530	8.2686	8.3803
1.4	*v*	0.1430	0.1635	0.1823	0.2003	0.2178	0.2351	0.2522	0.2691	0.2860	0.3028	0.3195	0.3362	0.3529
(195.05)	*h*	2803.0	2927.9	3041.0	3150.1	3258.0	3366.0	3474.7	3584.4	3695.4	3807.8	3921.8	4037.2	4154.3
	s	6.4975	6.7488	6.9553	7.1378	7.3044	7.4591	7.6045	7.7420	7.8729	7.9981	8.1183	8.2340	8.3457
1.5	*v*	0.1324	0.1520	0.1697	0.1866	0.2030	0.2192	0.2352	0.2510	0.2668	0.2825	0.2981	0.3137	0.3293
(198.30)	*h*	2796.0	2924.0	3038.3	3148.0	3256.4	3364.7	3473.6	3583.5	3694.6	3807.2	3921.2	4036.7	4153.9
	s	6.4537	6.7111	6.9199	7.1035	7.2708	7.4259	7.5716	7.7093	7.8404	7.9657	8.0860	8.2018	8.3135
1.6	*v*		0.1419	0.1587	0.1746	0.1901	0.2053	0.2203	0.2352	0.2500	0.2647	0.2794	0.2940	0.3087
(201.38)	*h*		2919.9	3035.5	3146.0	3254.7	3363.3	3472.5	3582.6	3693.9	3806.5	3920.6	4036.2	4153.4
	s		6.6754	6.8865	7.0713	7.2392	7.3948	7.5407	7.6787	7.8099	7.9354	8.0557	8.1716	8.2834
1.7	*v*		0.1330	0.1489	0.1640	0.1786	0.1930	0.2072	0.2212	0.2352	0.2491	0.2629	0.2767	0.2904
(204.31)	*h*		2915.9	3032.7	3143.9	3253.1	3362.0	3471.4	3581.6	3693.1	3805.8	3920.0	4035.7	4153.0
	s		6.6413	6.8548	7.0408	7.2094	7.3654	7.5117	7.6499	7.7813	7.9068	8.0273	8.1432	8.2551
1.8	*v*		0.1250	0.1402	0.1546	0.1685	0.1821	0.1955	0.2088	0.2220	0.2351	0.2482	0.2612	0.2743
(207.12)	*h*		2911.7	3029.9	3141.8	3251.5	3360.7	3470.3	3580.7	3692.3	3805.1	3919.4	4035.2	4152.5
	s		6.6087	6.8247	7.0119	7.1812	7.3377	7.4842	7.6226	7.7542	7.8799	8.0004	8.1164	8.2284
2.0	*v*		0.1115	0.1255	0.1386	0.1512	0.1635	0.1757	0.1877	0.1996	0.2115	0.2233	0.2350	0.2467
(212.38)	*h*		2903.2	3024.3	3137.6	3248.2	3358.1	3468.1	3578.9	3690.7	3803.8	3918.2	4034.2	4151.6
	s		6.5474	6.7685	6.9582	7.1290	7.2863	7.4335	7.5723	7.7042	7.8301	7.9509	8.0670	8.1791

v = specific volume, m³/kg *h* = enthalpy, kJ/kg *s* = entropy, kJ/(kg·K)

Table 6 (continued). Superheated Steam – SI Units

Pressure MPa (Sat. T)		Temperature—Degrees Celsius 225	250	300	350	400	450	500	550	600	650	700	750	800
2.2	*v*	0.0931	0.1004	0.1134	0.1255	0.1371	0.1484	0.1595	0.1704	0.1813	0.1921	0.2028	0.2136	0.2242
(217.26)	*h*	2824.5	2894.5	3018.5	3133.4	3244.9	3355.4	3465.9	3577.0	3689.1	3802.4	3917.1	4033.1	4150.7
	s	6.3531	6.4903	6.7168	6.9091	7.0813	7.2396	7.3873	7.5266	7.6588	7.7850	7.9059	8.0222	8.1344
2.4	*v*	0.0842	0.0911	0.1034	0.1146	0.1253	0.1357	0.1459	0.1560	0.1660	0.1760	0.1858	0.1957	0.2055
(221.80)	*h*	2812.1	2885.5	3012.6	3129.1	3241.6	3352.7	3463.7	3575.2	3687.6	3801.1	3915.9	4032.1	4149.8
	s	6.2926	6.4365	6.6688	6.8638	7.0375	7.1967	7.3450	7.4848	7.6173	7.7437	7.8648	7.9813	8.0936
2.6	*v*		0.0833	0.0948	0.1053	0.1153	0.1250	0.1345	0.1439	0.1531	0.1623	0.1714	0.1805	0.1896
(226.05)	*h*		2876.2	3006.6	3124.8	3238.3	3350.0	3461.5	3573.3	3686.0	3799.7	3914.7	4031.1	4148.9
	s		6.3854	6.6238	6.8216	6.9968	7.1570	7.3060	7.4461	7.5790	7.7056	7.8269	7.9435	8.0559
2.8	*v*		0.0765	0.0875	0.0974	0.1068	0.1158	0.1247	0.1334	0.1420	0.1506	0.1591	0.1676	0.1760
(230.06)	*h*		2866.5	3000.5	3120.5	3234.9	3347.4	3459.3	3571.5	3684.4	3798.4	3913.5	4030.0	4148.0
	s		6.3365	6.5814	6.7821	6.9589	7.1200	7.2696	7.4102	7.5434	7.6703	7.7918	7.9085	8.0210
3.0	*v*		0.0706	0.0812	0.0906	0.0994	0.1079	0.1162	0.1244	0.1324	0.1405	0.1484	0.1563	0.1642
(233.86)	*h*		2856.5	2994.3	3116.1	3231.6	3344.7	3457.0	3569.6	3682.8	3797.0	3912.3	4029.0	4147.0
	s		6.2893	6.5412	6.7449	6.9233	7.0853	7.2356	7.3767	7.5102	7.6373	7.7590	7.8759	7.9885
3.2	*v*		0.0655	0.0756	0.0845	0.0929	0.1009	0.1088	0.1165	0.1240	0.1316	0.1390	0.1465	0.1539
(237.46)	*h*		2846.2	2988.0	3111.6	3228.2	3341.9	3454.8	3567.7	3681.2	3795.6	3911.2	4028.0	4146.1
	s		6.2434	6.5029	6.7097	6.8897	7.0527	7.2036	7.3451	7.4790	7.6064	7.7283	7.8453	7.9581
3.4	*v*		0.0609	0.0707	0.0792	0.0872	0.0948	0.1022	0.1095	0.1166	0.1237	0.1308	0.1378	0.1448
(240.90)	*h*		2835.3	2981.6	3107.1	3224.8	3339.2	3452.6	3565.9	3679.6	3794.3	3910.0	4026.9	4145.2
	s		6.1986	6.4662	6.6762	6.8579	7.0219	7.1735	7.3154	7.4496	7.5773	7.6993	7.8165	7.9294
3.6	*v*		0.0568	0.0663	0.0745	0.0821	0.0893	0.0964	0.1033	0.1101	0.1168	0.1234	0.1301	0.1367
(244.19)	*h*		2824.0	2975.1	3102.6	3221.3	3336.5	3450.3	3564.0	3678.0	3792.9	3908.8	4025.9	4144.3
	s		6.1545	6.4309	6.6443	6.8276	6.9927	7.1449	7.2873	7.4219	7.5498	7.6720	7.7893	7.9023
3.8	*v*		0.0531	0.0624	0.0703	0.0775	0.0844	0.0911	0.0977	0.1042	0.1105	0.1169	0.1232	0.1294
(247.33)	*h*		2812.1	2968.4	3098.0	3217.9	3333.7	3448.1	3562.1	3676.4	3791.5	3907.6	4024.8	4143.4
	s		6.1107	6.3968	6.6137	6.7988	6.9649	7.1178	7.2607	7.3955	7.5237	7.6461	7.7636	7.8767
4.0	*v*			0.0589	0.0665	0.0734	0.0800	0.0864	0.0927	0.0989	0.1049	0.1110	0.1170	0.1229
(250.36)	*h*			2961.7	3093.3	3214.4	3331.0	3445.8	3560.2	3674.8	3790.2	3906.4	4023.8	4142.5
	s			6.3638	6.5843	6.7712	6.9383	7.0919	7.2353	7.3704	7.4989	7.6215	7.7391	7.8523
4.5	*v*			0.0514	0.0584	0.0648	0.0708	0.0765	0.0821	0.0877	0.0931	0.0985	0.1038	0.1092
(257.44)	*h*			2944.1	3081.5	3205.6	3324.0	3440.2	3555.5	3670.8	3786.7	3903.4	4021.2	4140.2
	s			6.2852	6.5153	6.7069	6.8767	7.0320	7.1765	7.3126	7.4416	7.5647	7.6827	7.7962
5.0	*v*			0.0453	0.0520	0.0578	0.0633	0.0686	0.0737	0.0787	0.0836	0.0885	0.0933	0.0982
(263.94)	*h*			2925.6	3069.3	3196.6	3317.0	3434.5	3550.8	3666.8	3783.3	3900.5	4018.6	4137.9
	s			6.2109	6.4515	6.6481	6.8208	6.9778	7.1235	7.2604	7.3901	7.5137	7.6321	7.7459
5.5	*v*			0.0404	0.0467	0.0522	0.0572	0.0621	0.0668	0.0714	0.0759	0.0803	0.0848	0.0891
(269.97)	*h*			2906.2	3056.8	3187.5	3309.9	3428.7	3546.0	3662.8	3779.8	3897.5	4016.0	4135.6
	s			6.1396	6.3919	6.5938	6.7693	6.9282	7.0751	7.2129	7.3432	7.4673	7.5861	7.7002
6.0	*v*			0.0362	0.0423	0.0474	0.0522	0.0567	0.0610	0.0653	0.0694	0.0735	0.0776	0.0816
(275.59)	*h*			2885.5	3043.9	3178.2	3302.8	3422.9	3541.2	3658.8	3776.4	3894.5	4013.4	4133.3
	s			6.0702	6.3356	6.5431	6.7216	6.8824	7.0306	7.1692	7.3002	7.4248	7.5439	7.6583
6.5	*v*			0.0326	0.0385	0.0434	0.0479	0.0521	0.0561	0.0601	0.0640	0.0678	0.0716	0.0753
(280.86)	*h*			2863.5	3030.6	3168.7	3295.5	3417.1	3536.4	3654.7	3772.9	3891.5	4010.7	4131.0
	s			6.0018	6.2819	6.4953	6.6771	6.8397	6.9892	7.1287	7.2603	7.3854	7.5050	7.6196

v = specific volume, m^3/kg   *h* = enthalpy, kJ/kg   *s* = entropy, kJ/(kg·K)

Table 6 (continued). Superheated Steam - SI Units

Pressure MPa (Sat. T)		Temperature—Degrees Celsius 300	325	350	375	400	450	500	550	600	650	700	750	800
7.0	*v*	0.0295	0.0326	0.0353	0.0377	0.0400	0.0442	0.0482	0.0520	0.0557	0.0593	0.0628	0.0664	0.0698
(285.83)	*h*	2839.8	2935.5	3016.8	3090.4	3159.1	3288.2	3411.3	3531.5	3650.6	3769.4	3888.5	4008.1	4128.6
	s	5.9335	6.0970	6.2303	6.3460	6.4501	6.6351	6.7997	6.9505	7.0909	7.2232	7.3488	7.4687	7.5837
7.5	*v*	0.0267	0.0298	0.0325	0.0348	0.0370	0.0410	0.0448	0.0483	0.0518	0.0552	0.0586	0.0619	0.0651
(290.54)	*h*	2814.3	2917.4	3002.7	3078.8	3149.3	3280.7	3405.3	3526.7	3646.5	3765.9	3885.4	4005.5	4126.3
	s	5.8644	6.0407	6.1805	6.3002	6.4070	6.5954	6.7620	6.9141	7.0555	7.1885	7.3145	7.4348	7.5501
8.0	*v*	0.0243	0.0274	0.0300	0.0323	0.0343	0.0382	0.0418	0.0452	0.0485	0.0517	0.0548	0.0579	0.0610
(295.01)	*h*	2786.4	2898.3	2988.1	3066.9	3139.3	3273.2	3399.4	3521.8	3642.4	3762.4	3882.4	4002.9	4124.0
	s	5.7935	5.9849	6.1319	6.2560	6.3657	6.5577	6.7264	6.8798	7.0221	7.1557	7.2823	7.4030	7.5186
8.5	*v*	0.0220	0.0252	0.0278	0.0300	0.0320	0.0357	0.0391	0.0424	0.0455	0.0485	0.0515	0.0545	0.0574
(299.27)	*h*	2755.4	2878.3	2972.9	3054.7	3129.1	3265.6	3393.4	3516.9	3638.3	3758.9	3879.4	4000.2	4121.7
	s	5.7193	5.9294	6.0845	6.2132	6.3259	6.5216	6.6925	6.8473	6.9905	7.1248	7.2519	7.3730	7.4889
9.0	*v*		0.0233	0.0258	0.0280	0.0300	0.0335	0.0368	0.0399	0.0429	0.0458	0.0486	0.0514	0.0541
(303.35)	*h*		2857.0	2957.2	3042.2	3118.8	3257.9	3387.3	3511.9	3634.2	3755.4	3876.4	3997.6	4119.4
	s		5.8736	6.0378	6.1716	6.2875	6.4871	6.6601	6.8163	6.9605	7.0955	7.2231	7.3446	7.4608
9.5	*v*		0.0215	0.0240	0.0262	0.0281	0.0316	0.0347	0.0377	0.0405	0.0433	0.0460	0.0486	0.0512
(307.25)	*h*		2834.4	2940.9	3029.4	3108.2	3250.2	3381.2	3506.9	3630.0	3751.9	3873.3	3994.9	4117.0
	s		5.8170	5.9917	6.1309	6.2502	6.4538	6.6291	6.7867	6.9319	7.0676	7.1957	7.3176	7.4341
10.0	*v*		0.0199	0.0224	0.0246	0.0264	0.0298	0.0328	0.0357	0.0384	0.0410	0.0436	0.0461	0.0486
(311.00)	*h*		2810.2	2924.0	3016.2	3097.4	3242.3	3375.1	3501.9	3625.8	3748.3	3870.3	3992.3	4114.7
	s		5.7593	5.9458	6.0910	6.2139	6.4217	6.5993	6.7584	6.9045	7.0409	7.1696	7.2918	7.4086
11.0	*v*		0.0170	0.0196	0.0217	0.0235	0.0267	0.0296	0.0322	0.0347	0.0371	0.0395	0.0418	0.0441
(318.08)	*h*		2755.6	2887.8	2988.7	3075.1	3226.2	3362.6	3491.9	3617.5	3741.2	3864.2	3987.0	4110.1
	s		5.6373	5.8541	6.0129	6.1438	6.3605	6.5430	6.7050	6.8531	6.9910	7.1207	7.2437	7.3612
12.0	*v*		0.0143	0.0172	0.0193	0.0211	0.0242	0.0268	0.0293	0.0317	0.0339	0.0361	0.0383	0.0404
(324.68)	*h*		2688.4	2848.0	2959.5	3051.9	3209.8	3350.0	3481.7	3609.0	3734.1	3858.0	3981.6	4105.4
	s		5.4988	5.7607	5.9362	6.0762	6.3027	6.4902	6.6553	6.8055	6.9448	7.0756	7.1994	7.3175
13.0	*v*			0.0151	0.0173	0.0190	0.0220	0.0245	0.0269	0.0291	0.0312	0.0332	0.0352	0.0372
(330.86)	*h*			2803.6	2928.3	3027.6	3192.9	3337.1	3471.4	3600.5	3726.9	3851.9	3976.3	4100.7
	s			5.6635	5.8600	6.0104	6.2475	6.4404	6.6087	6.7610	6.9018	7.0336	7.1583	7.2771
14.0	*v*			0.0132	0.0155	0.0172	0.0201	0.0225	0.0248	0.0268	0.0288	0.0308	0.0326	0.0345
(336.67)	*h*			2752.9	2894.9	3002.2	3175.6	3324.1	3461.0	3591.9	3719.7	3845.7	3970.9	4096.0
	s			5.5595	5.7832	5.9457	6.1945	6.3931	6.5648	6.7192	6.8615	6.9944	7.1200	7.2393
15.0	*v*			0.0115	0.0139	0.0157	0.0185	0.0208	0.0229	0.0249	0.0268	0.0286	0.0304	0.0321
(342.16)	*h*			2693.0	2858.9	2975.5	3157.8	3310.8	3450.5	3583.3	3712.4	3839.5	3965.6	4091.3
	s			5.4435	5.7049	5.8817	6.1433	6.3479	6.5230	6.6797	6.8235	6.9576	7.0839	7.2039
16.0	*v*			0.0098	0.0125	0.0143	0.0170	0.0193	0.0214	0.0232	0.0250	0.0267	0.0284	0.0301
(347.36)	*h*			2617.0	2819.5	2947.5	3139.6	3297.3	3439.8	3574.6	3705.1	3833.3	3960.2	4086.6
	s			5.3045	5.6238	5.8177	6.0935	6.3045	6.4832	6.6422	6.7876	6.9228	7.0499	7.1706
17.0	*v*				0.0112	0.0130	0.0158	0.0180	0.0199	0.0218	0.0235	0.0251	0.0267	0.0282
(352.29)	*h*				2775.9	2917.8	3120.9	3283.6	3429.1	3565.9	3697.8	3827.0	3954.8	4081.9
	s				5.5384	5.7533	6.0449	6.2627	6.4451	6.6064	6.7534	6.8897	7.0178	7.1391
18.0	*v*				0.0100	0.0119	0.0147	0.0168	0.0187	0.0204	0.0221	0.0236	0.0251	0.0266
(356.99)	*h*				2726.9	2886.3	3101.7	3269.7	3418.3	3557.0	3690.4	3820.7	3949.4	4077.2
	s				5.4465	5.6881	5.9973	6.2222	6.4085	6.5722	6.7208	6.8583	6.9872	7.1091

v = specific volume, m³/kg *h* = enthalpy, kJ/kg *s* = entropy, kJ/(kg·K)

Table 6 (continued). Superheated Steam – SI Units

Pressure MPa (Sat. T)		Temperature—Degrees Celsius 375	400	425	450	475	500	550	600	650	700	750	800
20	*v*	0.00768	0.00995	0.0115	0.0127	0.0138	0.0148	0.0166	0.0182	0.0197	0.0211	0.0225	0.0239
(365.75)	*h*	2602.4	2816.8	2952.9	3061.5	3155.8	3241.2	3396.2	3539.2	3675.6	3808.2	3938.5	4067.7
	s	5.2272	5.5525	5.7510	5.9041	6.0322	6.1445	6.3390	6.5077	6.6596	6.7994	6.9301	7.0534
22	*v*	0.00490	0.00826	0.00987	0.0111	0.0122	0.0131	0.0148	0.0163	0.0178	0.0191	0.0204	0.0216
(373.71)	*h*	2354.0	2735.8	2897.8	3019.0	3121.0	3211.8	3373.8	3521.2	3660.6	3795.5	3927.6	4058.2
	s	4.8240	5.4050	5.6417	5.8124	5.9511	6.0704	6.2736	6.4475	6.6029	6.7451	6.8776	7.0022
24	*v*	0.00206	0.00673	0.00850	0.00977	0.0108	0.0118	0.0134	0.0148	0.0161	0.0174	0.0186	0.0197
	h	1872.5	2637.4	2837.4	2974.0	3084.8	3181.4	3350.9	3502.9	3645.6	3782.8	3916.7	4048.8
	s	4.0731	5.2366	5.5289	5.7212	5.8720	5.9991	6.2116	6.3910	6.5499	6.6946	6.8289	6.9549
26	*v*	0.00192	0.00529	0.00731	0.00862	0.00967	0.0106	0.0121	0.0135	0.0148	0.0160	0.0171	0.0182
	h	1832.8	2510.6	2770.6	2926.1	3047.0	3150.2	3327.6	3484.4	3630.4	3770.0	3905.8	4039.3
	s	4.0059	5.0304	5.4106	5.6296	5.7942	5.9298	6.1523	6.3374	6.5000	6.6473	6.7833	6.9107
28	*v*	0.00185	0.00385	0.00625	0.00762	0.00867	0.00957	0.0111	0.0124	0.0136	0.0147	0.0158	0.0168
	h	1809.1	2334.4	2695.8	2875.1	3007.7	3117.9	3303.9	3465.7	3615.1	3757.1	3894.8	4029.7
	s	3.9635	4.7552	5.2841	5.5367	5.7170	5.8621	6.0953	6.2863	6.4527	6.6026	6.7405	6.8693
30	*v*	0.00179	0.00280	0.00530	0.00674	0.00780	0.00869	0.0102	0.0114	0.0126	0.0137	0.0147	0.0156
	h	1792.0	2152.4	2611.9	2820.9	2966.7	3084.8	3279.8	3446.9	3599.7	3744.2	3883.8	4020.2
	s	3.9314	4.4750	5.1473	5.4419	5.6402	5.7956	6.0403	6.2374	6.4077	6.5602	6.7000	6.8303
35	*v*	0.00170	0.00211	0.00344	0.00496	0.00606	0.00693	0.00835	0.00952	0.0106	0.0115	0.0124	0.0133
	h	1762.5	1988.4	2373.5	2671.0	2857.3	2998.0	3218.1	3399.0	3560.9	3711.9	3856.3	3996.5
	s	3.8725	4.2140	4.7752	5.1945	5.4480	5.6331	5.9093	6.1229	6.3032	6.4625	6.6072	6.7411
40	*v*	0.00164	0.00191	0.00254	0.00369	0.00476	0.00562	0.00699	0.00809	0.00905	0.00993	0.0107	0.0115
	h	1742.7	1931.1	2198.6	2511.8	2740.1	2906.7	3154.6	3350.4	3521.8	3679.4	3828.8	3972.8
	s	3.8290	4.1141	4.5037	4.9447	5.2555	5.4746	5.7859	6.0170	6.2079	6.3743	6.5239	6.6614
45	*v*	0.00160	0.00180	0.00219	0.00292	0.00382	0.00463	0.00594	0.00698	0.00788	0.00870	0.00945	0.0102
	h	1728.0	1897.6	2110.8	2377.3	2623.4	2813.4	3090.2	3301.5	3482.5	3647.0	3801.3	3949.3
	s	3.7939	4.0505	4.3612	4.7362	5.0710	5.3209	5.6685	5.9179	6.1197	6.2932	6.4479	6.5891
50	*v*	0.00156	0.00173	0.00201	0.00249	0.00317	0.00389	0.00512	0.00611	0.00696	0.00772	0.00842	0.00907
	h	1716.6	1874.3	2060.2	2284.4	2520.0	2722.5	3025.7	3252.6	3443.5	3614.8	3774.1	3926.0
	s	3.7642	4.0028	4.2738	4.5892	4.9096	5.1759	5.5566	5.8245	6.0372	6.2180	6.3777	6.5226
60	*v*	0.00150	0.00163	0.00182	0.00208	0.00247	0.00295	0.00395	0.00483	0.00559	0.00627	0.00688	0.00746
	h	1699.9	1843.1	2001.6	2179.8	2375.2	2570.4	2902.1	3157.0	3366.8	3551.4	3720.6	3880.2
	s	3.7148	3.9316	4.1626	4.4134	4.6790	4.9356	5.3519	5.6528	5.8867	6.0815	6.2512	6.4034
70	*v*	0.00146	0.00157	0.00171	0.00189	0.00214	0.00246	0.00322	0.00397	0.00465	0.00525	0.00580	0.00632
	h	1688.4	1822.9	1967.1	2123.4	2291.7	2466.2	2795.0	3067.5	3293.6	3490.5	3669.0	3835.8
	s	3.6743	3.8778	4.0880	4.3080	4.5368	4.7662	5.1786	5.5003	5.7522	5.9600	6.1390	6.2982
80	*v*	0.00143	0.00152	0.00163	0.00177	0.00196	0.00219	0.00276	0.00338	0.00398	0.00452	0.00501	0.00548
	h	1680.4	1808.8	1944.0	2087.6	2239.6	2397.6	2709.9	2988.1	3225.7	3432.9	3619.7	3793.3
	s	3.6395	3.8339	4.0311	4.2331	4.4398	4.6474	5.0391	5.3674	5.6321	5.8509	6.0382	6.2039
90	*v*	0.00140	0.00148	0.00157	0.00169	0.00184	0.00201	0.00246	0.00297	0.00348	0.00397	0.00442	0.00484
	h	1674.6	1798.6	1927.6	2062.7	2204.0	2350.3	2645.2	2920.8	3164.4	3379.5	3573.5	3753.0
	s	3.6089	3.7965	3.9847	4.1747	4.3669	4.5593	4.9288	5.2540	5.5255	5.7526	5.9470	6.1184
100	*v*	0.00137	0.00144	0.00153	0.00163	0.00175	0.00189	0.00225	0.00267	0.00311	0.00355	0.00395	0.00434
	h	1670.7	1791.1	1915.5	2044.5	2178.3	2316.2	2596.1	2865.1	3110.6	3330.8	3530.7	3715.2
	s	3.5815	3.7638	3.9452	4.1267	4.3086	4.4899	4.8407	5.1580	5.4316	5.6640	5.8644	6.0405

v = specific volume, m^3/kg *h* = enthalpy, kJ/kg *s* = entropy, kJ/(kg·K)

Appendix C

COMMON UNITS AND UNIT CONVERSION FACTORS

Table Conv-1. Conversion Factors for Pressure (Force/Area)

To obtain → / multiply by ↓	atm	bar	psia (lb_f/in^2)	in Hg (conventional)	mm Hg (conventional)	ft H_2O (conventional)	kPa	MPa
atm	1	**1.013 25**	14.695 95	**760/25.4** = 29.921 26	**760**	33.898 54	**101.325**	**0.101 325**
bar	**1/(1.013 25)** = $9.869\ 233\times10^{-1}$	**1**	14.503 77	29.529 99	**760/(1.013 25)** = 750.0617	33.455 26	**100**	**0.1**
psia (lb_f/in^2)	$6.804\ 596\times10^{-2}$	$6.894\ 757\times10^{-2}$	1	2.036 021	51.714 93	2.306 659	6.894 757	$6.894\ 757\times10^{-3}$
in Hg (conventional)	**25.4/760** = $3.342\ 105\times10^{-2}$	$3.386\ 388\times10^{-2}$	$4.911\ 541\times10^{-1}$	1	**25.4**	1.132 925	3.386 388	$3.386\ 388\times10^{-3}$
mm Hg (conventional)	**1/760** = $1.315\ 789\times10^{-3}$	**(1.013 25)/760** = $1.333\ 224\times10^{-3}$	$1.933\ 677\times10^{-2}$	**1/(25.4)** = $3.937\ 008\times10^{-2}$	1	$4.460\ 334\times10^{-2}$	**101.325/760** = $1.333\ 224\times10^{-1}$	**(0.101 325)/760** = $1.333\ 224\times10^{-4}$
ft H_2O (conventional)	$2.949\ 980\times10^{-2}$	$2.989\ 067\times10^{-2}$	$4.335\ 275\times10^{-1}$	$8.826\ 711\times10^{-1}$	22.419 85	1	2.989 067	$2.989\ 067\times10^{-3}$
kPa	**1/101.325** = $9.869\ 233\times10^{-3}$	**0.01**	$1.450\ 377\times10^{-1}$	$2.952\ 999\times10^{-1}$	**760/101.325** = 7.500 617	$3.345\ 526\times10^{-1}$	1	**0.001**
MPa	**1/(0.101 325)** = 9.869 233	**10**	145.0377	295.2999	**760/(0.101 325)** = $7.500\ 617\times10^{3}$	334.5526	**1000**	1

Table Conv-2. Conversion Factors for Specific Volume (Volume/Mass)

To obtain → / multiply by ↓	ft^3/lb_m	in^3/lb_m	US gal/lb_m	liter/kg (cm^3/g)	m^3/kg
ft^3/lb_m	**1**	**1728**	**1728/231** = 7.480 519	$6.242\ 796\times10^{1}$	$6.242\ 796\times10^{-2}$
in^3/lb_m	**1/1728** = $5.787\ 037\times10^{-4}$	**1**	**1/231** = $4.329\ 004\times10^{-3}$	$3.612\ 729\times10^{-2}$	$3.612\ 729\times10^{-5}$
US gal/lb_m	**231/1728** = $1.336\ 806\times10^{-1}$	**231**	**1**	8.345 404	$8.345\ 404\times10^{-3}$
liter/kg (cm^3/g)	$1.601\ 846\times10^{-2}$	$2.767\ 990\times10^{1}$	$1.198\ 264\times10^{-1}$	**1**	**0.001**
m^3/kg	$1.601\ 846\times10^{1}$	$2.767\ 990\times10^{4}$	$1.198\ 264\times10^{2}$	**1000**	**1**

Table Conv-3. Conversion Factors for Specific Enthalpy and Specific Energy (Energy/Mass)

To obtain → / multiply by ↓	Btu/lb$_m$	ft·lb$_f$/lb$_m$	hp·h/lb$_m$	kW·h/lb$_m$	psia/(lb$_m$/ft^3)	cal/g	kJ/kg
Btu/lb$_m$	**1**	$7.781\ 693\times10^{2}$	$3.930\ 148\times10^{-4}$	$2.930\ 711\times10^{-4}$	5.403 953	**1/1.8** $= 5.555\ 556\times10^{-1}$	**4.1868/1.8** = **2.326**
ft·lb$_f$/lb$_m$	$1.285\ 067\times10^{-3}$	**1**	**1/(1.98×10^6)** $= 5.050\ 505\times10^{-7}$	$3.766\ 161\times10^{-7}$	**1/144** $= 6.944\ 444\times10^{-3}$	$7.139\ 264\times10^{-4}$	$2.989\ 067\times10^{-3}$
hp·h/lb$_m$	$2.544\ 434\times10^{3}$	**1.98×10^6**	**1**	$7.456\ 999\times10^{-1}$	**(1.98×10^6)/144** = **1.375×10^4**	$1.413\ 574\times10^{3}$	$5.918\ 353\times10^{3}$
kW·h/lb$_m$	$3.412\ 142\times10^{3}$	$2.655\ 224\times10^{6}$	1.341 022	**1**	$1.843\ 905\times10^{4}$	$1.895\ 634\times10^{3}$	$7.936\ 641\times10^{3}$
psia/(lb$_m$/ft^3)	$1.850\ 497\times10^{-1}$	**144**	**144/(1.98×10^6)** $= 7.272\ 727\times10^{-5}$	$5.423\ 272\times10^{-5}$	**1**	$1.028\ 054\times10^{-1}$	$4.304\ 256\times10^{-1}$
cal/g	**1.8**	$1.400\ 705\times10^{3}$	$7.074\ 266\times10^{-4}$	$5.275\ 279\times10^{-4}$	9.727 116	**1**	**4.1868**
kJ/kg	**1/2.326** $= 4.299\ 226\times10^{-1}$	$3.345\ 526\times10^{2}$	$1.689\ 659\times10^{-4}$	$1.259\ 979\times10^{-4}$	2.323 282	**1/4.1868** $= 2.388\ 459\times10^{-1}$	**1**

Table Conv-4.
Conversion Factors for Specific Entropy, Heat Capacity, and Gas Constant (Energy/(Mass·Temperature))

To obtain → / multiply by ↓	Btu/(lb$_m$·°R)	ft·lb$_f$/(lb$_m$·°R)	kW·h/(lb$_m$·°R)	psia·ft^3/(lb$_m$·°R)	bar·cm^3/(g·K)	cal/(g·K)	kJ/(kg·K)
Btu/(lb$_m$·°R)	**1**	$7.781\,693\times10^{2}$	$2.930\,711\times10^{-4}$	5.403 953	**41.868**	**1**	**4.1868**
ft·lb$_f$/(lb$_m$·°R)	$1.285\,067\times10^{-3}$	**1**	$3.766\,161\times10^{-7}$	**1/144** = $6.944\,444\times10^{-3}$	$5.380\,320\times10^{-2}$	$1.285\,067\times10^{-3}$	$5.380\,320\times10^{-3}$
kW·h/(lb$_m$·°R)	$3.412\,142\times10^{3}$	$2.655\,224\times10^{6}$	**1**	$1.843\,905\times10^{4}$	$1.428\,595\times10^{5}$	$3.412\,142\times10^{3}$	$1.428\,595\times10^{4}$
psia·ft^3/(lb$_m$·°R)	$1.850\,497\times10^{-1}$	**144**	$5.423\,272\times10^{-5}$	**1**	7.747 661	$1.850\,497\times10^{-1}$	$7.747\,661\times10^{-1}$
bar·cm^3/(g·K)	**0.1/4.1868** = $2.388\,459\times10^{-2}$	$1.858\,625\times10^{1}$	$6.999\,882\times10^{-6}$	$1.290\,712\times10^{-1}$	**1**	**0.1/4.1868** = $2.388\,459\times10^{-2}$	**0.1**
cal/(g·K)	**1**	$7.781\,693\times10^{2}$	$2.930\,711\times10^{-4}$	5.403 953	**41.868**	**1**	**4.1868**
kJ/(kg·K)	**1/4.1868** = $2.388\,459\times10^{-1}$	$1.858\,625\times10^{2}$	$6.999\,882\times10^{-5}$	1.290 712	**10**	**1/4.1868** = $2.388\,459\times10^{-1}$	**1**

Appendix D

COMMON SYMBOLS

Greek Alphabet			
Αα	Alpha	Νν	Nu
Ββ	Beta	Ξξ	Xi
Γγ	Gamma	Οο	Omicron
Δδ	Delta	Ππ	Pi
Εε	Epsilon	Ρρ	Rho
Ζζ	Zeta	Σσς	Sigma
Ηη	Eta	Ττ	Tau
Θθ	Theta	Υυ	Upsilon
Ιι	Iota	Φφ	Phi
Κκ	Kappa	Χχ	Chi
Λλ	Lambda	Ψψ	Psi
Μμ	Mu	Ωω	Omega

Index

S